新工科建设之路·数据科学与大数据系列教材

# Hadoop 大数据平台技术与应用

孙风栋　主编

李绪成　褚　娜　陈明华　副主编

电子工业出版社.
Publishing House of Electronics Industry
北京·BEIJING

## 内 容 简 介

本书以案例为驱动，系统介绍 Hadoop 大数据平台技术及其应用，Hadoop 生态圈组件的工作机制、管理与开发，以及基于 Hadoop 大数据平台的大数据业务解决方案。全书共 10 章，包括 Hadoop 分布式集群、伪分布式集群的安装与配置，Hadoop 平台开发工具的安装与配置，大数据存储与管理技术（分布式文件系统、分布式数据库 HBase）、大数据分析技术（MapReduce 计算框架、数据仓库 Hive）、大数据迁移工具（Sqoop）、大数据采集工具（Flume），并利用 Hadoop 生态圈组件构建一个网站日志分析项目的解决方案，完成整个大数据业务处理流程。

本书适合作为高等院校大数据相关专业的教材，也适合作为 Hadoop 大数据平台技术的初学者，以及初、中级 Hadoop 大数据平台管理与开发人员的培训教材。

未经许可，不得以任何方式复制或抄袭本书之部分或全部内容。
版权所有，侵权必究。

**图书在版编目（CIP）数据**

Hadoop 大数据平台技术与应用 / 孙风栋主编. — 北京：电子工业出版社，2021.6
ISBN 978-7-121-41365-0

Ⅰ. ①H… Ⅱ. ①孙… Ⅲ. ①数据处理软件 Ⅳ. ①TP274

中国版本图书馆 CIP 数据核字(2021)第 117510 号

责任编辑：凌　毅
印　　刷：北京盛通数码印刷有限公司
装　　订：北京盛通数码印刷有限公司
出版发行：电子工业出版社
　　　　　北京市海淀区万寿路 173 信箱　邮编：100036
开　　本：787×1 092　1/16　印张：18.25　字数：490 千字
版　　次：2021 年 6 月第 1 版
印　　次：2025 年 1 月第 7 次印刷
定　　价：59.00 元

凡所购买电子工业出版社图书有缺损问题，请向购买书店调换。若书店售缺，请与本社发行部联系。联系及邮购电话：(010)88254888，88258888。
质量投诉请发邮件至 zlts@phei.com.cn，盗版侵权举报请发邮件至 dbqq@phei.com.cn。
本书咨询联系方式：(010)88254528，lingyi@phei.com.cn。

# 前 言

### 1. 编写背景

随着移动互联网、物联网、社交媒体等的发展,以及数据产生方式的变革,数据量每年增长约50%。云计算为大数据技术的发展提供了存储、计算等技术支撑,促进人类社会进入大数据时代。大数据技术已经广泛应用于金融、医疗、交通、教育、电信、政府、工业等各行各业,数据量也由B、KB、MB、GB级,发展到TB、PB、EB、ZB、YB级。

在国家"十四五"规划中,大数据中心建设与5G、工业互联网一同被提及,成为未来国家战略中的一个重要衡量指标。大数据中心为应对5G、人工智能、工业互联网等大数据需求应运而生,构成了新基建的"基础"。大数据技术的发展关键在于人才。2016年2月,北京大学、对外经济贸易大学和中南大学成为第一批申报并获批开设数据科学与大数据专业的高校。到2021年,全国共有近700所高校开设大数据相关专业。

Hadoop作为大数据开源软件的行业标准,是大数据生态系统的核心框架,为海量数据提供了分布式文件系统(HDFS)和分布式并行计算架构MapReduce。当前,Hadoop已经应用于各大互联网公司的核心业务,如阿里巴巴的云计算平台、淘宝的推荐系统、百度的搜索引擎等,只要是涉及大数据的相关领域,都能看到Hadoop的身影。

### 2. 本书特色

本书是作者根据多年教学经验及项目实践经验总结而成的,是一本面向应用型人才培养的教材。本书以实践为主、理论为辅,案例丰富,实用性强,引导读者步步深入,掌握Hadoop生态圈组件的应用。本书反映了最新的混合式教育教学改革的思想,以学生为中心,提供丰富的资源,包括电子课件、案例、习题、题库、试卷等,学生在实践过程中发现问题、分析问题、解决问题,教师精讲、指导,学与教有机融为一体。本书介绍Hadoop架构相关技术的最新版本,包括Hadoop 3.1.2 + HBase 2.2.6 + Hive 3.1.2 + Sqoop 1.4.7 + Flume 1.9.0。

### 3. 内容构成

本书包括10章。

第1章:介绍大数据技术发展应用及Hadoop生态圈组件与架构。

第2章:介绍Hadoop伪分布式平台从无到有的整个安装与配置过程,以及Hadoop开发环境的配置。

第3章:介绍分布式文件系统的基本架构、工作机制、管理与开发。

第4章:介绍MapReduce的工作机制、编程模型与应用开发。

第5章:介绍分布式数据库HBase的工作机制、架构、管理与应用开发。

第6章:介绍数据仓库Hive的架构、工作机制、HiveQL语句应用及Hive应用开发。

第7章:介绍数据迁移工具Sqoop的应用,包括Hadoop与MySQL数据库之间数据的导入与导出。

第8章:介绍数据采集工具Flume的工作机制、组件及Flume在不同场景中的应用。

第9章:介绍如何利用Hadoop生态圈组件构建一个网站日志分析项目的解决方案,并完成整个大数据业务处理流程。

第 10 章：介绍生产环境中 Hadoop 分布式集群、HBase 分布式集群的安装与配置。

每个章节都配有丰富的案例，通过案例介绍 Hadoop 生态圈组件在实际中的应用。读者可以根据需要采用 Hadoop 分布式集群环境或伪分布式集群环境，两者对用户是透明的，采用统一的访问 URI：hdfs://master:9000，读者可以根据学习环境进行选择，所有案例和项目都可以正常运行。

### 4．教材资源

本书配有电子课件、程序源代码、习题解答等教学资源，读者可以登录华信教育资源网（www.hxedu.com.cn），注册后免费下载。

### 5．致谢

首先感谢我的合作者们为本书出版所付出的努力。本书第 1～2 章由李绪成副教授编写，第 3～4 章由陈明华副教授编写，第 5～6 章由褚娜副教授编写，第 7～10 章由孙风栋教授编写。全书由孙风栋主持编写并统稿，褚娜主审。

此外，贾宁副教授、周伊佳博士、叶剑锋高级工程师、周慧教授、郑纯军教授、才晶晶老师等对本书的编写工作提出了宝贵意见，在此向他们表示感谢。

还要特别感谢电子工业出版社凌毅编辑为本书的编写和出版提供的帮助与支持！

Hadoop 大数据平台技术繁杂，作者水平有限，再加上编写时间仓促，本书错误或不妥之处在所难免，敬请读者批评指正！欢迎加入微信交流群"Hadoop 大数据平台技术与应用"。

孙风栋

2021 年 6 月于大连

E-mail：sunfengdong@neusoft.edu.cn

# 目 录

## 第 1 章 大数据技术与 Hadoop 概述 ·········· 1
- 1.1 大数据技术概述 ·········· 1
  - 1.1.1 大数据技术发展 ·········· 1
  - 1.1.2 大数据基本特征 ·········· 1
  - 1.1.3 大数据关键技术 ·········· 2
  - 1.1.4 大数据技术应用 ·········· 3
- 1.2 Hadoop 概述 ·········· 4
  - 1.2.1 Hadoop 简介 ·········· 4
  - 1.2.2 Hadoop 核心组件 ·········· 5
  - 1.2.3 Hadoop 生态圈组件 ·········· 6
  - 1.2.4 Hadoop 架构 ·········· 7
- 本章小结 ·········· 8
- 思考题与习题 ·········· 9

## 第 2 章 Hadoop 平台和开发环境的安装与配置 ·········· 10
- 2.1 Hadoop 平台安装准备 ·········· 10
  - 2.1.1 VMware 安装与配置 ·········· 10
  - 2.1.2 Ubuntu 安装与配置 ·········· 12
  - 2.1.3 网络配置 ·········· 19
- 2.2 Hadoop 伪分布式集群安装与配置 ·········· 24
  - 2.2.1 创建用户 hadoop ·········· 24
  - 2.2.2 修改主机名与域名映射 ·········· 25
  - 2.2.3 SSH 免密码登录设置 ·········· 25
  - 2.2.4 安装 Java 环境 ·········· 26
  - 2.2.5 伪分布式集群安装与配置 ·········· 27
- 2.3 Eclipse 开发环境安装与配置 ·········· 30
  - 2.3.1 Maven 安装与配置 ·········· 30
  - 2.3.2 Eclipse 安装与配置 ·········· 31
  - 2.3.3 Eclipse 中 Maven 设置 ·········· 31
- 本章小结 ·········· 32
- 思考题与习题 ·········· 33

## 第 3 章 分布式文件系统 ·········· 34
- 3.1 HDFS 概述 ·········· 34
  - 3.1.1 HDFS 架构 ·········· 34
  - 3.1.2 HDFS 设计目标 ·········· 36
  - 3.1.3 HDFS 高可用架构 ·········· 36
  - 3.1.4 HDFS 架构的优劣性 ·········· 38
- 3.2 HDFS 工作机制 ·········· 38
  - 3.2.1 HDFS 数据存储策略 ·········· 38
  - 3.2.2 HDFS 数据读取策略 ·········· 39
  - 3.2.3 HDFS 数据错误与恢复 ·········· 39
  - 3.2.4 HDFS 数据读写过程 ·········· 40
- 3.3 HDFS Shell 管理 ·········· 41
  - 3.3.1 HDFS 文件操作命令 ·········· 41
  - 3.3.2 HDFS 系统管理命令 ·········· 47
  - 3.3.3 HDFS Shell 操作实例 ·········· 50
- 3.4 HDFS Java 开发 ·········· 53
  - 3.4.1 HDFS Java 程序设计基础 ·········· 53
  - 3.4.2 HDFS 程序设计流程 ·········· 54
  - 3.4.3 常用 HDFS Java API ·········· 55
  - 3.4.4 HDFS 开发实例 ·········· 61
- 本章小结 ·········· 66
- 思考题与习题 ·········· 67

## 第 4 章 MapReduce ·········· 69
- 4.1 MapReduce 概述 ·········· 69
  - 4.1.1 MapReduce 简介 ·········· 69
  - 4.1.2 MapReduce 计算模型 ·········· 70
  - 4.1.3 MapReduce 编程模型 ·········· 70
- 4.2 MapReduce 架构 ·········· 71
  - 4.2.1 MapReduce V1 架构 ·········· 71
  - 4.2.2 MapReduce V2 架构 ·········· 73
- 4.3 MapReduce 编程组件 ·········· 76
  - 4.3.1 MapReduce 编程流程 ·········· 76
  - 4.3.2 InputFormat ·········· 77
  - 4.3.3 InputSplit ·········· 78
  - 4.3.4 RecordReader ·········· 79
  - 4.3.5 Mapper ·········· 79
  - 4.3.6 Shuffle ·········· 80
  - 4.3.7 Reducer ·········· 83
  - 4.3.8 OutputFormat ·········· 84

   4.3.9 序列化与反序列化 ················ 84
 4.4 WordCount 程序设计实例 ················ 87
   4.4.1 准备输入文件 ················ 87
   4.4.2 创建 Maven 工程 ················ 87
   4.4.3 配置 Maven 工程 ················ 87
   4.4.4 程序设计 ················ 88
   4.4.5 工程打包、部署与运行 ················ 92
   4.4.6 定制 WordCount 程序设计 ················ 92
 4.5 MapReduce 开发典型案例 ················ 96
   4.5.1 数据去重 ················ 96
   4.5.2 数据排序 ················ 98
   4.5.3 计算平均值 ················ 101
 4.6 网站浏览量统计分析 ················ 104
 本章小结 ················ 106
 思考题与习题 ················ 106

## 第 5 章 分布式数据库 HBase ················ 109

 5.1 HBase 概述 ················ 109
   5.1.1 HBase 简介 ················ 109
   5.1.2 HBase 特性 ················ 109
   5.1.3 HBase 适用场景 ················ 110
 5.2 HBase 数据模型 ················ 110
   5.2.1 HBase 基本概念 ················ 110
   5.2.2 概念视图 ················ 112
   5.2.3 物理视图 ················ 112
 5.3 HBase 体系结构 ················ 113
 5.4 HBase 安装与配置 ················ 116
   5.4.1 HBase 运行模式 ················ 116
   5.4.2 HBase 安装准备 ················ 117
   5.4.3 HBase 伪分布式集群安装与配置 ················ 118
 5.5 HBase Shell ················ 120
   5.5.1 HBase Shell 简介 ················ 120
   5.5.2 General 命令组 ················ 121
   5.5.3 DDL 命令组 ················ 121
   5.5.4 DML 命令组 ················ 124
   5.5.5 查询过滤器 ················ 127
 5.6 HBase 程序设计 ················ 131
   5.6.1 HBase Java API 简介 ················ 131
   5.6.2 Hbase 表管理程序设计 ················ 138
   5.6.3 HBase 数据操作程序设计 ················ 144
   5.6.4 HBase Filter API ················ 150

 5.7 HBase 与 MapReduce 融合 ················ 154
   5.7.1 HBase 与 MapReduce 融合概述 ················ 154
   5.7.2 HBase MapReduce Java API ················ 155
   5.7.3 HBase MapReduce 程序设计 ················ 156
 5.8 HBase 学生成绩分析 ················ 159
   5.8.1 任务描述 ················ 159
   5.8.2 导入原始数据到 HBase ················ 160
   5.8.3 统计学生平均成绩 ················ 161
 本章小结 ················ 163
 思考题与习题 ················ 163

## 第 6 章 数据仓库 Hive ················ 165

 6.1 Hive 基础 ················ 165
   6.1.1 Hive 简介 ················ 165
   6.1.2 Hive 系统架构 ················ 166
   6.1.3 Hive 工作原理 ················ 167
   6.1.4 Hive 数据存储模型 ················ 168
   6.1.5 Hive 数据类型 ················ 169
   6.1.6 Hive 数据存储格式 ················ 170
 6.2 Hive 安装与配置 ················ 170
   6.2.1 安装 MySQL ················ 170
   6.2.2 Hive 安装与配置过程 ················ 171
 6.3 Beeline ················ 174
   6.3.1 Beeline 简介 ················ 174
   6.3.2 Beeline 基本操作 ················ 174
 6.4 Hive DDL 操作 ················ 176
   6.4.1 Hive 数据库管理 ················ 176
   6.4.2 Hive 表管理 ················ 178
   6.4.3 视图管理 ················ 184
 6.5 Hive DML 操作 ················ 184
 6.6 Hive 数据查询 ················ 188
   6.6.1 Hive SELECT 基本语法 ················ 188
   6.6.2 无条件查询 ················ 189
   6.6.3 有条件查询 ················ 190
   6.6.4 查询统计 ················ 191
   6.6.5 分组查询 ················ 192
   6.6.6 子查询 ················ 193
   6.6.7 连接查询 ················ 194
   6.6.8 排序 ················ 195
   6.6.9 合并操作 ················ 196
   6.6.10 复合类型数据查询 ················ 196

6.7 Hive 内置函数 197
    6.7.1 数学函数 197
    6.7.2 集合函数 199
    6.7.3 类型转换函数 200
    6.7.4 日期函数 200
    6.7.5 条件函数 201
    6.7.6 字符串函数 201
    6.7.7 内置聚合函数 203
    6.7.8 内置表生成函数 204
    6.7.9 窗口函数 205
    6.7.10 其他函数 207
    6.7.11 词频统计实例 208
6.8 Hive 高级应用 208
    6.8.1 用户自定义函数 208
    6.8.2 Hive 与 HBase 整合 210
6.9 Hive 程序设计 212
本章小结 213
思考题与习题 213

## 第 7 章 数据迁移工具 Sqoop 216
7.1 Sqoop 概述 216
7.2 Sqoop 安装与配置 217
7.3 Sqoop 常用命令 218
7.4 Sqoop 数据导入 220
    7.4.1 Sqoop 命令参数 220
    7.4.2 数据从 MySQL 导入 HDFS 221
    7.4.3 数据从 MySQL 导入 Hive 224
    7.4.4 数据从 MySQL 导入 HBase 225
7.5 Sqoop 数据导出 226
    7.5.1 Sqoop export 命令参数 226
    7.5.2 从 HDFS 导出数据到 MySQL 227
    7.5.3 从 Hive 导出数据到 MySQL 228
    7.5.4 中文乱码问题 229
本章小结 230
思考题与习题 230

## 第 8 章 数据采集工具 Flume 232
8.1 Flume 概述 232
    8.1.1 Flume 简介 232
    8.1.2 Flume 架构 232
8.2 Flume 安装与配置 235
8.3 Flume 组件 237
    8.3.1 Source 组件 237
    8.3.2 Channel 组件 240
    8.3.3 Sink 组件 243
    8.3.4 Interceptor 组件 246
    8.3.5 Selector 组件 250
    8.3.6 Sink Processor 251
8.4 Flume 数据采集案例与实施 252
    8.4.1 实时采集本地文件到 HDFS 252
    8.4.2 多源与多目的地数据采集 254
本章小结 257
思考题与习题 257

## 第 9 章 网站日志分析 259
9.1 需求分析 259
    9.1.1 网站日志分析的必要性 259
    9.1.2 网站日志数据说明 259
    9.1.3 网站日志分析 KPI 指标 260
9.2 方案设计 260
9.3 数据采集 261
9.4 数据预处理 262
9.5 数据分析 266
9.6 数据分析结果导出及可视化 268
本章小结 272
思考题与习题 272

## 第 10 章 Hadoop 与 HBase 分布式集群安装与配置 273
10.1 Hadoop 分布式集群安装与配置 273
10.2 HBase 分布式集群安装与配置 280
本章小结 283
思考题与习题 283

## 参考文献 284

# 第1章 大数据技术与 Hadoop 概述

大数据是信息时代的重要标志，大数据技术是海量数据采集技术、存储技术、分析与处理技术、可视化技术的集成，能帮助人们从海量数据集合中挖掘出数据价值并用于决策支持。Hadoop 平台为大数据提供了分布式存储与分布式并行计算功能，是大数据的行业标准开源软件。本章将介绍大数据技术与 Hadoop 平台。

- 大数据技术概述：大数据技术发展、大数据基本特征、大数据关键技术及大数据技术应用。
- Hadoop 概述：Hadoop 简介、Hadoop 核心组件、Hadoop 生态圈组件及 Hadoop 架构。

## 1.1 大数据技术概述

### 1.1.1 大数据技术发展

"大数据"一词最早诞生于 1983 年出版的托夫勒的著作《第三次浪潮》中。作者将社会发展分为三次浪潮，农业时代归为第一次浪潮，工业时代归为第二次浪潮，信息时代归为第三次浪潮。大数据在信息时代中扮演着非常重要的角色。

2011 年美国著名的咨询公司麦肯锡的一份研究报告指出，各个国家的数据量呈现出一种爆炸式增长的趋势，这也标志着大数据时代的到来。

2012 年著名的数据科学家舍恩伯格在其著作《大数据时代》中对大数据的商业应用展开了讨论，人们开始关心大数据的价值。然而，传统算力与存储能力无法满足大数据的要求，幸运的是大数据的发展遇到了云计算的支持，很多公司都通过开发云计算技术来支持大数据的发展，比如谷歌云与阿里云。正是有了云计算的助力，大数据才能够进入爆炸式的增长通道。

大数据技术的发展与信息化紧密相关，正是信息化浪潮的涌动，开启了人类社会的大数据时代。信息科技的发展，使得存储设备容量不断增加、CPU 处理能力不断提升、网络带宽不断增加，这些为大数据的发展奠定了技术基础。同时，数据的产生经历了传统的运营式系统阶段、Web2.0 时代的用户原创内容阶段和当前的感知式系统阶段，数据量爆炸式增长，这为大数据的发展提供了数据基础。

### 1.1.2 大数据基本特征

对于大数据的概念，不同研究机构、不同公司有不同的诠释。2011 年，麦肯锡在研究报告中给出大数据的定义为：大数据是指大小超出典型数据库软件工具收集、存储、管理和分析能力的数据集。维基百科给出大数据的定义为：大数据或称巨量数据、海量数据、大资料，是指所涉及的数据量规模巨大到无法通过人工在合理时间内实现提取、管理、处理并整理成人类所能解读的信息。百度百科给出大数据的定义为：大数据或称巨量资料，是指所涉及的资料量规模巨大到无法通过目前主流软件工具在合理时间内实现提取、管理、处理并整理成为帮助企业经营决策的信息。

实际上，所有关于大数据概念的定义都是围绕大数据的特征进行的。关于大数据的特征，学术界和业界比较认可"4V"说法，即数据量大（Volume）、速度快（Velocity）、数据类型多样（Variety）和价值密度低（Value）。

**1. 数据量大**

大数据的处理量级已经由传统数据的 GB、TB 级达到了 PB 级甚至更高量级。根据 IDC 做出的估测，数据量每年增长约 50%，也就是说每两年就增长一倍。人类在最近两年产生的数据量相当于之前产生的全部数据量，到 2020 年，全球共拥有 35ZB 的数据量，相较于 2010 年，数据量增长了近 30 倍。

**2. 速度快**

大数据是一种以实时数据处理、实时结果导向为特征的解决方案，它的"快"有两个层面。

一是数据产生得快。有的数据是爆发式产生的，例如，欧洲核子研究中心的大型强子对撞机在工作状态下每秒产生 PB 级的数据；有的数据是涓涓细流式产生的，但是由于用户众多，短时间内产生的数据量依然非常庞大，例如，点击量、日志、射频识别数据、GPS（全球定位系统）位置信息等。

二是数据处理得快。大数据有批处理（"静止数据"转变为"正使用数据"）和流处理（"动态数据"转变为"正使用数据"）两种范式，以实现快速的数据处理。在数据处理速度方面，有一个著名的"1 秒定律"，即要在秒级时间范围内给出分析结果，超出这个时间，数据就失去了价值。大数据时代的很多应用，都需要基于快速生成的数据给出实时分析结果。从数据的生成到消耗，时间窗口非常小，可用于生成决策的时间非常短，这与传统的数据挖掘技术有着本质的不同。

**3. 数据类型多样**

大数据是由结构化、非结构化数据组成的。其中，结构化数据占 10%左右，主要指存储在关系型数据库中的数据；而非结构化数据占 90%左右，包括邮件、音频、视频、微信、微博、位置信息、网络日志、手机通话记录及传感器网络数据等。

**4. 价值密度低**

大数据的价值密度远远低于传统关系数据库中的数据，很多有价值的信息都分散在海量数据中。因此，大数据的本质是获取数据价值，关键是商业价值，即如何有效利用好这些数据。以视频为例，在连续不间断的监控过程中，可能有用的数据仅仅有一两秒，但是具有很高的商业价值。

### 1.1.3 大数据关键技术

按大数据分析的处理流程，大数据关键技术主要包括数据采集、数据预处理、数据存储与管理、数据处理与分析、数据可视化、数据隐私和安全等几个层面的内容，如图 1-1 所示。

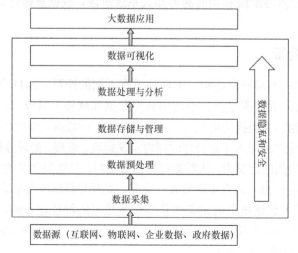

图 1-1 大数据关键技术

① 数据采集是指利用 ETL 工具将分布的、异构的数据源中的数据提取到临时中间层。

② 数据预处理是指对采集得到的数据进行清洗、转换、集成，最后加载到数据仓库或数据集市中，成为联机分析处理、数据挖掘的基础；或者把实时采集的数据作为流计算系统的输入，进行实时分析处理。

③ 数据存储与管理是指利用分布式文件系统、数据仓库、关系数据库、NoSQL 数据库、云数据库等，对结构化、半结构化和非结构化海量数据进行存储和管理。

④ 数据处理与分析是指利用分布式并行编程模型和计算框架，结合机器学习和数据挖掘算法，实现对海量数据的处理和分析。

⑤ 数据可视化是指对分析结果进行可视化呈现，从而更好地理解数据、分析数据。

⑥ 数据隐私和安全是指从大数据中挖掘潜在的商业价值和学术价值的同时，构建隐私数据保护体系和数据安全体系，有效保护个人隐私和数据安全。

在大数据的各种关键技术中，其核心技术为分布式存储和分布式并行计算。

分布式存储技术与数据存储介质的类型和数据的组织管理形式直接相关。目前主要的数据存储介质类型包括内存、磁盘、磁带等；主要的数据组织管理形式包括按行组织、按列组织、按键值组织和按关系组织。大数据分布式存储结构主要包括分布式文件系统、文档存储、列式存储、键值存储、图形存储、关系数据库存储、内存存储等。

大数据采用分布式并行计算框架，针对不同的业务场景采用不同的计算模式，主要包括以下几个方面。

● 批处理计算：主要解决针对大规模数据的批量处理。具有代表性的批处理计算产品包括 MapReduce、Spark 等。

● 流计算：可以实时处理来自不同数据源的、连续到达的流数据，经过实时分析处理，得到有价值的分析结果。具有代表性的流计算产品包括 IBM InfoSphere Streams、Twitter Storm、Spark Streaming 等。

● 图计算：针对大规模图结构数据的处理。代表性的图计算产品包括 Pregel、Spark GraphX 等。

● 查询分析计算：针对大规模数据的存储管理和查询分析，需要提供实时或准时的响应，才能更好地满足企业经营管理的需求。代表性产品包括 Dremel、Hive 等。

## 1.1.4 大数据技术应用

大数据的应用领域非常广泛，在商业、教育、工业、交通、医疗等领域中发挥着非常重要的作用。

**1. 商业领域**

目前，大数据在商业领域的应用最为成熟，一个原因是商业领域的大数据能够快速、直接地体现出来价值，另一个原因是商业领域产生的数据量非常庞大，消费者的行为都会成为对企业非常有价值的数据来源，这也让大数据在商业领域落地有了非常扎实的基础。商品推荐是大数据在商业领域的重要应用，根据用户的基础信息、消费信息、社交信息构建用户画像，精准地对用户进行价值判断，从而进行精准营销、精准匹配，提供个性化服务等。

**2. 教育领域**

大数据与教育的深度融合已成为必然趋势。基于大数据的精确学情诊断、个性化学习分析和智能决策支持，大大提升了教育品质，对促进教育公平、提高教育质量、优化教育治理都具有重要作用，已成为实现教育现代化必不可少的重要支撑。例如，通过对学生行为、课程资源、教师教学、课程评价等数据的实时精准采集，可以对教师授课质量进行精准评价、构建学生个人画像、进行学生成绩预测、发现学生异常行为、实现个性化的教学与学习等，为不同的利益相关者提供精准的教育服务。

**3. 工业领域**

工业大数据是工业领域产品和服务全生命周期数据的总称，包括工业企业在研发设计、生产制造、经营管理、运维服务等环节中生成和使用的数据，以及工业互联网平台中的数据等。工业大数据是未来工业在全球市场竞争中发挥优势的关键，各国制造业创新战略的实施基础都是工业大数据的搜集和特征分析，以及以此为未来制造系统搭建的无忧环境。

**4. 交通领域**

交通大数据主要应用在两个方面：一方面可以利用大量网络传感器数据来了解车辆通行密度，合理进行道路规划（包括单行线路规划）；另一方面，可以利用大数据来实现即时信号灯调度，提高已有道路的通行能力。科学安排信号灯是一项复杂的系统工程，必须利用大数据计算平台计算出一个较为合理的方案。科学的信号灯安排将会提高30%左右已有道路的通行能力。机场的航班起降依靠大数据将会提高航班管理的效率，航空公司利用大数据可以提高上座率，降低运行成本。铁路利用大数据可以有效安排客运和货运列车，提高效率、降低成本。

**5. 医疗领域**

目前大数据与人工智能在医疗诊断领域有了很多的突破，结合一些病例数据与医疗诊断图像，可以得到比专业医生更精准的诊断结论。例如，针对早期癌症诊断，基于大数据建立一个医疗数据与患有癌症概率的关联模型，收集患者血常规、血生化和尿常规等数据，就可以预测一个患者患癌症的概率。大数据预测早期癌症有3个亮点，一是早期发现的准确率高，检测非常准确；二是预测方法简便，只需要基于现有指标，不需要再采样；三是预测费用低。

借助大数据的技术优势，治疗方案的制定变得更加科学合理，治疗效果也得到了进一步提升。与此同时，医务工作者通过对远程监控系统产生的数据进行综合分析，可以及时采取措施做好疾病防控工作，进而避免大规模恶性疾病的爆发。此外，医务工作者通过全面分析患者的各项医疗数据，对比多种干预措施的有效性，可以找到针对特定病人的较适宜的治疗途径，并降低过度治疗及治疗不足等情况出现的可能性。

## 1.2 Hadoop 概述

### 1.2.1 Hadoop 简介

Hadoop 是 Apache 软件基金会旗下的一个开源分布式计算平台，为用户提供了系统底层透明的分布式基础架构。Hadoop 基于 Java 语言开发，具有很好的跨平台特性，并且可以部署在廉价的计算机集群中。Hadoop 被公认为行业大数据标准开源软件，在分布式环境下提供了海量数据的处理能力。目前，几乎所有主流厂商都围绕 Hadoop 提供开发工具、开源软件、商业化工具和技术服务，如谷歌、雅虎、微软、思科、阿里等。

**1. Hadoop 的发展史**

① Hadoop 源于 2002 年的一个开源网络搜索引擎项目 Apache Nutch。

② 2004 年，Nutch 开源实现了谷歌的 GFS，开发了自己的分布式文件系统 NDFS（Nutch Distributed File System），也就是 HDFS 的前身。

③ 2005 年，Nutch 开源实现了谷歌的 MapReduce。

④ 2006 年 2 月，Nutch 中的 NDFS 和 MapReduce 开始独立出来，称为 Hadoop。

⑤ 2008 年 1 月，Hadoop 正式成为 Apache 顶级项目。2008 年 4 月，Hadoop 打破世界纪录，采用一个由 910 个节点构成的集群、用时 209s 实现对 1TB 数据进行排序。2009 年 5 月，Hadoop 把 1TB 数据的排序时间缩短到 62s。Hadoop 从此名声大震，迅速发展成为大数据时代最具影响力的开源分布式开发平台，并成为事实上的大数据处理标准。

⑥ 2011 年 12 月，Hadoop 1.0.0 版本发布。

⑦ 2013 年 10 月，Hadoop 2.2.0 版本发布，Hadoop 正式进入 Hadoop 2.x 时代。

⑧ 2016 年，Hadoop 3.x-alpha 版本发布，预示着进入 Hadoop 3.x 时代。

**2．Hadoop 特性**

Hadoop 是一个能够对大量数据进行分布式处理的软件框架，并且是以一种可靠、高效、可伸缩的方式进行处理的，它具有以下几个方面的特性。

① 高可靠性：Hadoop 采用冗余数据存储方式，即使一个副本发生故障，其他副本也可以保证正常对外提供服务。

② 高效性：Hadoop 能够在节点之间动态地移动数据，并保证各个节点的动态平衡，因此处理速度非常快。

③ 高扩展性：Hadoop 是在可用的计算机集群间分配数据并完成计算任务的，集群可以方便地扩展到数以千计的节点中。

④ 高容错性：Hadoop 能够自动保存数据的多个副本，并且能够自动将失败的任务重新分配。

⑤ 低成本：Hadoop 采用廉价的计算机集群，成本比较低，普通用户也很容易用自己的 PC 搭建 Hadoop 运行环境。

⑥ 运行在 Linux 平台上：Hadoop 是基于 Java 语言开发的，可以较好地运行在 Linux 平台上。

⑦ 支持多种编程语言：Hadoop 上的应用程序也可以使用多种语言编写，如 C++、Python、Java 等。

## 1.2.2　Hadoop 核心组件

在 Hadoop 1.0 中，核心组件主要为 HDFS 和 MapReduce。由于存在单点故障问题、采用单一命名空间无法实现资源隔离、MapReduce 负载过重等原因，导致运行效率低下。在 Hadoop 2.0 中，引入了 YARN（Yet Another Resource Negotiator），专门负责资源调度与管理，MapReduce 专门用于计算，如图 1-2 所示。

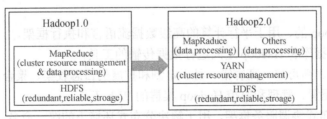

图 1-2　Hadoop 1.0 与 Hadoop 2.0 异同

从 Hadoop 2.0 开始，Hadoop 包含 Hadoop Common、HDFS、YARN 和 MapReduce 共 4 个核心组件。

（1）Hadoop Common（公共组件）

Hadoop Common 是为 Hadoop 其他组件提供支持的实用程序，是整个 Hadoop 项目的核心。主要包括一组分布式文件系统、通用 I/O 组件与接口（序列化、远程过程调用 RPC、持久化数据结构）。

（2）HDFS（分布式文件系统）

HDFS 提供对应用程序数据的高吞吐量的访问，是 Google GFS 的开源实现，将大小不一的众多数据文件切分成相同大小的数据块，以多副本的方式存储在不同的集群节点中。

（3）YARN（资源调度与管理系统）

YARN 的主要作用是负责集群的资源管理和统一调度，使多种计算框架可以运行在一个集群中，从而充分利用集群的资源，维护和管理也变得更加便捷。

（4）MapReduce（分布式计算框架）

从 Hadoop 2.0 开始，MapReduce 成为基于 YARN 的大型数据集并行处理系统。从 HDFS 中取出数据块并进行分片，每个分片对应一个 Map 任务，Map 计算结果通过 Reduce 任务进行汇总计算，得到最终结果并输出。

### 1.2.3 Hadoop 生态圈组件

除 4 个核心组件外，Hadoop 还包括众多生态圈组件。核心组件和众多生态圈组件共同形成一个丰富的 Hadoop 生态系统，如图 1-3 所示。

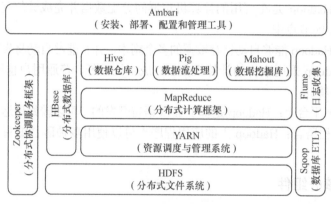

图 1-3 Hadoop 生态系统

**Hive**：Hadoop 中的数据仓库，定义了一种类似 SQL 的查询语言（HQL），将 SQL 转化为 MapReduce 任务在 Hadoop 上执行。

**HBase**：建立在 HDFS 之上，可伸缩、高可靠、高性能、分布式和面向列的动态模式数据库。

**Pig**：基于 Hadoop 的、用于平行计算的高级数据流语言和执行框架。

**Sqoop**：传统数据库和 Hadoop 之间进行数据传输的工具。

**Flume**：分布式、高可靠、高容错、易于定制和扩展的日志收集、聚合、移动框架。

**Ambari**：用于配置、管理和监控 Hadoop 集群的 Web 工具。

**Zookeeper**：分布式协调服务框架，用于解决分布式环境下的统一命名、状态同步、集群管理、配置同步等问题。

**Mahout**：机器学习和数据挖掘库，提供的 MapReduce 包含很多实现，包括聚类算法、回归测试、统计建模等。

另外，Tez 为基于 YARN 的通用数据流编程框架；Spark 为基于内存的通用并行编程框架；Storm 为大型数据流的实时处理开源框架，可以实时处理 Hadoop 的批量任务；Oozie 是一个工作流调度引擎，可以执行 MapReduce、Hive、Spark 等不同类型的单一或具有依赖性的作业；Kafka 为分布式发布订阅消息系统。

## 1.2.4 Hadoop 架构

**1. Hadoop 1.x 架构**

Hadoop 1.x 架构如图 1-4 所示。在该架构中，核心组件主要包括 HDFS 和 MapReduce。其中，HDFS 包括一个 NameNode 主节点，用于管理集群中的各种数据；一个 SecondaryNameNode 节点，用于 NameNode 节点的备份；多个 DataNode 从节点，用于存储集群中的各种数据。MapReduce 包括一个 JobTracker 主节点，用于接收用户的计算请求任务，并分配任务给从节点；多个 TaskTracker 从节点，负责执行 JobTracker 分配的任务。

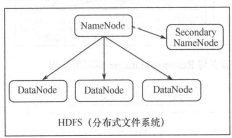
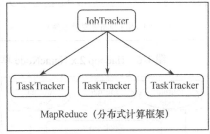

图 1-4　Hadoop 1.x 架构

**2. Hadoop 2.x NameNode 与 ResourceManager 单节点架构**

Hadoop 2.x NameNode 与 ResourceManager 单节点架构如图 1-5 所示。在该架构中，HDFS 包括一个 Active NameNode 主节点、一个 SecondaryNameNode 节点和多个 DataNode 从节点。YARN 包括一个 ResourceManager 主节点，用于接收用户的计算请求任务，并负责集群的资源分配及计算任务的划分；多个 NodeManager 从节点，负责执行 ResourceManager 分配的任务。MapReduce 分离出来，专门进行数据分析与计算。

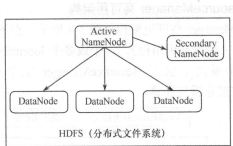

图 1-5　Hadoop 2.x NameNode 与 ResourceManager 单节点架构

**3. Hadoop 2.x NameNode 单节点与 ResourceManager 高可用架构**

Hadoop 2.x NameNode 单节点与 ResourceManager 高可用架构如图 1-6 所示。在该架构中，HDFS 包括一个 Active NameNode 主节点、一个 SecondaryNameNode 节点和多个 DataNode 从节点。YARN 包括两个或两个以上 ResourceManager 主节点，用于接收用户的计算请求任务，并负责集群的资源分配及计算任务的划分，通过 Zookeeper 实现 ResourceManager 的高可用；多个 NodeManager 从节点，负责执行主节点 ResourceManager 分配的任务。MapReduce 专门进行数据分析与计算。

**4. Hadoop 2.x NameNode 高可用与 ResourceManager 单节点架构**

Hadoop 2.x NameNode 高可用与 ResourceManager 单节点架构如图 1-7 所示。在该架构中，HDFS 包括两个或两个以上 NameNode 主节点（形成高可用状态），多个 DataNode 从节点，以及一个或多个用于文件系统元数据信息管理的 JournalNode 节点。YARN 包括一个 ResourceManager 主节点和多个 NodeManager 从节点。MapReduce 专门进行数据分析与计算。

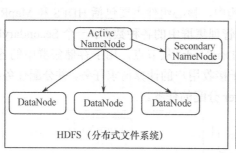 

图 1-6　Hadoop 2.x NameNode 单节点与 ResourceManager 高可用架构

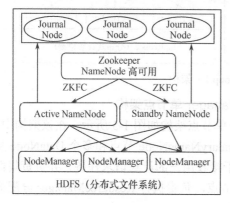 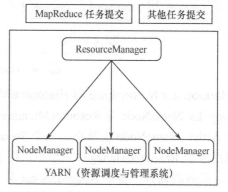

图 1-7　Hadoop 2.x NameNode 高可用与 ResourceManager 单节点架构

**5. Hadoop 2.x NameNode 高可用与 ResourceManager 高可用架构**

Hadoop 2.x NameNode 高可用与 ResourceManager 高可用架构如图 1-8 所示。在该架构中，HDFS 包括两个或两个以上 NameNode 主节点（实现高可用性），一个或多个 JournalNode 节点和多个 DataNode 从节点。YARN 包括两个或两个以上 ResourceManager 主节点和多个 NodeManager 从节点。MapReduce 专门进行数据分析与计算。

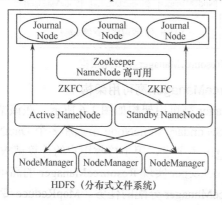 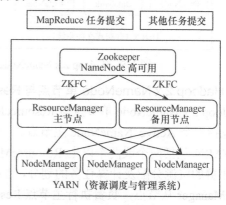

图 1-8　Hadoop 2.x NameNode 高可用与 ResourceManager 高可用架构

# 本 章 小 结

本章首先介绍了大数据技术的发展、大数据基本特征、大数据关键技术及大数据技术在各个领域的应用，然后介绍了 Hadoop 的发展历程、Hadoop 特性、Hadoop 核心组件、Hadoop 生态圈组件及 Hadoop 架构。通过本章学习，读者可以了解大数据基本特征、大数据业务处理历程、大数据关键技术及大数据技术应用，了解 Hadoop 大数据平台的发展、特性、核心组件、生态圈及 Hadoop 架构，为后续课程的学习奠定基础。

## 思考题与习题

### 1. 简答题
（1）简述大数据业务处理的基本流程与关键技术。
（2）简述大数据的特征。
（3）简述大数据在不同领域的应用。
（4）简述 Hadoop 核心组件、生态圈组件及其作用。
（5）简述 Hadoop 架构模型。

### 2. 选择题
（1）下列（　　）不是大数据发展的技术支撑。
A. 存储设备容量不断增加　　　　　B. 网络带宽不断增加
C. CUP 处理能力大幅提升　　　　　D. 互联网数据量增加

（2）关于大数据特征描述不正确的是（　　）。
A. 数据量大　　B. 数据价值密度高　　C. 数据类型多样　　D. 数据产生处理速度快

（3）关于大数据应用，描述不正确的是（　　）。
A. 大数据可以应用于交通、医疗、金融等行业
B. 利用大数据技术可以进行预测
C. 利用大数据技术可以分析疾病发展
D. 利用大数据技术可以研究事物的因果关系

（4）大数据技术的核心技术是（　　）。
A. 分布式存储与分布式计算　　　　B. 分布式数据采集与分布式计算
C. 并行计算与并行可视化　　　　　D. 分布式计算与分布式控制

（5）下列不属于分布式存储技术的是（　　）。
A. HDFS　　　　B. Oracle　　　　C. GFS　　　　D. HBase

（6）Hadoop 2.x 中，为了实现高可用性，必须首先配置（　　）。
A. YARN　　　　B. MapReduce　　　　C. Zookeeper　　　　D. HDFS

（7）在 Hadoop 架构中，用于日志信息采集的组件为（　　）。
A. Flume　　　　B. Kafka　　　　C. Sqoop　　　　D. Pig

（8）在 Hadoop 架构中，用于数据流处理的组件为（　　）。
A. MapReduce　　B. Storm　　　　C. Pig　　　　D. NoSQL

（9）下列不是 Hadoop 核心组件的是（　　）。
A. YARN　　　　B. MapReduce　　　　C. HBase　　　　D. Hadoop Common

（10）Hadoop 架构中，负责分布式计算的组件为（　　）。
A. HDFS　　　　B. YARN　　　　C. Storm　　　　D. MapReduce

# 第 2 章  Hadoop 平台和开发环境的安装与配置

Hadoop 平台的安装与配置是进行 Hadoop 大数据平台学习的基础，Eclipse 开发环境的安装与配置是 Hadoop 大数据开发的利器。本章将介绍 Hadoop 平台和开发环境的安装与配置。
- VMware 安装与配置：介绍 VMware 虚拟机的安装与配置。
- Ubuntu 安装与配置：介绍在 VMware 中安装 Ubuntu 操作系统。
- 伪分布式 Hadoop 平台的安装与配置：介绍在虚拟机中配置伪分布式 Hadoop 平台。
- Eclipse 开发环境的安装与配置：介绍 Maven 安装与配置、Eclipse 安装与配置。

## 2.1  Hadoop 平台安装准备

本书中 Hadoop 平台相关软件版本如下。
- Windows 操作系统：Windows 10。
- 虚拟机软件：VMware Workstation 15.5.6。
- Linux 系统：Ubuntu-18.04.5。
- Hadoop：Hadoop 3.1.2。

首先在 Windows 操作系统中安装 VMware Workstation 15.5.6，然后在 VMware Workstation 15.5.6 中创建 Ubuntu-18.04.5 虚拟机环境，然后在虚拟机中安装、配置 Hadoop 3.1.2，构建一个 Hadoop 平台。

### 2.1.1  VMware 安装与配置

VMware 安装与配置的基本步骤为：

（1）右键单击下载的安装程序，在弹出菜单中选择"以管理员身份运行"，打开 VMware 的安装向导，如图 2-1 所示。单击"下一步"按钮，进入图 2-2 所示的"最终用户许可协议"界面。

图 2-1  VMware 安装向导

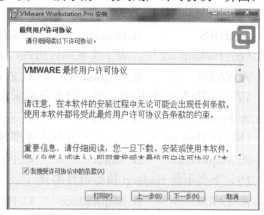

图 2-2  "最终用户许可协议"界面

（2）选中"我接受许可协议中的条款"选项后，单击"下一步"按钮，进入图 2-3 所示的"自定义安装"界面。

（3）单击"更改"按钮，选择 VMware 安装位置后，单击"下一步"按钮，进入图 2-4 所示的"用户体验设置"界面。

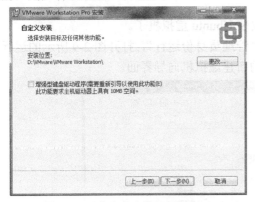 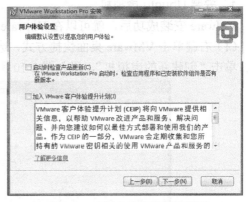

图 2-3 "自定义安装"界面　　　　　图 2-4 "用户体验设置"界面

（4）不勾选"启动时检查产品更新"和"加入 VMware 客户体验提升计划"选项，单击"下一步"按钮，进入图 2-5 所示的"快捷方式"界面。

（5）选中"桌面"和"开始菜单程序文件夹"选项后，单击"下一步"按钮，进入图 2-6 所示的"已准备好安装 VMware Workstation Pro"界面。

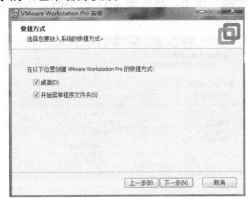 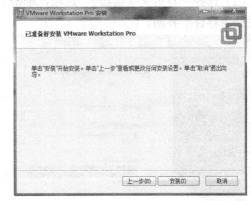

图 2-5 "快捷方式"界面　　　　　图 2-6 "已准备好安装 VMware Workstation Pro"界面

（6）单击"安装"按钮，进行 VMware 安装。安装结束后，可以退出安装，或进行许可认证，如图 2-7 所示。单击"许可证"按钮，进入图 2-8 所示的"输入许可证密钥"界面。

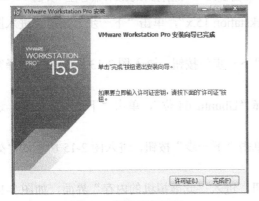 

图 2-7 安装结束界面　　　　　图 2-8 "输入许可证密钥"界面

（7）输入许可证密钥后，单击"输入"按钮完成 VMware 产品注册。

## 2.1.2　Ubuntu 安装与配置

### 1．安装 Ubuntu 虚拟机

VMware 安装成功后，就可以在 Vmware 中安装 Ubuntu 虚拟机了。

（1）右键单击 VMware 桌面快捷方式，选择"管理员身份运行"，打开图 2-9 所示的运行界面。单击"创建新的虚拟机"，进入图 2-10 所示的新建虚拟机向导界面。

图 2-9　VMware 运行界面

（2）选择"自定义（高级）"选项，单击"下一步"按钮，进入图 2-11 所示的"选择虚拟机硬件兼容性"界面。

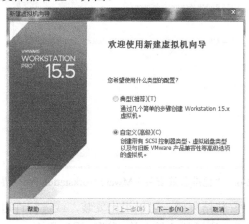

图 2-10　新建虚拟机向导界面

图 2-11　"选择虚拟机硬件兼容性"界面

（3）在"硬件兼容性"下拉菜单中选择"Workstation 15.x"，单击"下一步"按钮，进入图 2-12 所示的"安装客户机操作系统"界面。

（4）选择"稍后安装操作系统"选项，单击"下一步"按钮，进入图 2-13 所示的"选择客户机操作系统"界面。

（5）客户机操作系统选择"Linux"，版本选择"Ubuntu 64 位"，单击"下一步"按钮，进入图 2-14 所示的"命名虚拟机"界面。

（6）设置虚拟机名称和虚拟机安装位置后，单击"下一步"按钮，进入图 2-15 所示的"处理器配置"界面。

（7）完成处理器配置后，单击"下一步"按钮，进入"此虚拟机的内存"界面，如图 2-16 所示。

（8）完成虚拟机内存设置后，单击"下一步"按钮，进入图 2-17 所示的"网络类型"界面。

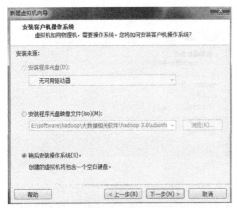
图2-12 "安装客户机操作系统"界面

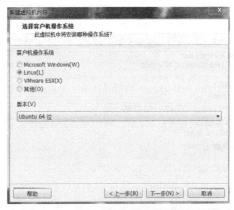
图2-13 "选择客户机操作系统"界面

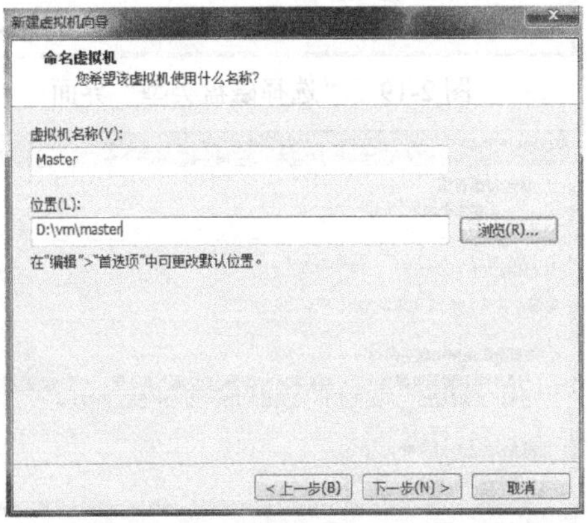

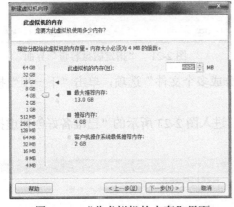
图2-14 "命名虚拟机"界面

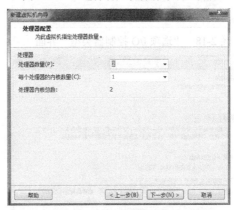
图2-15 "处理器配置"界面

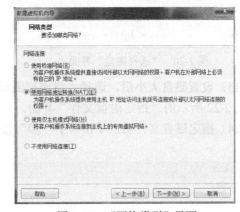
图2-16 "此虚拟机的内存"界面

图2-17 "网络类型"界面

（9）选择"使用网络地址转换（NAT）"选项，单击"下一步"按钮，进入图2-18所示的"选项I/O控制器类型"界面。

（10）选择"LSI Logic"选项，单击"下一步"按钮，进入图2-19所示的"选择磁盘类型"界面。

（11）选择"SCSI"选项，单击"下一步"按钮，进入图2-20所示的"选择磁盘"界面。

（12）选择"创建新虚拟磁盘"选项，单击"下一步"按钮，进入图2-21所示的"指定磁盘容量"界面。

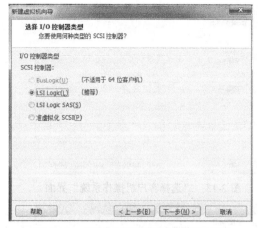
图 2-18 "选项 I/O 控制器类型"界面

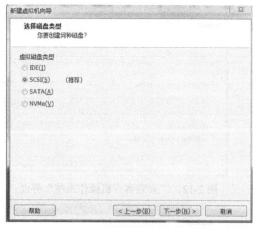
图 2-19 "选择磁盘类型"界面

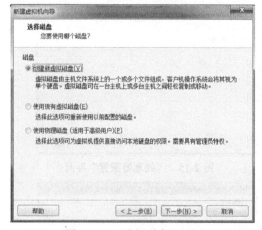
图 2-20 "选择磁盘"界面

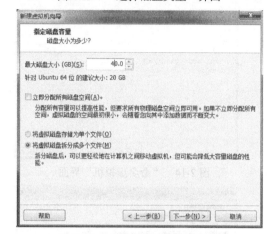
图 2-21 "指定磁盘容量"界面

（13）设置磁盘大小后，选择"将虚拟磁盘拆分成多个文件"选项，单击"下一步"按钮，进入图 2-22 所示的"指定磁盘文件"界面。

（14）指定磁盘文件后，单击"下一步"按钮，进入图 2-23 所示的"已准备好创建虚拟机"界面。

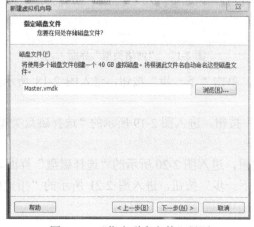
图 2-22 "指定磁盘文件"界面

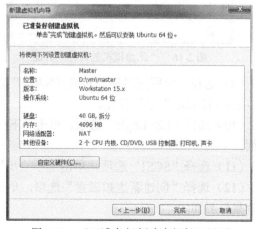
图 2-23 "已准备好创建虚拟机"界面

（15）单击"完成"按钮，将在 VMware 界面中显示创建好的虚拟机 Master，如图 2-24 所示。单击"编辑虚拟机设置"选项，进入图 2-25 所示的"虚拟机设置"界面。

图 2-24　VMware 虚拟机界面

图 2-25　"虚拟机设置"界面

（16）选择图 2-25 左侧列表框中的"处理器"，在右侧"虚拟化引擎"中选择"虚拟化 Intel VT-x/EPT 或 AMD-V/RVI"选项，如图 2-26 所示。

（17）选择图 2-26 左侧列表框中的"CD/DVD(SATA)"，在右侧"连接"中选择"使用 IOS 映像文件"选项，单击"浏览"按钮，选择下载的 Ubuntu 镜像文件 ubuntu-18.04.5-desktop-amd64.iso，如图 2-27 所示。

图 2-26　虚拟机处理器设置界面

图 2-27　虚拟机 CD/DVD 设置界面

（18）单击"确定"按钮，完成虚拟机的设置。

（19）单击图 2-24 中的"开启此虚拟机"选项，准备安装系统。等待一段时间后，进入图 2-28 所示的"欢迎"界面。

（20）在图 2-28 左侧列表框中选择"中文（简体）"后，单击"安装 Ubuntu"按钮，进入图 2-29 所示的"键盘布局"界面。

（21）选择"汉语"后，单击"继续"按钮，进入图 2-30 所示的"更新和其他软件"界面。

（22）选中"正常安装"选项，勾选"为图形或无线硬件，以及其他媒体格式安装第三方软件"选项后，单击"继续"按钮，进入图 2-31 所示的"安装类型"界面。

（23）选中"清除整个磁盘并安装 Ubuntu"选项后，弹出图 2-32 所示的"将改动写入磁盘吗？"提示界面，单击"继续"按钮，回到图 2-31。单击"现在安装"按钮，进入"您在什么地方？"的安装界面。输入"Shanghai"后，单击"继续"按钮，进入图 2-33 所示的"您是谁？"界面。

图 2-28 "欢迎"界面

图 2-29 "键盘布局"界面

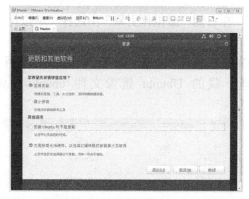

图 2-30 "更新和其他软件"界面

图 2-31 "安装类型"界面

图 2-32 "将改动写入磁盘吗？"提示界面

图 2-33 "您是谁？"界面

（24）设置好用户名、计算机名及密码后，选中"登录时需要密码"选项，单击"继续"按钮，开始 Ubuntu 虚拟机的安装。

（25）安装时间比较长，需要耐心等候。安装结束后，提示重启虚拟机。重启虚拟机后，输入用户密码，打开图 2-34 所示的 Ubuntu 系统主界面。

## 2．安装 VMware Tools 工具

为了方便实现 Windows 主机和 Ubuntu 虚拟机之间进行文件拖拽、文件共享，以及使用剪贴板进行信息的复制与粘贴，需要在 Ubuntu 虚拟机中安装 VMware Tools 工具。

图 2-34　Ubuntu 系统主界面

（1）虚拟机启动后，在图 2-35 所示 Ubuntu 系统主界面菜单中单击"虚拟机"，选择"安装 VMware Tools"选项，进入图 2-36 所示界面，加载虚拟光驱。

图 2-35　安装 VMware Tools 工具

图 2-36　加载虚拟光驱

（2）双击桌面上的"VMware Tools"快捷图标，打开图 2-37 所示的虚拟光驱文件。

（3）右键单击文件"VMwareTools-10.3.21-14772444.tar.gz"，在弹出菜单中选择"复制到"选项，如图 2-38 所示，选择复制目的地为"桌面"文件夹。

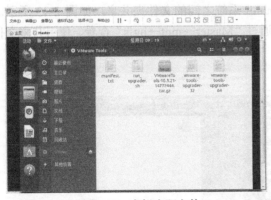

图 2-37　虚拟光驱文件

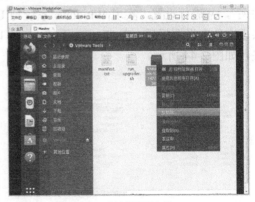

图 2-38　复制压缩文件

（4）打开"桌面"文件夹，可以看到文件"VMwareTools-10.3.21-14772444.tar.gz"，如图 2-39 所示。右键单击该文件，在弹出菜单中选择"提取到此处"选项，将文件解压。

（5）解压后的结果如图 2-40 所示。展开解压后的文件夹，如图 2-41 所示。

图 2-39　解压文件　　　　　　　　　图 2-40　解压后结果

（6）右键单击窗口空白位置，在弹出菜单中选择"在终端打开"选项，打开一个终端，如图 2-42 所示，执行"sudo ./ vmware-install.pl"命令，安装 VMware Tools 工具。除第一次交互输入 yes 外，其他都采用默认设置即可。

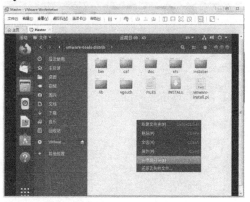

图 2-41　展开解压后的文件夹　　　　　图 2-42　安装 VMware Tools 工具

安装结束后，重新启动虚拟机，就可以从 Windows 系统拖拽文件到 Ubuntu 系统中，也可以共享剪贴板，实现 Windows 系统与 Ubuntu 系统的信息交换。

此外，还可以在"虚拟机设置"界面（见图 2-25）中，选中"选项"标签页中的"共享文件夹"，并进行设置，实现 Windows 系统与 Ubuntu 系统之间的文件夹共享，如图 2-43 所示。共享文件夹设置好后，在 Ubuntu 系统中，可以通过"Master→mnt→hgfs"查看、使用共享文件夹，如图 2-44 所示。

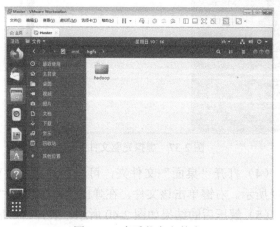

图 2-43　共享文件夹设置　　　　　　　图 2-44　查看共享文件夹

## 2.1.3 网络配置

VMware 提供了 3 种网络工作模式：桥接模式（Bridged）、网络地址转换模式（Network Address Translation，NAT）和仅主机模式（Host-Only）。

打开 VMware 虚拟机，在"编辑"菜单中选择"虚拟网络编辑器"选项，打开"虚拟网络编辑器"界面，可以看到 VMnet0（桥接模式）、VMnet1（仅主机模式）、VMnet8（NAT 模式），如图 2-45 所示。其实，VMnet0 表示的是用于桥接模式下的虚拟交换机，VMnet1 表示的是用于仅主机模式下的虚拟交换机，VMnet8 表示的是用于 NAT 模式下的虚拟交换机。

同时，在主机上对应有 VMware Network Adapter VMnet1 和 VMware Network Adapter VMnet8 两块虚拟网卡，如图 2-46 所示，分别作用于仅主机模式与 NAT 模式。在"网络连接"中，如果将这两块网卡卸载了，可以在 VMware 的"虚拟网络编辑器"界面中单击"还原默认设置"选项，重新将虚拟网卡还原。

图 2-45 "虚拟网络编辑器"界面

图 2-46 网络连接中的虚拟网卡

### 1. 桥接模式

桥接模式是指主机网卡与虚拟机的虚拟网卡利用虚拟网桥进行通信。通过虚拟网桥，将主机网卡与虚拟机的虚拟网卡都连接到虚拟交换机 VMnet0 上，因此桥接模式的虚拟机与主机之间可以相互通信。虚拟机 IP 地址必须与主机在同一个网段，且子网掩码相同。如果虚拟机需要联网，则网关和 DNS 需要与主机网卡一致。其网络结构如图 2-47 所示。

桥接模式设置的步骤如下。

（1）设置虚拟机连接模式。虚拟机安装完成后，在开启虚拟机之前，单击"编辑虚拟机设置"，打开"虚拟机设置"界面，选择"硬件"标签页中的"网络适配器"，在"网络连接"中选择"桥接模式（B）：直接连接物理网络"选项，然后单击"确定"按钮。如图 2-48 所示。

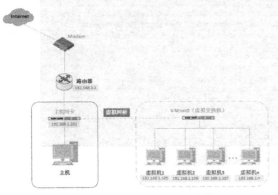

图 2-47 桥接模式网络结构图

图 2-48 桥接模式网络连接设置

（2）确认主机 IP 地址、网关、DNS 等信息。在主机的"网络连接"界面中，右键单击当前的网络连接，选择"状态"选项，在弹出的状态对话框中单击"详细信息"按钮，可以查看主机网络连接详细信息，如图 2-49 所示。

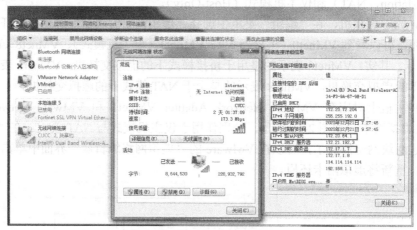

图 2-49　主机网络连接详细信息

（3）设置虚拟机网络。启动虚拟机后，编辑虚拟机网卡配置文件/etc/netplan/01-network-manager-all.yaml。

$ sudo gedit /etc/netplan/01-network-manager-all.yaml

写入下列内容：

network:
　version: 2
　renderer: NetworkManager
　ethernets:
　　ens33:　　　　　　#配置的网卡名称
　　　dhcp4: no　　　#关闭dhcp4
　　　dhcp6: no　　　#关闭dhcp6
　　　addresses: [172.20.72.205/24]　　#设置网卡IP地址和掩码
　　　gateway4 : 172.20.64.1#设置网关
　　　nameservers:　　　　　　#设置域名服务器DNS
　　　　addresses: [114.114.114.114,172.17.1.7]

（4）配置完成后，保存并退出，执行 sudo netplan apply 命令，可以让配置直接生效。测试与主机的连接，如图 2-50 所示。

```
sfd@Master:~$ sudo netplan apply
sfd@Master:~$ ifconfig
ens33: flags=4163<UP,BROADCAST,RUNNING,MULTICAST>  mtu 1500
        inet 172.20.72.205  netmask 255.255.255.0  broadcast 172.20.72.255
        inet6 fe80::20c:29ff:fe2c:63fb  prefixlen 64  scopeid 0x20<link>
        ether 00:0c:29:2c:63:fb  txqueuelen 1000  (以太网)
        RX packets 729  bytes 112617 (112.6 KB)
        RX errors 0  dropped 0  overruns 0  frame 0
        TX packets 740  bytes 115908 (115.9 KB)
        TX errors 0  dropped 0 overruns 0  carrier 0  collisions 0

sfd@Master:~$ ping 172.20.72.204
PING 172.20.72.204 (172.20.72.204) 56(84) bytes of data.
64 bytes from 172.20.72.204: icmp_seq=1 ttl=128 time=0.370 ms
64 bytes from 172.20.72.204: icmp_seq=2 ttl=128 time=0.361 ms
```

图 2-50　桥接模式测试

## 2．NAT 模式

NAT 模式是指虚拟机借助虚拟 NAT 设备和虚拟 DHCP 服务器，通过主机所在的网络访问外网，其网络结构如图 2-51 所示。采用 NAT 模式最大的优势是，虚拟机接入互联网非常简单，不需要进行任何其他配置，只需要主机能访问互联网即可。

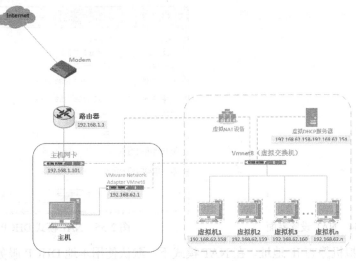

图 2-51　NAT 模式网络结构图

在 NAT 模式中，虚拟机将虚拟 NAT 设备和虚拟 DHCP 服务器连接到 VMnet8 虚拟交换机上，同时将主机上的虚拟网卡 VMware Network Adapter VMnet8 连接到 VMnet8 虚拟交换机上。需要注意的是，虚拟网卡 VMware Network Adapter VMnet8 只是作为主机与虚拟机通信的接口，虚拟机并不是依靠虚拟网卡 VMware Network Adapter VMnet8 来访问外网的。

NAT 模式的设置步骤如下。

（1）设置虚拟机连接模式。虚拟机安装完成后，在开启虚拟机之前，单击"编辑虚拟机设置"，打开"虚拟机设置"界面，选择"硬件"标签页中的"网络适配器"，在"网络连接"中选择"NAT 模式（N）：用于共享主机的 IP 地址"选项，然后单击"确定"按钮，如图 2-52 所示。

（2）设置 NAT 参数及 DHCP 参数。启动虚拟机后，选择"编辑"菜单中的"虚拟网络编辑器"选项，打开"虚拟网络编辑器"界面，选择"VMnet8"，如图 2-53 所示。

图 2-52　NAT 模式网络连接设置

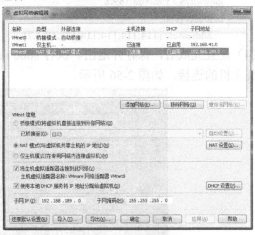

图 2-53　NAT 模式

单击"NAT 设置"按钮，打开 NAT 设置界面，如图 2-54 所示。可以进行 NAT 设置，查看虚拟机的子网 IP、子网掩码、网关等信息。

在图 2-53 所示界面中，单击"DHCP 设置"按钮，打开 DHCP 设置界面，如图 2-55 所示。可以进行 DHCP 设置，设置虚拟机 IP 地址的有效范围。

图 2-54　NAT 设置界面 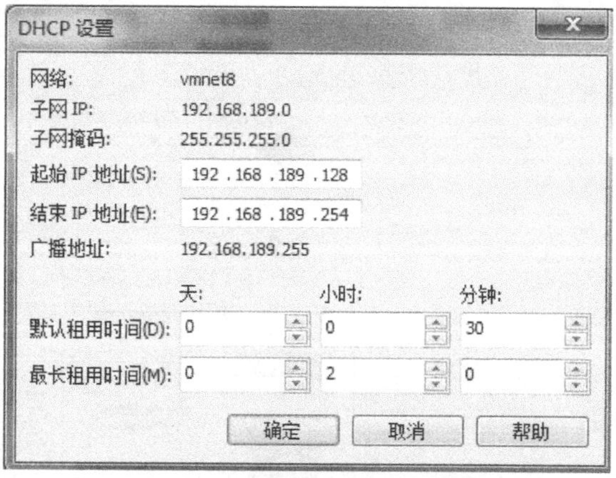　　　　　　　图 2-55　NAT 模式 DHCP 设置界面

（3）设置虚拟机静态 IP 地址。在 NAT 模式下，默认使用本地 DHCP 服务将 IP 地址分配给虚拟机，不需要额外设置。如果要为虚拟机设置静态 IP 地址，首先取消勾选图 2-53 中"使用本地 DHCP 服务将 IP 地址分配给虚拟机"选项。然后，编辑虚拟机网卡配置文件/etc/netplan/01-network-manager-all.yaml。

$ sudo gedit /etc/netplan/01-network-manager-all.yaml

写入下列内容：
network:
　version: 2
　renderer: NetworkManager
　ethernets:
　　ens33:
　　　dhcp4: no
　　　dhcp6: no
　　　addresses: [192.168.189.128/24]
　　　gateway4 : 192.168.189.2
　　　nameservers:
　　　　addresses: [114.114.114.114,172.17.1.7]

（4）配置完成后，保存并退出。执行 sudo netplan apply 命令，可以让配置直接生效。测试与外部主机的连接，如图 2-56 所示。

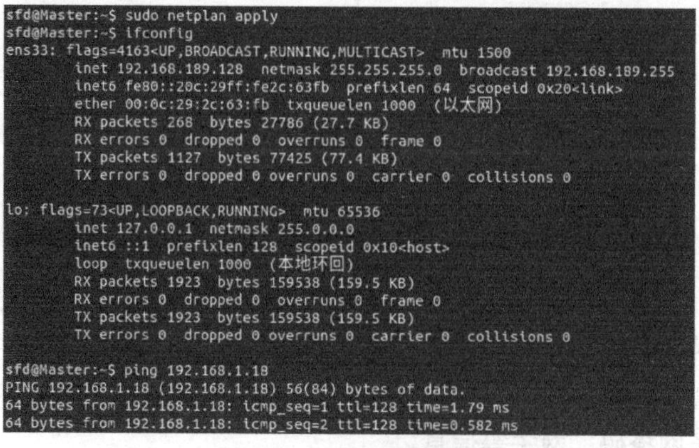

图 2-56　NAT 模式测试

## 3. 仅主机模式

仅主机模式其实就是 NAT 模式去除虚拟 NAT 设备，然后使用 VMware Network Adapter VMnet1 虚拟网卡连接 VMnet1 虚拟交换机来与虚拟机通信。所有虚拟机系统与主机都可以相互通信，但虚拟机系统与真实的网络是隔离的。仅主机模式网络结构如图 2-57 所示。

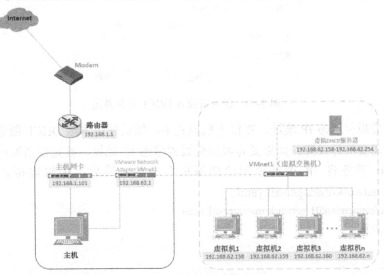

图 2-57　仅主机模式网络结构图

仅主机模式的设置步骤如下。

（1）设置虚拟机连接模式。虚拟机安装完成后，在开启虚拟机之前，单击"编辑虚拟机设置"，打开"虚拟机设置"界面，选择"硬件"标签页中的"网络适配器"，在"网络连接"中选择"仅主机模式（H）：与主机共享的专用网络"选项，然后单击"确定"按钮，如图 2-58 所示。

（2）设置 DHCP 参数。启动虚拟机后，选择"编辑"菜单中的"虚拟网络编辑器"选项，打开"虚拟网络编辑器"窗口，选择"VMnet1"，如图 2-59 所示。

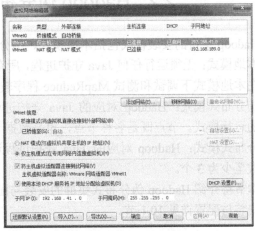

图 2-58　仅主机模式网络连接设置　　　　图 2-59　仅主机模式

单击"DHCP 设置"按钮，打开 DHCP 设置界面，如图 2-60 所示。可以进行 DHCP 设置，设置虚拟机 IP 地址的有效范围。

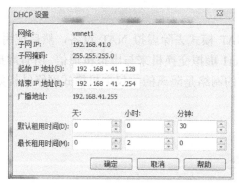

图 2-60 仅主机模式 DHCP 设置界面

（3）设置虚拟机静态 IP 地址。在仅主机模式下，默认使用本地 DHCP 服务将 IP 地址分配给虚拟机，不需要额外设置。如果要为虚拟机设置静态 IP 地址，首先取消勾选图 2-59 中"使用本地 DHCP 服务将 IP 地址分配给虚拟机"选项。然后，编辑虚拟机网卡配置文件 /etc/netplan/01-network-manager-all.yaml。

```
$ sudo gedit /etc/netplan/01-network-manager-all.yaml
```

写入下列内容：

```
network:
  version: 2
  renderer: NetworkManager
  ethernets:
    ens33:
      dhcp4: no
      dhcp6: no
      addresses: [192.168.41.128/24]
      gateway4: 192.168.41.1
```

（4）配置完成后，保存并退出。执行 sudo netplan apply 命令，可以让配置直接生效。

## 2.2　Hadoop 伪分布式集群安装与配置

Hadoop 支持的安装模式包括本地模式、伪分布式模式和分布式模式。

**本地模式**：无须运行任何 Java 守护进程，所有程序都在同一个 JVM（Java 虚拟机）上执行。由于在本地模式下调试和测试 MapReduce 程序较为方便，因此，这种模式适合用于开发阶段。

**伪分布式模式**：Hadoop 对应的 Java 守护进程都运行在一个物理机器上，模拟一个小规模集群的运行模式。节点既是名称节点，也是数据节点，读取的是 HDFS。

**分布式模式**：Hadoop 对应的 Java 守护进程运行在多个节点构成的集群环境中。通常集群节点数至少为 3 个。

本节将介绍 Hadoop 伪分布式集群的安装与配置。关于基于 4 个虚拟机的 Hadoop 分布式集群的安装与配置详见 10.1 节。

### 2.2.1　创建用户 hadoop

（1）如果安装 Ubuntu 18 时没有创建用户 hadoop，那么需要创建一个名为 hadoop 的用户，专门用于 Hadoop 集群操作。可以执行下列命令查看用户 hadoop 是否存在：

```
$ cat  /etc/passwd |grep hadoop
```

（2）如果用户 hadoop 不存在，则创建用户 hadoop，继续执行步骤（3）；如果用户 hadoop 存在，则执行步骤（4）。

（3）创建用户 hadoop，并设置密码为 hadoop。

$ sudo useradd -m hadoop -s /bin/bash

设置用户 hadoop 的密码为 hadoop（注意：密码不回显）：

$ sudo passwd hadoop

（4）将用户 hadoop 添加到 sudo 组中，为用户授权。

$ sudo adduser   hadoop sudo

### 2.2.2 修改主机名与域名映射

（1）修改虚拟机主机名称为 master。

编辑/etc/hostname 文件，写入虚拟机主机名称"master"：

$ sudo gedit /etc/hostname

（2）将虚拟机的 IP 地址与主机名称写入/etc/hosts 中，完成域名映射的添加。

$ sudo   gedit   /etc/hosts

在文件末尾添加一行下列信息：

192.168.189.128 master

其中，192.168.189.128 为虚拟机 IP 地址，master 为主机名称。修改完后，保存文件。

（3）重新启动虚拟机，以用户 hadoop 登录系统。

$ sudo reboot

### 2.2.3 SSH 免密码登录设置

为了实现 Hadoop 集群节点之间的 SSH 免密码登录，需要进行 SSH 免密码登录设置。

（1）在虚拟机上安装 SSH。

执行下列命令安装 SSH：

$ sudo apt-get update

$ sudo   apt-get   installopenssh-server

（2）在虚拟机上生成公钥和私钥。

$ ssh-keygen   -t   rsa

执行过程如图 2-61 所示（确认环节按回车键即可）。

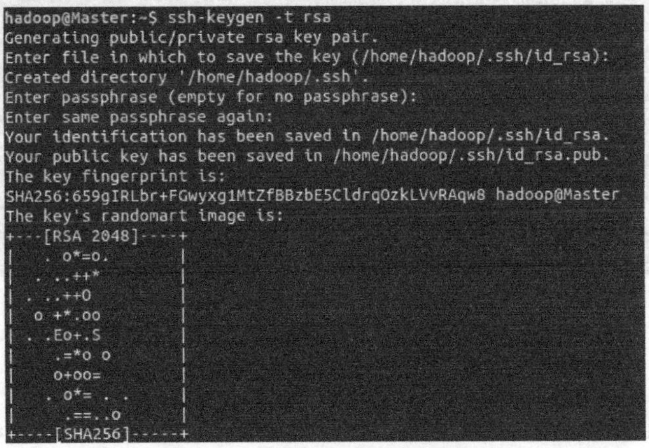

图 2-61  虚拟机密钥生成过程

执行完成后，在~/目录下（/home/hadoop）自动创建目录.ssh，内部包含 id_rsa（私钥）、id_rsa.pub（公钥）两个文件。

（3）将虚拟机的公钥发送到要登录节点的.ssh/authorized_keys 文件中。

```
$ cd  ~/.ssh
$ ssh-copy-id  -i  id_rsa.pub  hadoop@master
```

执行过程如图 2-62 所示。

```
hadoop@Master:~$ cd .ssh
hadoop@Master:~/.ssh$ ssh-copy-id -i id_rsa.pub hadoop@Master
/usr/bin/ssh-copy-id: INFO: Source of key(s) to be installed: "id_rsa.pub"
The authenticity of host 'master (127.0.1.1)' can't be established.
ECDSA key fingerprint is SHA256:cou80nm2aydhbwXB/BhGiPBBwYfHGf6YQ4wdlqQQ7Iw.
Are you sure you want to continue connecting (yes/no)? yes
/usr/bin/ssh-copy-id: INFO: attempting to log in with the new key(s), to filter out any that are already installed
/usr/bin/ssh-copy-id: INFO: 1 key(s) remain to be installed -- if you are prompted now it is to install the new keys
hadoop@master's password:

Number of key(s) added: 1

Now try logging into the machine, with:   "ssh 'hadoop@Master'"
and check to make sure that only the key(s) you wanted were added.
```

图 2-62　公钥发送到登录节点

此时，查看登录节点的~/.ssh/authorized_keys 文件，可以看到相应的客户端节点信息，如图 2-63 所示。

```
hadoop@Master:~/.ssh$ cat authorized_keys
ssh-rsa AAAAB3NzaC1yc2EAAAADAQABAAABAQCxt3TIv+AZmaFPxbw0E0yFazVdfxHndliWFnq7CP0PfOk3493zb8UU8TWfCce
8OVrG+HX8OOXe8imYu/J7FpyNHhx+gDWDgOlRCJ4YIJBze2UQjdazkH1th6Z1zu08XL1l/Fieez80daQQve94Kr89PWYEL0mipG
N6KCReeoB2utIXhbslwQ4JDBzKS40LWRs5lFfrimh+/k6pOQXRKBg118tCpPnjJ6js0Mv72YAfVvwmpBRD3IBCCEepOLve+bnV+
xKX8LLyFz2qVECjI+P/gGv7HxfOJ7zfxKYDeuCfMd1NRSl8kwpZ4Xl9XTanR1b/kmr5t27PP/u3km2HygSn hadoop@Master
```

图 2-63　authorized_keys 文件中的客户端节点信息

（4）测试 SSH 免密码登录。

```
$ ssh master
```

执行过程如图 2-64 所示，测试成功后，必须执行 exit 命令结束远程登录。

```
hadoop@Master:~/.ssh$ ssh hadoop@Master
Welcome to Ubuntu 18.04.5 LTS (GNU/Linux 5.4.0-58-generic x86_64)

 * Documentation:  https://help.ubuntu.com
 * Management:     https://landscape.canonical.com
 * Support:        https://ubuntu.com/advantage


 * Canonical Livepatch is available for installation.
   - Reduce system reboots and improve kernel security. Activate at:
     https://ubuntu.com/livepatch

46 个可升级软件包。
0 个安全更新。

New release '20.04.1 LTS' available.
Run 'do-release-upgrade' to upgrade to it.

Your Hardware Enablement Stack (HWE) is supported until April 2023.
hadoop@Master:~$ exit
注销
Connection to master closed.
```

图 2-64　SSH 免密登录测试

## 2.2.4　安装 Java 环境

由于 Hadoop 是使用 Java 语言开发的，因此 Hadoop 平台运行需要 Java 环境。

（1）在目录/usr/lib 中创建 jvm 目录，并将目录所有者修改为用户 hadoop。

```
$ sudo mkdir /usr/lib/jvm/
```

$ sudo chown-R hadoop /usr/lib/jvm

（2）使用 tar 命令解压安装 jdk-8u121-linux-x64.tar.gz 文件到目录/usr/lib/jvm 中。

$ cd ~/Downloads       #进入dk-8u121-linux-x64.tar.gz文件所在目录
$ sudo tar -zxvf jdk-8u231-linux-x64.tar.gz -C /usr/lib/jvm   #解压文件到/usr/lib/jvm中

（3）配置 JDK 环境变量，并使其生效。

① 使用 gedit 命令打开用户的配置文件.bashrc。

$ gedit ~/.bashrc

② 在文件中加入下列内容：

export JAVA_HOME=/usr/lib/jvm/jdk1.8.0_231
export JRE_HOME=$JAVA_HOME/jre
export PATH=$JAVA_HOME/bin:$JAVA_HOME/jre/bin:$PATH
export CLASSPATH=$CLASSPATH:.:$JAVA_HOME/lib:$JAVA_HOME/jre/lib

③ 使环境变量生效。

$ source ~/.bashrc

④ 验证 JDK 是否安装成功。

$ java    -version

如果打印出 Java 版本信息，则 Java 环境安装成功，如图 2-65 所示。

```
hadoop@Master:/usr/lib/jvm$ java -version
java version "1.8.0_231"
Java(TM) SE Runtime Environment (build 1.8.0_231-b11)
Java HotSpot(TM) 64-Bit Server VM (build 25.231-b11, mixed mode)
```

图 2-65   Java 版本信息

## 2.2.5   伪分布式集群安装与配置

（1）使用 tar 命令解压安装 hadoop-3.1.2.tar.gz 文件到目录/usr/local 中，并将文件夹重命名为 hadoop。

$ cd ~/Downloads          #进入源文件hadoop-3.1.2.tar.gz所在目录
$ sudo tar -zxvf hadoop-3.1.2.tar.gz-C /usr/local   #解压文件到/usr/local中
$ cd /usr/local
$ sudo mv hadoop-3.1.2 hadoop   #为简化操作，将文件夹重命名为hadoop

（2）将目录/usr/local/hadoop 的所有者修改为用户 hadoop。

$ sudo chown -R hadoop /usr/local/hadoop

（3）配置环境变量，并使其生效。

① 使用 gedit 命令打开用户的配置文件.bashrc。

$ gedit   ~/.bashrc

② 在文件中加入下列内容：

export HADOOP_HOME=/usr/local/hadoop
export PATH=$HADOOP_HONE/bin:$HADOOP_HOME/sbin:$PATH

③ 使环境变量生效。

$ source   ~/.bashrc

（4）配置 Hadoop 文件。

配置 Hadoop 文件由只读配置文件（Read-only default）和站点指定配置文件（Site-specific）组成。其中，只读配置文件包括 core-default.xml、hdfs-default.xml、yarn-default.xml 和 mapred-default.xml，里面记录了 Hadoop 平台的一些默认的属性配置。如果用户需要对其属性值进行修改，可在站点指定配置文件 core-site.xml、hdfs-site.xml、yarn-site.xml 和 mapred-site.xml

•27•

中对属性进行重新赋值。此外，可以通过在 hadoop-env.sh 和 yarn-env.sh 文件中设置指定的值来控制 bin 目录中的 Hadoop 脚本。

在伪分布式 Hadoop 集群配置中，需要配置/usr/local/hadoop/etc/hadoop 目录中的 3 个配置文件，分别为 hadoop-env.sh、core-site.xml 和 hdfs-site.xml，其他属性可以采用只读配置文件中的默认设置。

① 配置 hadoop-env.sh 文件。

用 gedit 命令打开 hadoop-env.sh 文件：

$ gedit /usr/local/hadoop/etc/hadoop/hadoop-env.sh

将第 37 行代码"# JAVA_HOME=/usr/java/testing hdfs dfs -ls"修改为"export JAVA_HOME=/usr/lib/jvm/jdk1.8.0_231"。

② 配置 core-site.xml 文件。

用 gedit 命令打开 core-site.xml 文件：

$ gedit /usr/local/hadoop/etc/hadoop/core-site.xml

在<configuration>和</configuration>标记之间写入下列内容：

```
<property>
<name>fs.defaultFS</name>
<value>hdfs://master:9000/</value>
</property>
<property>
<name>hadoop.tmp.dir</name>
<value>file:/usr/local/hadoop/tmp</value>
</property>
```

③ 配置 hdfs-site.xml 文件。

用 gedit 命令打开 hdfs-site.xml 文件：

$ gedit /usr/local/hadoop/etc/hadoop/hdfs-site.xml

在<configuration>和</configuration>标记之间写入下列内容：

```
<property>
<name>dfs.namenode.name.dir</name>
<value>file:/usr/local/hadoop/dfs/name</value>
</property>
<property>
<name>dfs.datanode.data.dir</name>
<value>file:/usr/local/hadoop/dfs/data</value>
</property>
<property>
<name>dfs.replication</name>
<value>1</value>
</property>
```

（5）格式化。在使用集群之前，需要先在主节点上进行格式化，在主节点中生成元数据。

$ hdfs namenode -format

注意：如果格式化之后，重新修改了系统配置，或者 Hadoop 启动不了，可能需要重新进行格式化操作。重新格式化，需要首先停止 Hadoop 运行，然后删除 tmp、dfs、logs 文件夹。例如：

$ stop-dfs.sh
$ stop-yarn.sh
$ cd /usr/local/hadoop

```
$ rm -r dfs/ logs/ tmp/
$ hdfs namenode -format
```

（6）启动 Hadoop 服务。可以一次性启动 HDFS 和 YARN：

```
$ start-all.sh
```

启动过程如图 2-66 所示。

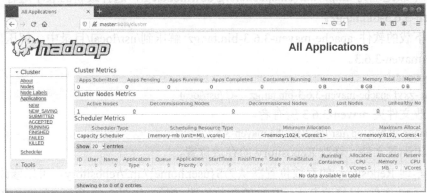

图 2-66　Hadoop 服务启动过程

可以分别启动 HDFS 和 YARN：

```
$ start-dfs.sh
$ start-yarn.sh
```

（7）查看 Hadoop 进程。Hadoop 集群启动后，可以使用 jps 命令查看 Hadoop 相关进程，如图 2-67 所示。

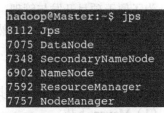

图 2-67　Hadoop 相关进程

（8）浏览器访问。通过 http://master:8088/cluster 访问 YARN 管理界面，如图 2-68 所示。

图 2-68　YARN 管理界面

通过 http://master:9870 查看 HDFS 管理界面，如图 2-69 所示。

图 2-69　HDFS 管理界面

（9）运行 MapReduce 程序。可以通过运行 Hadoop 自带的样例程序，验证 MapReduce 程序是否可以正常运行。

```
$ hadoop jar /usr/local/hadoop/share/hadoop/mapreduce/hadoop-mapreduce-examples-3.1.2.jar pi 2 1000
```

运行结果如图 2-70 所示。

```
hadoop@Master:~$ hadoop jar /usr/local/hadoop/share/hadoop/mapreduce/hadoop-mapreduce-examples-3.1
.2.jar pi 2 1000
Number of Maps  = 2
Samples per Map = 1000
Wrote input for Map #0
Wrote input for Map #1
Starting Job
2020-12-24 10:03:31,678 INFO impl.MetricsConfig: loaded properties from hadoop-metrics2.properties
Job Finished in 5.05 seconds
Estimated value of Pi is 3.14400000000000000000
```

图 2-70　MapReduce 程序运行测试

## 2.3　Eclipse 开发环境安装与配置

为了便于项目开发与管理，需要创建 Maven 工程。由于 Eclipse 内置的 Maven 版本较低，需要预先安装并配置 Maven，然后在 Eclipse 中进行 Maven 配置。

### 2.3.1　Maven 安装与配置

Maven 安装与配置的基本步骤如下。

（1）Maven 下载。

Maven 下载地址为 http://maven.apache.org/download.cgi，选择最新的版本进行下载，例如 apache-maven-3.6.3-bin.tar.gz。

（2）将下载的软件 apache-maven-3.6.3-bin.tar.gz 解压到/usr/local/目录中，解压后的文件夹名为 apache-maven-3.6.3。

```
$ cd Downloads/
$ sudo tar -zxvf apache-maven-3.6.3-bin.tar.gz   -C /usr/local
```

（3）修改文件夹 maven-3.6.3 所有者。

```
$ sudo chown -R hadoop /usr/local/apache-maven-3.6.3
```

（4）编辑配置文件，修改 Repository 的存放位置。

Maven 本地 Repository 的存放位置默认为${user.home}/.m2/repository，可以修改 conf 目录下的 settings.xml 文件，以修改 Repository 的存放位置。例如，如图 2-71 所示，在<localRepository>与</localRepository>标记中添加下列内容：

```
<localRepository>/usr/local/apache-maven-3.6.3/.m2/repository</localRepository>
```

```
<settings xmlns="http://maven.apache.org/SETTINGS/1.0.0"
          xmlns:xsi="http://www.w3.org/2001/XMLSchema-instance"
          xsi:schemaLocation="http://maven.apache.org/SETTINGS/1.0.0 http://maven.apache.org/xsd/settings-1.0.0.xsd">

  <localRepository>/usr/local/apache-maven-3.6.3/.m2/repository</localRepository>
```

图 2-71　修改 Repository 的存放位置

（5）配置环境变量，并使其生效。

① 打开.bashrc 文件：

```
$ gedit ~/.bashrc
```

将下列内容添加到.bashrc 文件中：

```
export MAVEN_HOME=/usr/local/ apache-maven-3.6.3
export PATH=$MAVEN_HOME/bin:$PATH
```

② 使配置文件生效。

保存.bashrc 文件后，执行下列命令使配置文件生效：

$ source ~/.bashrc

## 2.3.2　Eclipse 安装与配置

（1）Eclipse 下载。

Eclipse 下载网址为 https://www.eclipse.org/downloads/packages，选择最新的 Linux 版本的 Eclipse，例如 Eclipse IDE 2020-12 R Packages，下载后文件为 eclipse-inst-jre-linux64.tar.gz。

（2）将安装包解压到/usr/local 目录中。

$ cd Downloads/
$ sudo tar -zxvf eclipse-inst-jre-linux64.tar.gz -C /usr/local

（3）将解压后的文件夹 eclipse-installer 重命名为 eclipse。

$ sudo mv /usr/local/eclipse-installer/ /usr/local/eclipse

（4）修改文件夹/usr/local/eclipse 所有者为用户 hadoop。

$ sudo chown -R hadoop/usr/local/eclipse

（5）执行/usr/local/eclipse/eclipse-inst 命令，启动 eclipse installer。

$ /usr/local/eclipse/eclipse-inst

（6）在图 2-72 所示的"eclipse installer"界面中，双击"Eclipse IDE for Java Developers"选项，进入图 2-73 所示的 Eclipse 安装设置界面，设置完安装路径后，单击"INSTALL"按钮进行 Eclipse 安装。

图 2-72　"eclipse installer"界面

图 2-73　Eclipse 安装设置界面

（7）安装结束后，可以单击"LANCH"按钮启动 Eclipse。

（8）编辑配置文件.bashrc。

使用 gedit ~/.bashrc 命令打开配置文件，将下列内容添加到.bashrc 文件中：

export ECLIPSE_HOME=/usr/local/eclipse/java-2020-12/eclipse/
export PATH=$ECLIPSE_HOME:$PATH

保存文件后，执行下列命令使配置文件生效：

$ source ~/.bashrc

## 2.3.3　Eclipse 中 Maven 设置

启动 Eclipse 后，可以手动设置预先安装的 Maven 版本。

（1）执行下列命令，打开 Eclipse：

$ eclipse

（2）选择"Windows→Preferences→Maven→Installations"命令，打开"Preferences"界面，如图 2-74 所示。

(3)展开左侧的"Maven"选项,选择"Installations"选项,单击右侧的"Add"按钮,打开"New Maven Runtime"界面,如图 2-75 所示。单击"Directory"按钮,选择预先安装的 Maven 版本。最后单击"Finish"按钮,完成 Maven 版本的添加。

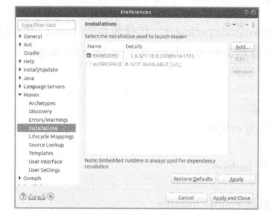
图 2-74 "Preferences"界面

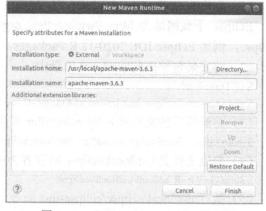
图 2-75 "New Maven Runtime"界面

(4)返回"Preferences"界面中,选择新添加的 Maven 版本,如图 2-76 所示。

(5)选择左侧的"User Settings"选项,进行用户设置,如图 2-77 所示。

图 2-76 选择新添加的 Maven 版本　　　　图 2-77 用户设置

(6)设置"Global Settings"为用户自己安装的 Maven 配置文件,设置"Local Repository"为用户自己安装的 Maven 本地库存放位置。

(7)完成用户设置后,单击"Apply and Close"按钮,完成 Maven 配置。

## 本 章 小 结

本章首先介绍了伪分布式 Hadoop 平台的安装与配置,包括 VMware 安装与配置、Ubuntu 的安装与配置、伪分布式 Hadoop 平台的完整配置过程。然后介绍了 Hadoop 平台中 Eclipse 开发环境的安装与配置,包括 Maven 安装与配置、Eclipse 安装与配置及 Eclipse 中 Maven 环境配置等。"工欲善其事,必先利其器",通过本章的学习,读者可以为后续学习奠定基础。

# 思考题与习题

## 1. 简答题
（1）Hadoop 平台运行模式有哪些？有何异同？
（2）VMware 网络连接模式有哪几种？分别适合什么场景？
（3）简述 Hadoop 伪分布式集群安装的基本步骤。
（4）简述 SSH 免密码登录的基本原理。

## 2. 选择题
（1）配置 Hadoop 时，JAVA_HOME 包含在哪一个配置文件中？（　　）

　　A. hadoop-default.xml　　　　　　B. hadoop-env.sh
　　C. hadoop-site.xml　　　　　　　 D. configuration.xsl

（2）下列不属于 Hadoop 集群的运行模式的是（　　）。

　　A. 单机模式　　B. 伪分布式模式　　C. 完全分布式模式　　D. 互联模式

（3）下列哪个程序通常与 NameNode 在同一个节点上启动？（　　）

　　A. SecondaryNameNode　　　　　　B. DataNode
　　C. Resoucemanager　　　　　　　 D. Nodemanager

（4）NameNode 在启动时自动进入安全模式，在安全模式阶段，说法错误的是（　　）。

　　A. 安全模式的目的是在系统启动时检查各个 DataNode 上数据块的有效性
　　B. 根据策略对数据块进行必要的复制或删除
　　C. 当数据块最小百分比满足最小副本数条件时，会自动退出安全模式
　　D. 文件系统允许有修改

（5）如果不同物理主机上的虚拟机需要构建集群，应采用 VMware 中的下列哪种网络配置模式？（　　）

　　A. NAT 模式　　B.仅主机模式　　C.桥接模式　　D.互联模式

## 3. 实训题
（1）安装 VMware 虚拟机软件。
（2）安装 Ubuntu 18 操作系统。
（3）进行伪分布式 Hadoop 平台安装与配置。
（4）检查伪分布式 Hadoop 平台安装是否成功（检查 Hadoop 进程及通过 Web 方式访问）。
（5）在 Ubuntu 虚拟机中配置开发环境，安装与配置 Maven 和 Eclipse。

# 第3章 分布式文件系统

分布式文件系统（Hadoop Distributed File System，HDFS）是 Hadoop 的核心组件，源于 2003 年 10 月 Google 公司的 GFS 论文，是 GFS 的开源实现，是分布式计算中数据存储管理的基础，是基于流数据模式访问和处理超大文件的需求而开发的。本章将介绍 HDFS 管理与开发的相关知识。

- HDFS 概述：介绍 HDFS 架构、设计目标、高可用架构及其优劣性。
- HDFS 工作机制：介绍 HDFS 数据存储策略、读取策略、错误与恢复及读写过程。
- HDFS Shell 管理：介绍 HDFS 文件操作命令、系统管理命令及 HDFS Shell 操作实例。
- HDFS Java 开发：介绍 HDFS Java 程序设计基础知识及 HDFS 常用的 Java API。
- HDFS 开发实例：介绍 Eclipse 中进行 HDFS 操作的 Maven 工程管理与程序设计。

## 3.1 HDFS 概述

### 3.1.1 HDFS 架构

作为 Hadoop 的核心存储技术，HDFS 在物理上由一系列的计算机节点构成，如图 3-1 所示。这些节点分为两类：一类为主节点（Master Node），又称名称节点（NameNode），用于存储 HDFS 的元数据，负责管理数据节点与数据块的映射关系；另一类为从节点（Slave Node），又称数据节点（DataNode），负责数据的存储和读取。

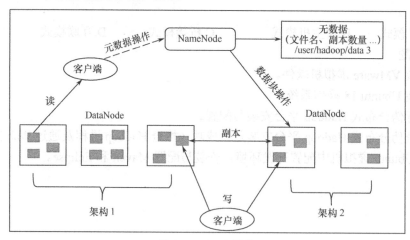

图 3-1 HDFS 架构

存储数据时，由名称节点分配文件的存储位置，然后客户端把数据直接写入相应的数据节点。读取数据时，客户端从名称节点获取数据块和数据节点的映射关系，然后到相应的数据节点访问数据块。数据块存储在数据节点本地的 Linux 系统中。

HDFS 架构包括下面几个核心概念。

（1）数据块（Block）

文件写入 HDFS 时会被切分成若干个数据块，数据块大小固定，默认大小为 128MB。数据块是 HDFS 的最小存储单元。若一个文件的大小小于数据块大小，则不会占用整个块的空间。由于数据块的大小远远大于普通文件系统的大小，因此可以最小化寻址开销。默认情况下，每个数据块有 3 个副本，这是数据可靠性的一个保障。

在 HDFS 中引入抽象数据块的逻辑结构，具有下列作用。

① 支持大规模文件存储。文件以块为单位进行存储，一个大规模文件可以被分拆成若干个数据块，不同的数据块可以被分发到不同的节点上，因此，一个文件的大小不会受到单个节点的存储容量的限制，可以远远大于网络中任意节点的存储容量。

② 简化系统设计：因为数据块大小是固定的，可以很容易计算出一个节点可以存储多少个数据块。此外，元数据不需要和数据块一起存储，可以由其他系统负责管理元数据。

③ 适合数据备份：每个数据块都可以冗余存储到多个节点上，大大提高了系统的容错性和可用性。在 HDFS 中，数据的冗余存储是以块为单位进行的。

（2）名称节点（NameNode）

名称节点用于管理 HDFS 的命名空间，维护文件目录及文件的元数据信息，记录每个文件中各个数据块所在的数据节点的位置信息，处理客户端的读写请求。

在 Hadoop 2.x 中，名称节点包括活跃名称节点（Active NameNode）和备用名称节点（Standby NameNode）两种。其中，活跃名称节点是主名称节点，负责所有客户端操作；而备用名称节点是活跃名称节点的热备节点，负责维护状态信息以便在需要时能快速切换。

名称节点主要包括 FsImage 和 EditLog 两种数据结构。

① FsImage 文件：用于维护系统文件树及文件树中所有文件和文件夹的元数据信息，如：一个文件夹下有哪些子文件夹、子文件是什么、文件名是什么，文件副本数有多少，文件由哪些块组成等。

② EditLog 文件：记录所有针对文件夹和文件的创建、删除、重命名等的操作日志。

名称节点的工作机制如图 3-2 所示。在名称节点启动时，将 FsImage 文件内容加载到内存中，然后执行 EditLog 文件中的各项操作，使得内存中的元数据保持最新。同时将最新的 FsImage 文件保存到磁盘中，并生成一个空的 EditLog 文件。名称节点进入正常运行状态后，对 HDFS 的更新操作会被写入新生成的 EditLog 文件中，而不是写入 FsImage 文件中，因为 FsImage 文件通常非常大，如果直接更新 FsImage 文件，会使系统变得非常缓慢。名称节点定期将新增的 EditLog 文件与内存中的 FsImage 文件合并保存到磁盘中。

图 3-2 名称节点的工作机制

（3）备用名称节点（Standby NameNode）

备用名称节点是 Hadoop 2.x 中引入的，是活跃名称节点的热备节点，周期性同步 EditLog 文件操作日志，定期合并 FsImage 文件与 EditLog 文件到本地磁盘中，与活跃名称节点的状态同步。当活跃名称节点出现故障时，备用名称节点切换为活跃名称节点，避免了系统的单点故障。

（4）数据节点（DataNode）

数据节点是 HDFS 中的存储节点，可以有成百上千个。数据节点存储数据块和数据校验和，执行客户端的读写请求操作，通过心跳机制定期向数据节点汇报运行状态和所有块列表信息。在集群启动时，数据节点向名称节点提供存储的数据块的信息，数据节点中的数据会被保存在各自节点的本地 Linux 文件系统中。

### 3.1.2 HDFS 设计目标

HDFS 是运行在大量廉价服务器上，为海量数据存储提供高容错性、高可靠性、高扩展性、高吞吐率的分布式文件系统。

HDFS 的设计目标为：

① HDFS 可以由成百上千个服务器构成，每个服务器上存储着文件系统的部分数据。任何一个服务器都可能失效，因此错误检测和快速、自动恢复是 HDFS 的最核心架构目标。

② HDFS 以支持大数据集合为目标。HDFS 上的一个典型文件大小一般都在 GB 至 TB 量级，一个单一的 HDFS 实例应能支撑数以千万计的文件。

③ 移动计算的代价比移动数据的代价低。一个应用请求的计算，离它操作的数据越近就越高效，这在数据达到海量级别时更是如此。将计算移动到数据附近，比之将数据移动到计算附近显然更好。HDFS 为应用提供了将计算移动到数据附近的接口的服务。

④ HDFS 应用以流式读为主进行数据的批量处理。HDFS 更关注数据访问的高吞吐量，而非关注数据访问的低延迟。

⑤ HDFS 应用对文件要求的是"一次写入，多次读取"的访问模型。一个文件经过创建、写入、关闭之后就不需要再改变。

⑥ 在异构的硬件和软件平台上具有良好的可移植性。

### 3.1.3 HDFS 高可用架构

在 HDFS 架构中，如果名称节点出现故障，整个系统将无法运行。为了解决该单点故障，在 Hadoop 2.x 中，引入了 HDFS 名称节点高可用架构。在该架构中，存在两个独立的名称节点。任何时刻，只有一个名称节点处于 Active 状态（主名称节点），另一个处于 Standby 状态（备用名称节点），如图 3-3 所示。

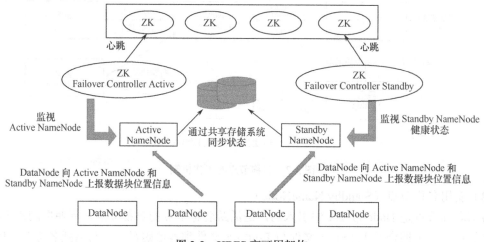

图 3-3　HDFS 高可用架构

HDFS 高可用架构中各个组件的作用如下。

① 主备切换控制器 Zookeeper（ZK）Failover Controller：作为独立进程运行，对 NameNode 的主备切换进行总体控制。主备切换控制器能及时检测到 NameNode 的健康状况，在 Active NameNode 发生故障时借助 Zookeeper 实现自动的主备选举和切换。NameNode 也支持不依赖于 Zookeeper 的手动主备切换。

② Zookeeper 集群：为主备切换控制器提供主备选举支持。

③ 共享存储系统：是实现 NameNode 高可用性最为关键的部分，共享存储系统保存了 NameNode 在运行过程中所产生的 HDFS 的元数据。Active NameNode 和 Standby NameNode 通过共享存储系统实现元数据同步。在进行主备切换时，新的 Active NameNode 在确认元数据完全同步之后才能继续对外提供服务。

④ DataNode：除通过共享存储系统共享 HDFS 的元数据信息外，Active NameNode 和 Standby NameNode 还需要共享 HDFS 的数据块与数据节点之间的映射关系。DataNode 会同时向 Active NameNode 和 Standby NameNode 上报数据块的位置信息。

在 HDFS 的高可用架构中，虽然提供了多个 NameNode，但在任一时刻只有一个 NameNode 处于活跃状态，从而整个集群的命名空间只保存在一个 Active NameNode 中，而 Active NameNode 在内存中存储了整个分布式文件系统中的元数据信息，这限制了集群中数据块、文件和文件夹的数目。从性能上来讲，单个 NameNode 资源有限，进而限制了文件操作过程的吞吐量。从业务的独立性上来讲，单个 NameNode 无法做到业务隔离，一个应用可能影响整个集群。为此，在 Hadoop 2.x 架构中引入了 HDFS Federation，即 HDFS 联邦机制，如图 3-4 所示。

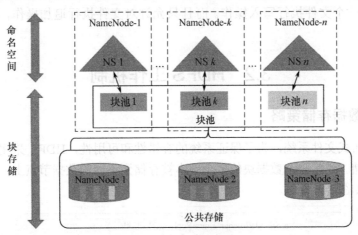

图 3-4 HDFS 联邦机制

在 HDFS Federation 中，设计了多个相互独立的 NameNode，这些 NameNode 分别进行各自命名空间和数据块的管理，相互之间是联盟（Federation）关系，不需要彼此协调。所有 NameNode 共享底层的 DataNode 存储资源，DataNode 向所有 NameNode 定时发送心跳，并处理所有 NameNode 发送的指令。

属于同一个命名空间（NameSpace，NS）的数据块构成一个"块池"（Block Pools），块池的管理是相互独立的。DataNode 保存了集群中所有的块池。

HDFS Federation 采用多名称节点机制，具有下列优点：
- 扩展了集群的命名空间，提高了集群的扩展性；
- 多个名称节点管理不同的数据，提高了集群的吞吐率；

- 用户可根据需要将不同业务数据交由不同名称节点管理，使得业务具有良好的隔离性。

### 3.1.4 HDFS架构的优劣性

通过HDFS设计目标及架构介绍，可以看出HDFS具有下列优点。

① 高容错性：数据自动保存多个副本，副本丢失后，可以自动恢复。

② 适合大量数据的批量处理：Hadoop架构以数据为中心，在进行计算时并不移动数据，而是将计算分配给数据，适合GB级、TB级甚至PB级的数据量，数据文件的数量可以达到百万级别，系统中节点数可以达到上万的规模。

③ 简单的数据模型：HDFS采用"一次写入，多次读取"的简单文件模型，文件一旦完成写入，关闭后就无法再次写入，只能被读取。

④ 构建成本低、安全可靠：HDFS采用成千上万的廉价服务器存储数据，显著降低了Hadoop集群的架构成本。

与HDFS的优点相对应，其缺点也非常明显。

① 不适合低延迟数据访问：由于HDFS面向大规模数据的批量处理，采用流式数据读取，具有很高的数据吞吐率，但也导致较高的延迟。

② 不适合大量小文件存储：HDFS使用名称节点管理文件系统的元数据，这些元数据被保存到内存中，过多的小文件会占用大量的内存空间，将导致元数据检索效率降低。此外，在多个节点中读取小文件，磁盘寻道时间将超过读取时间，严重影响系统性能。

③ 不支持多用户并发写入及任意修改文件：HDFS只允许一个文件有一个写入者，不允许多个用户同时对一个文件执行写入操作。而且只允许对文件执行追加操作，不能执行随机写操作。

## 3.2 HDFS工作机制

### 3.2.1 HDFS数据存储策略

作为一个分布式文件系统，为了保证系统的容错性和可用性，HDFS采用了多副本方式对数据进行冗余存储。通常一个数据块的多个副本被存储到不同的数据节点上，默认每个数据块保存3个副本，如图3-5所示。

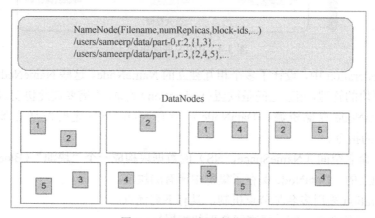

图3-5 HDFS多副本存储

在一个HDFS系统中，数据块副本数量在配置文件hdfs-site.xml中设置。通过属性

"dfs.replication"来设定副本数量。采用多副本存储可以加快数据传输速度,容易检查数据错误,保证数据可靠性。

数据的多个副本是如何存放到数据节点上的呢?

① 如果是在集群内提交文件,则文件的第一个副本存放在上传文件的数据节点上;如果是在集群外提交文件,则随机挑选一个数据节点存放文件第一个副本。

② 文件的第二个副本存放在与第一个副本不同机架的数据节点上。

③ 文件的第三个副本存放在与第一个副本相同机架的其他数据节点上。

④ 如果需要存储更多副本,则其余副本随机选择数据节点存储。

可以使用命令将文件上传到集群中,并查看文件的存储信息。

### 3.2.2 HDFS 数据读取策略

数据存储到 HDFS 中后,如何读取数据呢?

当客户端读取数据时,首先从名称节点获得数据块不同副本的存放位置列表,然后客户端可以调用 API 来确定客户端和这些数据节点所属的机架 ID。如果某个数据块副本对应的机架 ID 和客户端对应的机架 ID 相同,则优先选择该副本读取数据;否则,随机选择一个副本读取数据。

### 3.2.3 HDFS 数据错误与恢复

HDFS 具有较高的容错性,可以兼容廉价的硬件,它把硬件出错看作一种常态,而不是异常,并设计了相应的机制检测数据错误和进行自动恢复,主要包括以下几种情形:

- 名称节点出错;
- 数据节点出错;
- 数据出错。

(1)名称节点出错

名称节点保存了所有的元数据信息,最核心的两大数据结构是 FsImage 和 EditLog,如果这两个文件发生损坏,那么整个 HDFS 实例将失效。

如果进行了 HDFS 高可用性配置,那么当主名称节点出现故障时,Zookeeper 将启动选举机制,从备用名称节点中选取一个名称节点,并切换为激活状态的名称节点,从而保证了 HDFS 集群的持续运行。

如果没有进行 HDFS 高可用性配置,系统把名称节点的核心文件同步复制到备份名称节点(SecondaryNameNode)上,作为名称节点的冷备份。当名称节点出错时,就可以根据 SecondaryNameNode 中的 FsImage 和 EditLog 文件进行恢复。

(2)数据节点出错

每个数据节点会定期向名称节点发送"心跳"信息,向名称节点报告自己的状态。当数据节点发生故障,或者网络发生断网时,名称节点就无法收到来自一些数据节点的心跳信息,这时,这些数据节点就会被标记为"宕机",节点上面的所有数据都会被标记为"不可读",名称节点不会再给它们发送任何 I/O 请求。此时,有可能出现一些数据块的副本数量小于冗余因子(有些数据节点个数小于 3)的情况。

名称节点会定期检查这种情况,一旦发现某个数据块的副本数量小于冗余因子,就会启动数据冗余复制,在其他正常运行的数据节点上为该数据块生成新的副本。

（3）数据出错

网络传输和磁盘错误等因素，都会造成数据错误。

在文件被创建时，客户端就会对每一个数据块进行信息摘录，并把这些信息写入同一个路径的隐藏文件里。当客户端读取文件时，会先读取该信息文件，然后利用该信息文件对每个读取的数据块进行校验，如果校验出错，客户端就会请求到另一个数据节点读取该数据块，并且向名称节点报告这个数据块有错误，名称节点会定期检查并且重新复制这个数据块。

### 3.2.4 HDFS 数据读写过程

#### 1. HDFS 数据写入流程

HDFS 数据写入流程如图 3-6 所示。

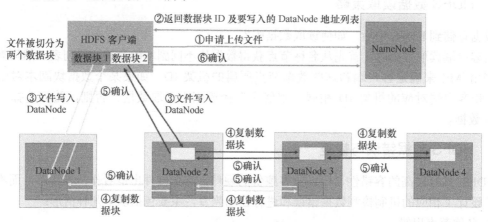

图 3-6　HDFS 数据写入流程

① 客户端向 NameNode 发送文件上传申请。

② NameNode 向客户端返回数据块 ID 及存储数据块的 DataNode 地址列表。

③ 客户端向 DataNode 中写入数据块。

④ 当客户端写入一个数据块后，在 DataNode 之间异步进行数据块复制。

⑤ 最后一个 DataNode 上数据块写入完成后，会发送一个确认信息给前一个 DataNode。第一个写入数据块的 DataNode 反馈确认信息给客户端，数据写入完毕。

⑥ 客户端向 NameNode 发送最终的确认信息。

#### 2. HDFS 数据读取的基本流程

HDFS 数据读取的基本流程如图 3-7 所示。

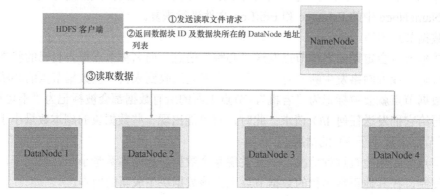

图 3-7　HDFS 数据读取流程

① 客户端向 NameNode 发送读取数据的请求。

② NameNode 返回数据块 ID 及存储数据块的 DataNode 地址列表。该列表按照存储起始数据块的 DataNode 与客户端的距离进行排序。

③ 客户端直接在最近的 DataNode 上读取数据。如果读取失败，则从另一个副本所在的 DataNode 上读取数据。

## 3.3 HDFS Shell 管理

Hadoop 提供了大量 HDFS Shell 命令用于 HDFS 集群管理、文件管理及问题调试，所有的 HDFS 命令都通过 bin/hdfs 脚本调用。HDFS Shell 命令与 Linux Shell 命令的不同之处在于，HDFS Shell 命令操作的对象是 Hadoop 集群中的文件，Linux 命令操作的对象是本地的文件。

可以通过 hdfs -help 命令查看 HDFS Shell 支持的所有命令，如图 3-8 所示。

```
hadoop@Master:/usr/local/hadoop$ ./bin/hdfs -help
Usage: hdfs [--config confdir] [--loglevel loglevel] COMMAND
       where COMMAND is one of:
  dfs                  run a filesystem command on the file systems supported in Hadoop.
  classpath            prints the classpath
  namenode -format     format the DFS filesystem
  secondarynamenode    run the DFS secondary namenode
  namenode             run the DFS namenode
  journalnode          run the DFS journalnode
  zkfc                 run the ZK Failover Controller daemon
  datanode             run a DFS datanode
  dfsadmin             run a DFS admin client
  haadmin              run a DFS HA admin client
  fsck                 run a DFS filesystem checking utility
  balancer             run a cluster balancing utility
  jmxget               get JMX exported values from NameNode or DataNode.
  mover                run a utility to move block replicas across
                       storage types
  oiv                  apply the offline fsimage viewer to an fsimage
  oiv_legacy           apply the offline fsimage viewer to an legacy fsimage
  oev                  apply the offline edits viewer to an edits file
  fetchdt              fetch a delegation token from the NameNode
  getconf              get config values from configuration
  groups               get the groups which users belong to
  snapshotDiff         diff two snapshots of a directory or diff the
                       current directory contents with a snapshot
  lsSnapshottableDir   list all snapshottable dirs owned by the current user
                                               Use -help to see options
  portmap              run a portmap service
  nfs3                 run an NFS version 3 gateway
  cacheadmin           configure the HDFS cache
  crypto               configure HDFS encryption zones
  storagepolicies      list/get/set block storage policies
  version              print the version

Most commands print help when invoked w/o parameters.
```

图 3-8　HDFS Shell 支持的所有命令

### 3.3.1 HDFS 文件操作命令

在 Hadoop 中，用于文件操作的命令形式为 hdfs dfs -command <args>。其中 command 为文件操作命令（见图 3-9），args 为命令参数。所有的文件操作命令都使用 URI 作为路径参数。URI 格式为 scheme://authority/path。对 HDFS，scheme 为 hdfs；对本地文件系统，scheme 为 file。其中，scheme 和 authority 是可选项，如果未加指定，就会使用配置中指定的默认 scheme。一个 HDFS 文件或目录，比如/parent/child，可以表示为 hdfs://namenode:port/parent/child（其中 hdfs://namenode:port 为配置文件 core-sitx.mxl 中属性“fs.defaultFS”的值，如 hdfs://master:9000），也可以简化为 /parent/child。

需要注意的是，在 HDFS 中，当前目录为/user/current_user，例如当前用户名为 hadoop 时，系统当前目录为/user/hadoop，可以用相对路径表示为./。

可以使用 hdfs dfs –help 命令查看所有的文件操作命令，如图 3-9 所示。

```
hadoop@Master:/usr/local/hadoop$ ./bin/hdfs dfs -help
Usage: hadoop fs [generic options]
        [-appendToFile <localsrc> ... <dst>]
        [-cat [-ignoreCrc] <src> ...]
        [-checksum <src> ...]
        [-chgrp [-R] GROUP PATH...]
        [-chmod [-R] <MODE[,MODE]... | OCTALMODE> PATH...]
        [-chown [-R] [OWNER][:[GROUP]] PATH...]
        [-copyFromLocal [-f] [-p] [-l] <localsrc> ... <dst>]
        [-copyToLocal [-p] [-ignoreCrc] [-crc] <src> ... <localdst>]
        [-count [-q] [-h] <path> ...]
        [-cp [-f] [-p | -p[topax]] <src> ... <dst>]
        [-createSnapshot <snapshotDir> [<snapshotName>]]
        [-deleteSnapshot <snapshotDir> <snapshotName>]
        [-df [-h] [<path> ...]]
        [-du [-s] [-h] <path> ...]
        [-expunge]
        [-find <path> ... <expression> ...]
        [-get [-p] [-ignoreCrc] [-crc] <src> ... <localdst>]
        [-getfacl [-R] <path>]
        [-getfattr [-R] {-n name | -d} [-e en] <path>]
        [-getmerge [-nl] <src> <localdst>]
        [-help [cmd ...]]
        [-ls [-d] [-h] [-R] [<path> ...]]
        [-mkdir [-p] <path> ...]
        [-moveFromLocal <localsrc> ... <dst>]
        [-moveToLocal <src> <localdst>]
        [-mv <src> ... <dst>]
        [-put [-f] [-p] [-l] <localsrc> ... <dst>]
        [-renameSnapshot <snapshotDir> <oldName> <newName>]
        [-rm [-f] [-r|-R] [-skipTrash] <src> ...]
        [-rmdir [--ignore-fail-on-non-empty] <dir> ...]
        [-setfacl [-R] [{-b|-k} {-m|-x <acl_spec>} <path>]|[--set <acl_spec> <path>]]
        [-setfattr {-n name [-v value] | -x name} <path>]
        [-setrep [-R] [-w] <rep> <path> ...]
        [-stat [format] <path> ...]
        [-tail [-f] <file>]
        [-test -[defsz] <path>]
        [-text [-ignoreCrc] <src> ...]
        [-touchz <path> ...]
        [-truncate [-w] <length> <path> ...]
        [-usage [cmd ...]]
```

图 3-9  文件操作命令

（1）appendToFile

功能：将本地一个或多个文件追加到目标文件中，或者从标准输入中读取输入并追加到目标文件中。

语法：hdfs dfs  -appendToFile <localsrc>…<dst>

示例：

```
$ hdfs dfs -appendToFile localfile /user/hadoop/hadoopfile
$ hdfs dfs -appendToFile localfile1 localfile2 /user/hadoop/hadoopfile
$ hdfs dfs -appendToFile - hdfs://master:9000/user/hadoop/hadoopfile
```

返回值：成功返回 0，失败返回-1。

注意：在 Hadoop 3.x 中，如果集群中 DataNode 的数量少于或等于 3，则执行 appendToFile 命令时，需要在集群配置文件 hdfs-site.xml 中添加下列配置信息，并重启集群。

```
<property>
<name>dfs.client.block.write.replace-datanode-on-failure.enable</name>
<value>true</value>
```

```
</property>
<property>
<name>dfs.client.block.write.replace-datanode-on-failure.policy</name>
<value>NEVER</value>
</property>
```

（2）cat

功能：将路径指定文件的内容输出到标准输出 stdout。

语法：hdfs dfs -cat URI [URI…]

示例：

```
$ hdfs dfs -cat hdfs://host1:port1/file1 hdfs://host2:port2/file2
$ hdfs dfs -cat file:///file3   /user/hadoop/file4
```

返回值：成功返回 0，失败返回-1。

（3）chgrp

功能：改变文件所属的组。用户必须为文件所有者或超级用户。

语法：hdfs dfs -chgrp [-R] group URI [URI…]

参数：

R：将使改变在目录结构下递归进行，即所有子目录所属组同时改变。

示例：

```
$ hdfs dfs -chgrp -R hadoop /user/hadoop/hadoopfile
```

（4）chmod

功能：改变文件的权限。用户必须为文件所有者或超级用户。

语法：hdfs dfs -chmod [-R] <MODE[,MODE]… | OCTALMODE> URI [URI…]

参数：

R：将使改变在目录结构下递归进行，即所有子目录权限同时改变。

示例：

```
$ hdfs dfs -chown -R hadoop /user/hadoop/hadoopfile
```

（5）chown

功能：改变文件的拥有者。用户必须为超级用户。

语法：hdfs dfs -chown [-R] [OWNER][:[GROUP]] URI [URI ]

参数：

R：使改变在目录结构下递归进行，即所有子目录所有者同时改变。

示例：

```
$ hdfs dfs -chown -R hadoop:hadoop /user/hadoop/hadoopfile
```

（6）copyFromLocal

功能：将本地源文件复制到指定的位置，与 put 命令类似。

使用方法：hdfs dfs -copyFromLocal [-f][-p]<localsrc> URI

参数：

p：保持文件访问时间、修改时间、所有者及相关权限。

f：如果目标文件存在，则覆盖目标文件。

示例：

```
$ hdfs dfs -copyFromLocallocalfile /user/hadoop/
```

（7）copyToLocal

功能：将文件复制到本地文件或本地文件夹中，与 get 命令类似。

语法：hdfs dfs -copyToLocal [-ignoreCrc] [-crc] URI<localdst>
参数：
ignoreCrc：复制 CRC 校验失败的文件。
crc：复制文件及其 CRC 信息。
示例：
$ hdfs dfs -copyToLocal /user/hadoop/hadoopfile ./

（8）count
功能：统计指定路径下的目录数、文件数、字节数，获取配额和使用情况。输入列为 DIR_COUNT、FILE_COUNT、CONTENT_SIZE、PATHNAME。
语法：hadoop dfs -count <path>
示例：
$ hdfs dfs -count -q   /user/hadoop
返回值：成功返回 0，失败返回-1。

（9）cp
功能：将文件从源路径复制到目标路径。允许有多个源路径，此时目标路径必须是一个目录。
语法：hdfs dfs -cp [-f] [-p] URI [URI…] <dst>
参数：
p：保持文件访问时间、修改时间、所有者及相关权限。
f：如果目标文件存在，则覆盖目标文件。
示例：
$ hdfs dfs -cp/user/hadoop/file1   /user/hadoop/file2
$ hdfs dfs -cp /user/hadoop/file1 /user/hadoop/file2 /user/hadoop/dir
返回值：成功返回 0，失败返回-1。

（10）du
功能：显示目录中所有文件的大小，或者当只指定一个文件时，显示此文件的大小。
语法：hdfs dfs  -du [-s] [-h] URI [URI…]
示例：
$ hdfs dfs -du /user/hadoop/dir1 /user/hadoop/file1
返回值：成功返回 0，失败返回-1。

（11）expunge
功能：清空回收站。
语法：hdfs dfs -expunge

（12）find
功能：从指定文件中查找符合条件的文件。
语法：hdfs dfs -fine <path>…[-name]<expression>
示例：
$ hdfs dfs -fine / -name test -print
返回值：成功返回 0，失败返回-1。

（13）get
功能：复制 HDFS 文件到本地文件系统中。
语法：hdfs dfs -get [-ignoreCrc] [-crc] [-p] [-f] <src><localdst>

参数：

p：保持文件访问时间、修改时间、所有者及相关权限。

f：如果目标文件存在，则覆盖目标文件。

ignoreCrc：复制 CRC 校验失败的文件。

crc：复制文件及 CRC 信息。

示例：

$ hdfs dfs -get /user/hadoop/file localfile
$ hdfs dfs -get hdfs://host:port/user/hadoop/file localfile

返回值：成功返回 0，失败返回-1。

（14）getmerge

功能：接受一个源目录和一个目标文件作为输入，并且将源目录中所有的文件合并成本地目标文件。

语法：hdfs dfs -getmerge [-nl] <src><localdst>

参数：

nl：用于指定在每个文件结尾添加一个换行符。

示例：

$ hadoop dfs -getmerge -nl /src /opt/output.txt
$ hadoop dfs -getmerge -nl /src/file1.txt /src/file2.txt /output.txt

返回值：成功返回 0，失败返回-1。

（15）ls

功能：返回文件或目录信息。如果是文件，则按照如下格式返回文件信息：文件名 <副本数> 文件大小 修改日期 修改时间 权限 用户 ID 组 ID。如果是目录，则返回其直接子目录的一个列表，目录返回列表的信息为：目录名 <dir> 修改日期 修改时间 权限 用户 ID 组 ID。

语法：hdfs dfs -ls [-C] [-d] [-h] [-R] [-t] <args>

参数：

C：仅显示文件和目录的路径信息。

d：显示目录信息。

h：以可读格式输出文件大小。

R：递归显示目录中所有子目录信息。

t：输出结果根据修改时间进行排序。

示例：

$ hdfs dfs -ls /user/hadoop/file1 /user/hadoop/file2

返回值：成功返回 0，失败返回-1。

（16）mkdir

功能：创建一个或多个目录。

语法：hdfs dfs -mkdir [-p] <path>

参数：

p：递归创建目录。

示例：

$ hdfs dfs -mkdir -p /user/hadoop/parent/child
$ hdfs dfs -mkdir /user/hadoop/dir1 /user/hadoop/dir2
$ hdfs dfs -mkdir hdfs://host1:port1/user/hadoop/dir hdfs://host2:port2/user/hadoop/dir

返回值：成功返回 0，失败返回-1。

（17）moveFromLocal

功能：将本地文件上传到指定位置，同时删除本地源文件。与 put 命令类似。

语法： hadoop dfs -moveFromLocal <localsrc><dst>

（18）mv

功能：将文件从源路径移动到目标路径。允许有多个源路径，此时目标路径必须是一个目录。不允许在不同的文件系统之间移动文件。

语法： hdfs dfs -mv URI [URI…] <dst>

示例：

$ hdfs dfs -mv /user/hadoop/file1 /user/hadoop/file2
$ hdfs dfs -mv /user/hadoop/file1    /user/hadoop/file2    /user/hadoop/dir1

返回值：成功返回 0，失败返回-1。

（19）put

功能：从本地文件系统中复制一个或多个文件到目标文件系统中。也支持从标准输入中读取输入并写入目标文件系统。

语法： hadoop dfs -put [-f] [-p] <localsrc> …<dst>

参数：

p：保持文件访问时间、修改时间、所有者和相关权限。

f：如果目标文件存在，则覆盖目标文件。

示例：

$ hdfs dfs -put localfile /user/hadoop/hadoopfile
$ hdfs dfs -put localfile1 localfile2 /user/hadoop/hadoopdir
$ hdfs dfs -put - hdfs://host:port/hadoop/hadoopfile

返回值：成功返回 0，失败返回-1。

（20）rm

功能：删除指定的文件或目录。如果启用了回收站，则文件系统会将删除的文件或目录移到回收站中。

语法： hadoop dfs -rm [-f] [-r |-R] [-skipTrash] [-safety] URI [URI…]

参数：

p：如果文件不存在，将不显示诊断消息或者修改退出状态以反映错误。

r，R：递归删除目录及其下的所有内容。

skipTrash：将绕过回收站（如果启用），并立即删除指定的文件。当需要从超过配额的目录中删除文件时，这会很有用。

safety：删除文件数大于预设值时需要进行安全确认。

示例：

$ hdfs dfs -rm hdfs://host:port/file /user/hadoop/emptydir

返回值：成功返回 0，失败返回-1。

（21）setrep

功能：改变一个文件的副本数量。如果指定的对象是目录，则修改目录中所有对象的副本数量。

语法： hdfs dfs -setrep [-R] [-w][numReplicas]<path>

参数：

R：为了后续兼容。
w：等待复制结束才返回。
numReplicas：副本数量。
示例：
$ hdfs dfs -setrep -w 3 /user/hadoop/dir1
返回值：成功返回 0，失败返回-1。

（22）stat
功能：按指定格式返回指定的文件或目录的统计信息。
语法：hdfs dfs -stat [-format]URI [URI…]
示例：
$ hdfs dfs -stat "type:%F perm:%a %u:%g size:%b mtime:%y atime:%x name:%n" /file
返回值：成功返回 0，失败返回-1。

（23）tail
功能：将文件尾部的 1KB 内容输出到 stdout。
语法：hdfs dfs -tail [-f] URI
参数：
f：持续检测新添加到文件中的内容。
示例：
$ hdfs dfs -tail pathname
返回值：成功返回 0，失败返回-1。

（24）test
功能：对文件或目录进行检测。
语法：hdfs dfs -test -[defsz] URI
参数：
d：如果指定路径为目录，则返回 0。
e：如果指定路径存在，则返回 0。
f：如果指定路径为文件，则返回 0。
s：如果指定路径为空，则返回 0。
z：如果指定文件长度为 0 字节，则返回 0。
示例：
$ hdfs dfs -test -e filename

（25）text
功能：将指定文件以文本格式输出。
语法：hdfs fs -text <src>

（26）touchz
功能：创建一个空文件。
语法：hdfs dfs -touchz <src>

## 3.3.2 HDFS 系统管理命令

Hadoop 提供了大量用于 HDFS 集群管理的 Shell 命令，如 balancer（平衡集群负载）、datanode（运行数据节点）、dfsadmin（运行 HDFS 管理客户端）、dfsrouter（HDFS 路由管理）、namenode

（运行名称节点）、fsck（HDFS 文件系统检查）等。

### 1. dfsadmin 命令

在 HDFS 集群管理过程中，管理员经常使用 hdfs dfsadmin 命令运行 HDFS 管理客户端。dfsadmin 命令如图 3-10 所示。

```
hadoop@Master:~$ hdfs dfsadmin -help
hdfs dfsadmin
        [-report [-live] [-dead] [-decommissioning]]
        [-safemode <enter | leave | get | wait>]
        [-saveNamespace]
        [-rollEdits]
        [-restoreFailedStorage true|false|check]
        [-refreshNodes]
        [-setQuota <quota> <dirname>...<dirname>]
        [-clrQuota <dirname>...<dirname>]
        [-setSpaceQuota <quota> [-storageType <storagetype>] <dirname>...<dirname>]
        [-clrSpaceQuota [-storageType <storagetype>] <dirname>...<dirname>]
        [-finalizeUpgrade]
        [-rollingUpgrade [<query|prepare|finalize>]]
        [-refreshServiceAcl]
        [-refreshUserToGroupsMappings]
        [-refreshSuperUserGroupsConfiguration]
        [-refreshCallQueue]
        [-refresh <host:ipc_port> <key> [arg1..argn]
        [-reconfig <datanode|...> <host:ipc_port> <start|status>]
        [-printTopology]
        [-refreshNamenodes datanode_host:ipc_port]
        [-deleteBlockPool datanode_host:ipc_port blockpoolId [force]]
        [-setBalancerBandwidth <bandwidth in bytes per second>]
        [-fetchImage <local directory>]
        [-allowSnapshot <snapshotDir>]
        [-disallowSnapshot <snapshotDir>]
        [-shutdownDatanode <datanode_host:ipc_port> [upgrade]]
        [-getDatanodeInfo <datanode_host:ipc_port>]
        [-metasave filename]
        [-triggerBlockReport [-incremental] <datanode_host:ipc_port>]
        [-help [cmd]]
```

图 3-10  dfsadmin 命令

dfsadmin 命令主要选项及其描述如表 3-1 所示。

表 3-1  dfsadmin 命令主要选项及其描述

| 命令选项 | 描述 |
| --- | --- |
| -report | 显示文件系统的基本信息和统计信息 |
| -safemode | 集群安全模式维护命令 |
| -saveNamespace | 在安全模式下保存当前名称空间到存储目录中，更新日志文件 |
| -rollEdits | 在主名称节点上回滚日志 |
| -refreshNodes | 重新读取主机和排除文件，以更新允许连接到 NameNode 上的 DataNode 集，以及应取消配置或重新调试的 DataNode 集 |
| -setQuota | 设置目录配额 |
| -finalizeUpgrade | 完成 HDFS 的升级。对 DataNode，删除其以前版本的工作目录；对 NameNode，执行相同操作 |
| -rollingUpgrade | 回滚升级操作 |
| -refreshNamenodes | 对给定的 DataNode，重新加载配置文件，停止为已删除的块池提供服务，并开始为新的块池提供服务 |
| -deleteBlockPool | 如果传递 force，将删除给定 DataNode 上指定的块池目录及其内容，否则仅当目录为空时删除该目录。如果 DataNode 仍在为块池提供服务，则该命令将失败 |
| -setBalancerBandwidth | 更改 HDFS 块平衡期间每个数据节点使用的网络带宽。bandwidth 是每个 DataNode 每秒将使用的最大字节数，此值将覆盖 dfs.datanode.balance.bandwidthPerSec 参数。注意：新值在 DataNode 上不是持久的 |
| -fetchImage | 从 NameNode 下载最新的 FsImage 文件并将其保存在指定的本地目录中 |

续表

| 命令选项 | 描述 |
|---|---|
| -shutdownDatanode | 向指定的 DataNode 提交 shutdown 请求 |
| -getDatanodeInfo | 获取指定的 DataNode 信息 |
| -metasave | 将 NameNode 的主要数据结构保存到 hadoop.log.dir 属性指定目录下的文件中。如果文件名存在，则覆盖该文件 |
| -triggerBlockReport | 触发一个块报告给指定的 DataNode |

在 HDFS 集群启动过程中，NameNode 会自动进入安全模式。主名称节点处于安全模式时，不允许修改命名空间，不能复制或删除数据块。安全模式的目的是在系统启动时检查各个 DataNode 上数据块的有效性，同时根据策略对数据块进行必要的复制或删除，当数据块副本数量满足最小副本数量条件时，会自动退出安全模式。

进行 HDFS 集群安全模式管理语法为：

hdfs dfsadmin [-safemode enter | leave | get | wait | forceExit]

- 集群进入安全模式：$ hdfs dfsadmin -safemode enter
- 集群退出安全模式：$ hdfs dfsadmin -safemode leave
- 强制集群退出安全模式：$ hdfs dfsadmin -safemode forceExit
- 查看集群是否进入安全模式：$ hdfs dfsadmin -safemode get

如果是手动方式进入安全模式的，只能采用手动方式关闭安全模式。

### 2. namenode 命令

运行 namenode 命令可以进行集群主名称节点格式化、升级回滚等操作。namenode 命令如图 3-11 所示。

图 3-11 namenode 命令

namenode 命令主要选项及其描述如表 3-2 所示。

表 3-2 namenode 命令主要选项及其描述

| 命令选项 | 描述 |
|---|---|
| -backup | 开始对节点进行备份操作 |
| -checkpoint | 开始对节点进行检查点操作 |
| -format | 格式化 NameNode。先启动 NameNode，再格式化，最后关闭 |
| -upgrade | NameNode 版本更新后，应以 upgrade 方式启动 |
| -upgradeOnly | 更新指定的 NameNode，然后关闭 NameNode |
| -rollback | 将 NameNode 回滚到前一个版本。必须先停止集群，并且分发旧版本 |

续表

| 命令选项 | 描述 |
|---|---|
| -importCheckpoint | 从检查点目录加载镜像，并保存为当前镜像 |
| -initializeSharedEdits | 格式化一个新的共享日志目录，并复制足够的日志段，以便备用 NameNode 可以启动 |
| -bootstrapStandby | 允许从活跃 NameNode 复制最新的命名空间快照，来引导备用 NameNode 的存储目录 |
| -recover | 在损坏的文件系统中恢复最后的元数据 |
| -metadataVersion | 验证配置的目录是否存在，然后打印软件和镜像的元数据版本 |

例如，在 Hadoop 集群安装配置过程中，需要在 NameNode 上执行格式命令：

```
$ hdfs namenode -format
```

### 3. fsck 命令

运行 fsck 命令可以检查 HDFS，如数据块分布等。fsck 命令如图 3-12 所示。

```
hadoop@master:~$ hdfs fsck -help
Usage: hdfs fsck <path> [-list-corruptfileblocks | [-move | -delete | -openforwrite] [-files
[-blocks [-locations | -racks | -replicaDetails | -upgradedomains]]] [-includeSnapshots] [
-storagepolicies] [-blockId <blk_Id>]
```

图 3-12 fsck 命令

fsck 命令主要选项及其描述如表 3-3 所示。

表 3-3 fsck 命令主要选项及其描述

| 命令选项 | 描述 |
|---|---|
| path | 检查指定路径中的文件 |
| -delete | 删除损坏的文件 |
| -files | 打印正在进行检查的文件名 |
| -blocks | 打印数据块的报告（与-files 选项一起使用） |
| -locations | 打印某个数据块的位置信息（与-blocks 选项一起使用） |
| -racks | 打印 DataNode 位置信息的网络拓扑结构（与-blocks 选项一起使用） |
| -replicaDetails | 打印每个副本的详细信息（与-blocks 选项一起使用） |
| -upgradedomains | 打印每个数据块的更新域（与-blocks 选项一起使用） |
| -list-corruptfileblocks | 打印丢失的数据块列表 |
| -move | 移动损坏的文件到/lost+found 中 |
| -openforwrite | 打印正在进行写操作的文件 |
| -blockId | 打印指定数据块的相应信息 |

例如，将文件 sale_user.csv 上传到 HDFS 中，可以看到文件被切分为一个数据块，该块的 3 个副本被保存在 3 个数据节点中，如图 3-13 所示。

### 3.3.3 HDFS Shell 操作实例

（1）查看 HDFS 根目录下的文件子目录信息。

```
$ hdfs dfs -ls hdfs://master:9000/
$ hdfs dfs -ls /
```

执行结果如图 3-14 所示。

```
hadoop@Master:/mnt/hgfs/hadoop$ hdfs dfs -put ./sale_user.csv /
hadoop@Master:/mnt/hgfs/hadoop$ hdfs fsck /sale_user.csv -files -blocks  -locations
Connecting to namenode via http://Master:50070/fsck?ugi=hadoop&files=1&blocks=1&locations=1&path=%2Fsale_user.csv
FSCK started by hadoop (auth:SIMPLE) from /192.168.216.128 for path /sale_user.csv at Tue Apr 09 12:56:01 CST 2019
/sale_user.csv 13530111 bytes, 1 block(s):  OK
0. BP-59132588-192.168.216.128-1520946293944:blk_1073741891_1071 len=13530111 repl=3 [DatanodeInfoWithStorage[192.168.216.128:5001
0,DS-033fc691-050c-4013-8b80-5ef822936ded,DISK], DatanodeInfoWithStorage[192.168.216.129:50010,DS-7c56234e-7c99-4c9f-a7ff-8617afa1
ccc7,DISK], DatanodeInfoWithStorage[192.168.216.130:50010,DS-561ea432-b091-4813-9180-db1e27fb651b,DISK]]

Status: HEALTHY
 Total size:    13530111 B
 Total dirs:    0
 Total files:   1
 Total symlinks:                0
 Total blocks (validated):      1 (avg. block size 13530111 B)
 Minimally replicated blocks:   1 (100.0 %)
 Over-replicated blocks:        0 (0.0 %)
 Under-replicated blocks:       0 (0.0 %)
 Mis-replicated blocks:         0 (0.0 %)
 Default replication factor:    3
 Average block replication:     3.0
 Corrupt blocks:                0
 Missing replicas:              0 (0.0 %)
 Number of data-nodes:          3
 Number of racks:               1
FSCK ended at Tue Apr 09 12:56:01 CST 2019 in 1 milliseconds


The filesystem under path '/sale_user.csv' is HEALTHY
```

图 3-13　查看文件存储信息

```
hadoop@Master:~$ hdfs dfs -ls hdfs://master:9000/
Found 3 items
drwxr-xr-x   - hadoop supergroup          0 2020-10-22 13:25 hdfs://localhost:9000/hbase
drwxr-xr-x   - hadoop supergroup          0 2020-08-04 15:42 hdfs://localhost:9000/home
drwxr-xr-x   - hadoop supergroup          0 2020-05-13 10:22 hdfs://localhost:9000/user
hadoop@Master:~$ hdfs dfs -ls /
Found 3 items
drwxr-xr-x   - hadoop supergroup          0 2020-10-22 13:25 /hbase
drwxr-xr-x   - hadoop supergroup          0 2020-08-04 15:42 /home
drwxr-xr-x   - hadoop supergroup          0 2020-05-13 10:22 /user
```

图 3-14　查看 HDFS 根目录

（2）查看 HDFS 当前目录（/user/hadoop）及其子目录中的文件信息。

```
$ hdfs dfs -ls -R hdfs://master:9000/user/hadoop
$ hdfd dfs -ls -R ./
```

执行结果如图 3-15 所示。

```
hadoop@Master:~$ hdfs dfs -ls -R ./
drwxr-xr-x   - hadoop supergroup          0 2020-10-28 10:48 emp
-rw-r--r--   1 hadoop supergroup       1657 2020-10-28 10:48 emp/emp_high
-rw-r--r--   1 hadoop supergroup       3811 2020-10-28 10:48 emp/emp_low
drwxr-xr-x   - hadoop supergroup          0 2020-10-29 14:42 emp_comp
-rw-r--r--   1 hadoop supergroup       1657 2020-10-29 14:42 emp_comp/emp_high
-rw-r--r--   1 hadoop supergroup       3811 2020-10-29 14:42 emp_comp/emp_low
drwxr-xr-x   - hadoop supergroup          0 2020-10-08 14:37 words
drwxr-xr-x   - hadoop supergroup          0 2020-10-08 14:35 words/input
```

图 3-15　查看 HDFS 当前目录

（3）在 HDFS 的当前目录中创建一个 bigdata/weblogs 子目录。

```
$ hdfs dfs -mkdir -p ./bigdata/weblogs
```

执行结果如图 3-16 所示。

```
hadoop@Master:~$ hdfs dfs -mkdir -p ./bigdata/weblogs
hadoop@Master:~$ hdfs dfs -ls ./bigdata
Found 1 items
drwxr-xr-x   - hadoop supergroup          0 2020-11-03 14:33 bigdata/weblogs
```

图 3-16　创建目录

（4）将本地宿主目录（/home/hadoop）中的文件 weblog1 和 weblog2 上传到集群的 bigdata/weblogs 目录中。

```
$ hdfs dfs -put /home/hadoop/weblog1 /home/hadoop/weblog2 bigdata/weblogs
```

执行结果如图 3-17 所示。

```
hadoop@Master:~$ hdfs dfs -put /home/hadoop/weblog1 /home/hadoop/weblog2 bigdata/weblogs
hadoop@Master:~$ hdfs dfs -ls bigdata/weblogs
Found 2 items
-rw-r--r--   1 hadoop supergroup     956192 2020-11-03 14:41 bigdata/weblogs/weblog1
-rw-r--r--   1 hadoop supergroup    1289692 2020-11-03 14:41 bigdata/weblogs/weblog2
```

图 3-17　上传文件

（5）将集群/user/hadoop/bigdata/weblogs 目录中的文件 weblog1 和 weblog2 合并下载到本地宿主目录中，目标文件命名为 weblogs。

$ hdfs dfs -getmerge ./bigdata/weblogs/weblog1 ./bigdata/weblogs/weblog2 /home/hadoop/weblogs

执行结果如图 3-18 所示。

```
hadoop@Master:~$ hdfs dfs -getmerge bigdata/weblogs/weblog1 bigdata/weblogs/weblog2 /home/hadoop/weblogs
hadoop@Master:~$ ls
apache-flume-1.7.0-bin  Documents    hadoop-2.7.3                metastore_db  Public          spark-warehouse  testd2   weblog1  workspace
derby.log               Downloads    hdfdtest                    IdeaProjects  Music           python           Templates venv    weblog2  workspace2
Desktop                 file:        hdfstest                    lib           Pictures        sbt              test     Videos  weblogs
```

图 3-18　合并文件

（6）将集群 bigdata/weblogs 目录中的文件下载到本地/home/hadoop/newweblogs 目录中。

$ hdfs dfs -get ./bigdata/weblogs /home/hadoop/newweblogs

执行结果如图 3-19 所示。

```
hadoop@Master:~$ hdfs dfs -get ./bigdata/weblogs /home/hadoop/newweblogs
hadoop@Master:~$ ls -l -R /home/hadoop/newweblogs
/home/hadoop/newweblogs:
total 2196
-rw-r--r-- 1 hadoop hadoop  956192 11月  3 16:40 weblog1
-rw-r--r-- 1 hadoop hadoop 1289692 11月  3 16:40 weblog2
```

图 3-19　下载文件

（7）查看集群 bigdata/weblogs 目录中各个文件的大小。

$ hdfs dfs -du ./bigdata/weblogs

执行结果如图 3-20 所示。

```
hadoop@Master:~$ hdfs dfs -du ./bigdata/weblogs
956192   bigdata/weblogs/weblog1
1289692  bigdata/weblogs/weblog2
```

图 3-20　查看文件大小

（8）将集群 bigdata/weblogs 目录中的文件 weblog1 和 weblog2 复制到 bigdata 目录中。

$ hdfs dfs -cp ./bigdata/weblogs/*　　./bigdata

执行结果如图 3-21 所示。

```
hadoop@Master:~$ hdfs dfs -cp ./bigdata/weblogs/*  ./bigdata
hadoop@Master:~$ hdfs dfs -ls ./bigdata
Found 3 items
-rw-r--r--   1 hadoop supergroup     956192 2020-11-03 16:53 bigdata/weblog1
-rw-r--r--   1 hadoop supergroup    1289692 2020-11-03 16:53 bigdata/weblog2
```

图 3-21　复制文件

（9）将集群/user/hadoop/bigdata/weblogs 目录中的所有文件移植到集群当前目录中。

$ hdfs dfs -mv ./bigdata/weblogs ./

执行结果如图 3-22 所示。

```
hadoop@Master:~$ hdfs dfs -mv ./bigdata/weblogs ./
hadoop@Master:~$ hdfs dfs -ls ./
Found 5 items
drwxr-xr-x   - hadoop supergroup          0 2020-11-03 16:56 bigdata
drwxr-xr-x   - hadoop supergroup          0 2020-10-28 10:48 emp
drwxr-xr-x   - hadoop supergroup          0 2020-10-29 14:42 emp_comp
drwxr-xr-x   - hadoop supergroup          0 2020-11-03 16:52 weblogs
drwxr-xr-x   - hadoop supergroup          0 2020-10-08 14:37 words
```

图 3-22　移植文件

（10）删除集群的/user/hadoop/weblogs 目录。

```
$ hdfs dfs -rm -R ./weblogs
```

执行结果如图 3-23 所示。

```
hadoop@Master:~$ hdfs dfs -rm -R ./weblogs
20/11/03 20:59:08 INFO fs.TrashPolicyDefault: Namenode trash configuration:
 Deletion interval = 0 minutes, Emptier interval = 0 minutes.
Deleted weblogs
```

图 3-23　删除目录

## 3.4　HDFS Java 开发

除可以使用 HDFS Shell 方式进行 HDFS 操作外，Hadoop 还提供了 Java API 的方式操作 HDFS。实际上，大数据应用开发都是通过程序设计完成的。

### 3.4.1　HDFS Java 程序设计基础

Hadoop 整合了众多文件系统，提供了一个高层的文件系统抽象类 org.apache.hadoop.fs. FileSystem，用来定义 Hadoop 中的文件系统接口。只要某个文件系统实现了这个接口，那么它就可以作为 Hadoop 支持的文件系统。HDFS 只是这个抽象的实现类 DistributionFileSystem 的一个实例。Hadoop 文件系统有不同的实现，如表 3-4 所示。

表 3-4　Hadoop 文件系统及其实现

| 文件系统 | URI 方案 | Java 实现 | 定义 |
| --- | --- | --- | --- |
| Local | file | fs.LocalFileSystem | 支持有客户端校验和的本地文件系统。带有校验和的本地文件系统在 fs.RawLocalFileSystem 中实现 |
| HDFS | hdfs | hdfs.DistributionFileSystem | Hadoop 的分布式文件系统 |
| HFTP | hftp | hdfs.HftpFileSystem | 支持通过 HTTP 方式以只读的方式访问 HDFS，distcp 经常用在不同的 HDFS 集群间复制数据 |
| HSFTP | hsftp | hdfs.HsftpFileSystem | 支持通过 HTTPS 方式以只读的方式访问 HDFS |
| HAR | har | fs.HarFileSystem | 构建在 Hadoop 文件系统之上，对文件进行归档。Hadoop 归档文件主要用来减少 NameNode 的内存使用 |
| KFS | kfs | fs.kfs.KosmosFileSystem | Cloudstore（其前身是 Kosmos 文件系统）文件系统类似于 HDFS 和 Google 的 GFS 文件系统，使用 C++编写 |
| FTP | ftp | fs.ftp.FtpFileSystem | 由 FTP 服务器支持的文件系统 |
| S3（本地） | s3n | fs.s3native.NativeS3FileSystem | 基于 Amazon S3 的文件系统 |
| S3（基于块） | s3 | fs.s3.NativeS3FileSystem | 基于 Amazon S3 的文件系统，以块格式存储解决了 S3 的 5GB 文件大小的限制 |

对 Hadoop 文件系统的操作都是从 FileSystem 的具体实例开始的。FileSystem 类提供了 3 个重载的 get 方法，用于获取具体的文件系统静态实例。

- static FileSystem get(Configuration conf)
- static FileSystem get(URI uri, Configuration conf)
- static FileSystem get(URI uri, Configuration conf, String user)

其中，Configuration 类（位于包 org.apache.hadoop.conf 中）对象封装了客户端或服务器的

配置。配置由资源指定，资源包含一组作为 XML 数据的名称/值对。每个资源都由字符串或路径命名。如果使用字符串命名，则检查环境变量 classpath 以查找具有该名称的文件；如果通过路径命名，则直接检查本地文件系统，而不引用环境变量 classpath。

除非显式关闭，否则 Hadoop 默认指定了两个资源，按顺序从 classpath 加载。

① core-default.xml：Hadoop 的只读默认值。

② core-site.xml：Hadoop 安装配置文件。

通过读取资源，创建 Configuration 对象。也可以通过 Configuration 对象的 get 和 set 方法访问、设置配置项，资源会在第一次使用时自动加载到对象中。例如：

Configuration conf=new Configuration();
conf.set("fs.defalutDF","hdfs://master:9000");
conf.set("yarn.resourcemanager.hostname","master");
conf.set("fs.hdfs.impl", "org.apache.hadoop.hdfs.DistributedFileSystem");

URI（Uniform Resource Identifiers）对象是通用资源标识符，按方案和权限确定用户可以使用的文件系统。如果给定 URI 中没有指定方案，则返回默认文件系统（在 core-site.xml 中指定）。URI 语法形式为：

[scheme:][//authority][path][?query][#fragment]

其中，scheme 指定方案；authority 设置权限，形式为[user-info@]host[:port]；path 设置资源路径。例如，HDFS 集群根目录下一个 example 文件可以表示为：

URI uri=new URI("hdfs://master:9000/example")

此时，就可以基于 URI 对象 uri 和 Configuration 对象 conf 创建一个文件系统实例：

FileSystem fs=FileSystem.get(uri,conf)

Hadoop 中的文件操作类基本都在 org.apache.hadoop.fs 中定义，这些操作类支持创建文件、打开文件、读写文件、查看文件状态等的操作。

### 3.4.2 HDFS 程序设计流程

#### 1. HDFS 数据写入流程

利用 Java API 进行 HDFS 数据写入流程如图 3-24 所示。

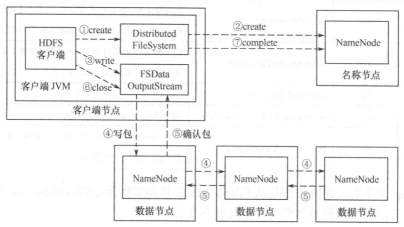

图 3-24　Java API 写入 HDFS 数据流程

① 客户端调用 DistributedFileSystem 对象的 create 方法，创建一个文件输出流 FSDataOutputStream 对象。

② 客户端通过 DistributedFileSystem 对象远程调用 NameNode，发送文件上传申请，返回

数据块 ID 和存储数据块的 DataNode 地址列表。

③ 客户端通过 FSDataOutputStream 对象向 DataNode 中写入数据。数据首先被写入 FSDataOutputStream 对象内部的缓冲区中，然后被分割成一个个 Packet 数据包。

④ 以 Packet 数据包为单位，在一组 DataNode（默认是 3 个）组成的管道上依次传输。

⑤ 在传输管道的反方向上，后一个 DataNode 向前一个 DataNode 发送确认信息，最终由传输管道中第一个 DataNode 将确认信息发送给客户端。

⑥ 完成数据写入后，客户端调用文件输出流 FSDataOutputStream 对象的 close 方法，关闭输出流对象。

⑦ 客户端调用 DistributedFileSystem 对象的 complete 方法，通知 NameNode 文件写入成功。

在 Java 程序设计中，只需要调用 FileSystem 对象的 get 方法创建一个 FileSystem 对象，然后调用该对象的 Create 方法创建 FSDataOutputStream 输出流对象，最后调用输出流对象的 write 方法向 HDFS 中写数据就可以了。

**2. HDFS 数据读取流程**

利用 Java API 进行 HDFS 数据读取流程如图 3-25 所示。

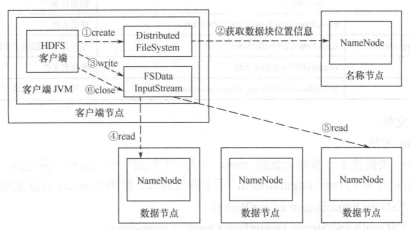

图 3-25　利用 Java API 进行 HDFS 数据读取流程

① 客户端调用 DistributedFileSystem 对象的 open 方法打开要读取的文件。

② 客户端通过 DistributedFileSystem 对象远程调用 NameNode，得到文件的数据块及保存数据块的 DataNode 地址列表。

③ DistributedFileSystem 对象返回输入流对象 FSDataInputStream 给客户端，用来读取数据。

④ 客户端调用 FSDataInputStream 对象的 read 方法读取数据。

⑤ 当客户端读取完所有数据后，调用 FSDataInputStream 的 close 方法关闭文件。

在 Java 程序设计中，只需要调用 FileSystem 对象的 get 方法创建一个 FileSystem 对象，然后调用该对象的 open 方法创建 FSDataInputStream 输入流对象，最后利用输入流对象读取 HDFS 中的数据。

### 3.4.3　常用 HDFS Java API

在 FileSystem 对象中定义了大量常用的文件操作 API，如表 3-5 所示。

表 3-5 常用文件操作 API

| 返回值类型 | API | 描述 |
|---|---|---|
| FSDataOutputStream | create(Path f, boolean overwrite, int bufferSize) | 创建文件，打开一个输入流 |
| FSDataOutputStream | append(Path f) | 向已有文件中追加数据 |
| void | copyFromLocalFile(Path src, Path dst) | 上传文件 |
| void | copyToLocalFile(Path src, Path dst) | 下载文件 |
| abstract boolean | delete(Path f, boolean recursive) | 删除文件 |
| boolean | abstract boolean | |
| BlockLocation[] | getFileBlockLocations(FileStatus file, long start, long len) | 返回指定文件所在主机名、偏移量、大小等信息 |
| abstract FileStatus | getFileStatus(Path f) | 获取指定路径下文件状态信息 |
| abstract FileStatus[] | listStatus(Path f) | 列举文件、记录状态 |
| FileStatus[] | listStatus(Path[] files) | 列举文件、记录状态 |
| boolean | mkdirs(Path f) | 创建目录 |
| void | moveFromLocalFile(Path src, Path dst) | 移植文件 |
| abstract FSDataInputStream | open(Path f, int bufferSize) | 打开文件，创建一个输入流 |
| abstract boolean | rename(Path src, Path dst) | 重命名文件 |
| boolean | setReplication(Path src, short replication) | 设置文件副本数量 |

### 1. 写入文件

（1） create 方法

FileSystem 类提供了一系列重载的 create 方法用于创建一个空文件，并返回一个 HDFS 文件系统的输出流对象 FSDataOutputStream，用于数据写入。常用的 create 方法原型包括：

- public FSDataOutputStream create(Path f)
- public FSDataOutputStream create(Path f,boolean overwrite)
- public FSDataOutputStream create(Path f,boolean overwrite,int bufferSize)
- public FSDataOutputStream create(Path f,Progressable progress)
- public FSDataOutputStream create(Path f,short replication)
- public FSDataOutputStream create(Path f,short replication,Progressable progress)

例 1：创建一个 HDFS 文件/hdfs/hello.txt，并将一个字符串写入该文件中。

```
public void createFile() throws Exception {
    String dst = "hdfs://master:9000/hdfs/hello.txt";
    Configuration conf = new Configuration();
    conf.set("fs.defaultFS","hdfs://master:9000");
    conf.set("fs.hdfs.impl","org.apache.hadoop.hdfs.DistributedFileSystem");
    FileSystem fs = FileSystem.get(URI.create(dst),conf);
    byte[] buff = "This is a example of output to HDFS file.".getBytes();
    FSDataOutputStream out = fs.create(new Path(dst));
    out.write(buff,0,buff.length);
    out.close();
}
```

注意：在创建工程时，将 HDFS 的配置文件 core-site.xml 和 hdfs-site.xml 复制到 eclipse 目

录下的 bin 目录中，此时就不需要执行 conf.set 方法进行配置项设置了。

可以利用 create 方法在 HDFS 中创建文件的同时，将本地 Linux 系统中的文件复制到新建的文件中。

**例 2**：在 HDFS 中新建一个文件/hdfs/word.txt，并将一个 Linux 本地文件/home/hadoop/data/word.txt 复制到新建文件中。

```
public void createFileCopy() throws Exception {
    String src= "/home/hadoop/data/word.txt";
    String dst = "hdfs://master:9000/hdfs/word.txt";
    Configuration conf = new Configuration();
    FileSystem fs = FileSystem.get(URI.create(dst), conf);
    InputStream in = new BufferedInputStream(new FileInputStream(src));
    FSDataOutputStream out = fs.create(new Path(dst));
    IOUtils.copyBytes(in, out, 4096, true);
    IOUtils.closeStream(in);
}
```

（2）append 方法

利用 FileSystem 类提供的一系列重载的 append 方法，可以将 Linux 本地文件追加到 HDFS 中已经存在的文件中。append 方法的原型包括：

- public FSDataOutputStream append(Path f)
- public FSDataOutputStream append(Path f,int bufferSize)
- public abstract FSDataOutputStream append(Path f,int bufferSize,Progressable progress)

**例 3**：将 Linux 本地文件/home/hadoop/data/score.txt 追加到 HDFS 的/hdfs/word.txt 文件中。

```
public void appendFile() throws Exception {
    String localSrc = "/home/hadoop/data/demo.txt";
    String dst = "hdfs://master:9000/hdfs/word.txt";
    BufferedInputStream in = null;
    in = new BufferedInputStream(new FileInputStream(localSrc));
    Configuration conf = new Configuration();
    conf.set("dfs.client.block.write.replace-datanode-on-failure","NEVER");
    conf.setBoolean("dfs.client.block.write.replace-datanode-on-failure.enable",true);
    FileSystem fs = FileSystem.get(URI.create(dst), conf);
    FSDataOutputStream out = fs.append(new Path(dst));
    IOUtils.copyBytes(in, out, 4096, false);
    IOUtils.closeStream(in);
}
```

**2. 读取文件**

为了读取 HDFS 文件中的数据，需要首先使用 open 方法打开文件，返回一个 HDFS 文件系统的输入流对象 FSDataInputStream，用于数据读取。open 方法原型包括：

- public FSDataInputStream open(Path f)
- public abstract FSDataInputStream open(Path f,int bufferSize)
- public FSDataInputStream open(PathHandle fd)
- public FSDataInputStream open(PathHandle fd,int bufferSize)

（1）整体读取文件

整体读取文件可以使用 IOUtils 工具类的 copyBytes 方法进行。

**例 4**：读取 HDFS 中的/hdfs/word.txt 文件，并输出到控制台。

```
public void readFile() throws Exception {
    String uri = "hdfs://master:9000/hdfs/word.txt";
    Configuration conf = new Configuration();
    FileSystem fs = FileSystem.get(URI.create(uri), conf);
    InputStream in = null;
    in = fs.open(new Path(uri));
    IOUtils.copyBytes(in, System.out, 4096, false);
    IOUtils.closeStream(in);
}
```

（2）随机读取文件

由于 FSDataInputStream 对象不是标准的 java.io.InputStream 类对象，而是继承了 java.io.DataInputStream 接口的一个特殊类，并支持数据的随机访问，因此可以从数据流的任意位置读取数据。

例5：读取 HDFS 中的/hdfs/word.txt 文件，每次读取 256 字节数据并输出到控制台。

```
public void randomReadFile() throws Exception {
    String uri = "hdfs://master:9000/hdfs/word.txt";
    Configuration conf = new Configuration();
    FileSystem fs = FileSystem.get(URI.create(uri), conf);
    FSDataInputStream in = null;
    in = fs.open(new Path(uri));
    byte buffer[] = new byte[256];
    int byteRead = 0;
    while ((byteRead = in.read(buffer)) > 0) {
        System.out.write(buffer, 0, byteRead);
    }
    IOUtils.closeStream(in);
}
```

### 3. 判断文件是否存在

在 HDFS 中进行文件读写操作时，经常需要判断文件或文件夹是否存在，可以使用 FileSystem 类提供的 exists 方法进行判断。

例6：判断/hdfs/word.txt 是否存在。

```
public void fileExists() throws Exception{
    String uri = "hdfs://master:9000/hdfs/word.txt";
    Configuration conf = new Configuration();
    FileSystem fs = FileSystem.get(URI.create(uri),conf);
    if(fs.exists(new Path(uri))){
        System.out.println("文件存在");
    }else{
        System.out.println("文件不存在");
    }
}
```

### 4. 上传本地 Linux 系统文件到 HDFS

如果要将本地 Linux 系统中的文件上传到 HDFS，可以使用 FileSystem 类提供的 copyFromLocalFile 方法。copyFromLocalFile 方法原型如下，可以根据需要进行选择。

● public void copyToLocalFile(Path src,Path dst)

● public void copyFromLocalFile(boolean delSrc,Path src,Path dst)

- public void copyFromLocalFile(boolean delSrc,boolean overwrite,Path[] srcs,Path dst)
- public void copyFromLocalFile(boolean delSrc,boolean overwrite,Path src,Path dst)

例7：将本地 Linux 系统中的文件/home/hadoop/data/score.txt 上传到 HDFS 的/hdfs 文件夹中。

```
public void uploadfile() throws Exception {
    String src = "/home/hadoop/data/demo.txt";
    String dst =" hdfs://master:9000/";
    Configuration conf = new Configuration();
    FileSystem fs = FileSystem.get(URI.create(dst),conf);
    fs.copyFromLocalFile(new Path(src),new Path(dst));
}
```

### 5．下载 HDFS 文件到本地 Linux 文件系统

如果要将 HDFS 文件下载到本地 Linux 系统中，可以使用 FileSystem 类提供的 copyToLocalFile 方法。copyToLocalFile 方法原型包括：

- public void copyToLocalFile(Path src,Path dst)
- public void copyFromLocalFile(boolean delSrc,Path src,Path dst)
- public void copyFromLocalFile(boolean delSrc,boolean overwrite,Path[] srcs,Path dst)
- public void copyFromLocalFile(boolean delSrc,boolean overwrite,Path src,Path dst)

例8：将 HDFS 文件/hdfs/word.txt 下载到本地 Linux 系统中，文件路径及名称为/home/hadoop/data/newWord.txt。

```
public void downloadFile() throws Exception {
    String src = "hdfs://master:9000/hdfs/word.txt";
    String dst = "/home/hadoop/data/newWord.txt";
    Configuration conf = new Configuration();
    FileSystem fs = FileSystem.get(URI.create(src),conf);
    Path hdfsPath=new Path(src);
    Path localPath=new Path(dst);
    fs.copyToLocalFile(hdfsPath, localPath);
}
```

### 6．创建文件夹

如果要在 HDFS 中创建文件夹，可以使用 FileSystem 类提供的 mkdirs 方法。mkdirs 方法原型包括：

- public boolean mkdirs(Path f)
- public abstract boolean mkdirs(Path f,FsPermission permission)
- public static boolean mkdirs(FileSystem fs,Path dir,FsPermission permission)

例9：在 HDFS 的/hdfs 目录下创建一个 course 的文件夹。

```
public void createDir() throws Exception {
    String uri = "hdfs://master:9000/hdfs/course";
    Configuration conf = new Configuration();
    FileSystem fs = FileSystem.get(URI.create(uri), conf);
    fs.mkdirs(new Path(uri));
}
```

### 7．重命名文件

如果要在 HDFS 中修改文件或文件夹的名称，可以使用 FileSystem 类提供的 rename 方法。rename 方法原型包括：

- public abstract boolean rename(Path src,Path dst)
- protected void rename(Path src,Path dst,org.apache.hadoop.fs.Options.Rename…options)

例 10：将 HDFS 的文件夹/hdfs/course 重命名为/hdfs/new_course。

```
public void renameDir() throws Exception {
    String oldName= "hdfs://master:9000/hdfs/course";
    String newName= "hdfs://master:9000/hdfs/new_course";
    Configuration conf = new Configuration();
    FileSystem fs = FileSystem.get(URI.create(oldName), conf);
    fs.rename(new Path(oldName),new Path(newName));
}
```

### 8. 查看文件状态

如果要查看文件信息，可以使用 FileSystem 类提供的 getFileStatus 方法。该方法返回 FileStatus 对象，封装了文件系统中文件和目录的元数据，包括文件长度、块大小、备份数、修改时间、所有者和文件权限信息等。getFileStatus 方法的原型为：

- public abstract FileStatus getFileStatus(Path f)

例 11：查看 HDFS 中文件/hdfs/word.txt 的路径、块大小、所有者、权限、长度、备份数和修改时间等。

```
public void fileStatus() throws Exception {
    String uri = "hdfs://master:9000/hdfs/word.txt";
    Configuration conf = new Configuration();
    FileSystem fs = FileSystem.get(URI.create(uri), conf);
    boolean flag = fs.exists(new Path(uri));
    if(flag){
        FileStatus fileStatus = fs.getFileStatus(new Path(uri));
        System.out.println("文件路径："+fileStatus.getPath());
        System.out.println("块的大小："+fileStatus.getBlockSize());
        System.out.println("文件所有者："+fileStatus.getOwner()+":"+fileStatus.getGroup());
        System.out.println("文件权限："+fileStatus.getPermission());
        System.out.println("文件长度："+fileStatus.getLen());
        System.out.println("备份数："+fileStatus.getReplication());
        System.out.println("修改时间："+fileStatus.getModificationTime());
    }
}
```

### 9. 查看文件夹中所有文件状态

如果要查看 HDFS 文件系统中某个文件夹下所有文件的状态信息，可以使用 FileSystem 类提供的 listStatus 方法，该方法返回一个 FileStatus 数组，包含指定文件夹中所有文件和子文件夹的状态信息。listStatus 方法的原型包括：

- public abstract FileStatus[] listStatus(Path f)
- public FileStatus[] listStatus(Path[] files)
- public FileStatus[] listStatus(Path[] files,PathFilter filter)
- public FileStatus[] listStatus(Path f,PathFilter filter)

例 12：查看 HDFS 中/hdfs 和/user/hadoop 两个文件夹中所有文件的路径信息。

```
public void dirStatus() throws Exception {
    String[] uris = {"hdfs://master:9000/hdfs","hdfs://master:9000/user/hadoop"};
    Configuration conf = new Configuration();
```

```
        FileSystem fs = FileSystem.get(URI.create(uris[0]), conf);
        Path[] paths = new Path[uris.length];
        for (int i = 0; i < paths.length; i++) {
            paths[i] = new Path(uris[i]);
        }
        FileStatus[] status = fs.listStatus(paths);
        Path[] listedPaths = FileUtil.stat2Paths(status);
        for (Path p : listedPaths) {
            System.out.println(p);
        }
    }
```

### 9. 删除文件

如果要在 HDFS 中删除文件或目录，可以使用 FileSystem 类提供的 delete 方法。delete 方法原型包括：

● public abstract boolean delete(Path f,boolean recursive)

**例 13**：删除 HDFS 中的/hdfs/hello.txt 文件。

```
public void delete_hdfs() throws Exception {
    String uri = "hdfs://master:9000/hdfs/hello.txt";
    Configuration conf = new Configuration();
    FileSystem fs = FileSystem.get(URI.create(uri), conf);
    fs.delete(new Path(uri),true);
}
```

## 3.4.4 HDFS 开发实例

### 1. 新建一个 Maven 工程

① 选择菜单"File→New→Project"，打开"New Project"界面，选择"Maven"列表下的"Maven Project"选项，如图 3-26 所示。

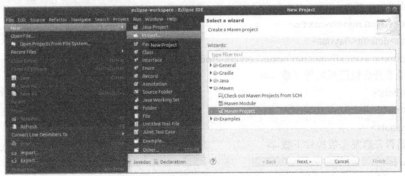

图 3-26 创建 Maven 工程

② 单击"Next"按钮，进入图 3-27 所示的"New Maven Project"界面，勾选"Create a simple project"和"Use default Workspace location"选项。

③ 单击"Next"按钮，进入图 3-28 所示的工程组与工程名称设置界面。

④ 单击"Finish"按钮，完成 Maven 工程的创建。

### 2. 编辑 pom.xml 文件，添加 Hadoop 依赖包

在 Maven 工程中，使用 pom.xml 文件定义工程的基本信息、描述工程如何构建、声明工程依赖包等。

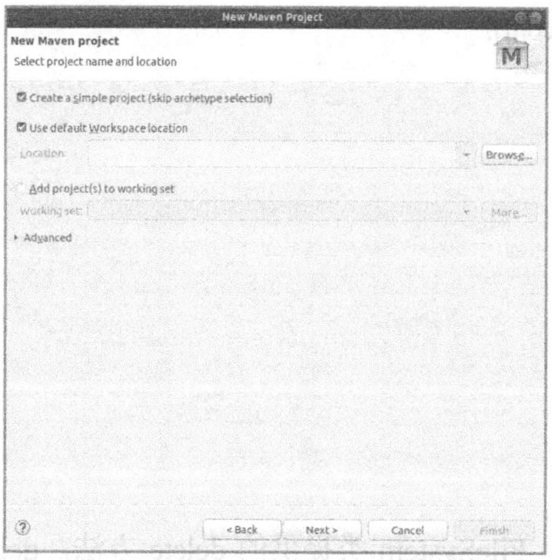

图 3-27 "New Maven Project"界面 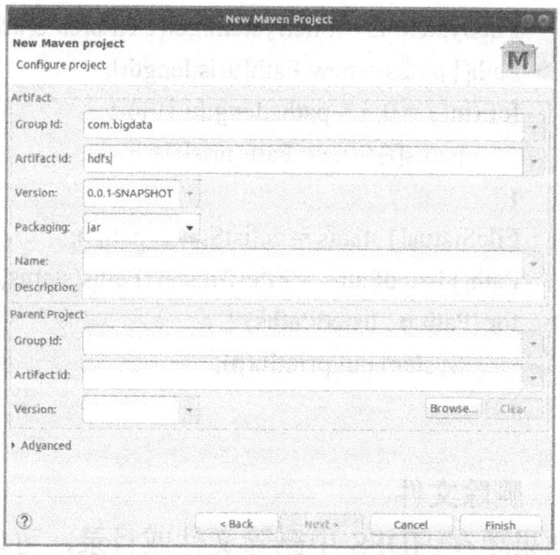　　　　图 3-28 工程组与工程名设置界面

为了实现对 HDFS 的操作，需要 hadoop-common、hadoop-client、hadoop-hdfs 等依赖包，以及用于测试的 junit 等依赖包。默认情况下，系统将从 Maven 远程仓库 http://repo1.maven.org/maven2 下载各种依赖包到本地仓库，但下载速度比较慢，可以设置国内的 Maven 仓库镜像，例如阿里 Maven 镜像仓库，以提高依赖包的下载速度。

可以在 Maven 配置文件 settings.xml 中设置全局镜像节点，例如：

```xml
<mirror>
    <id> alimaven </id>
    <mirrorOf>central</mirrorOf>
    <name>aliyun maven</name>
    <url>http://maven.aliyun.com/nexus/content/groups/public</url>
</mirror>
```

也可以之间在 pom.xml 文件中配置局部远程 Maven 仓库镜像，例如：

```xml
<repository>
    <id>alimaven</id>
    <name>aliyun maven</name>
    <layout>default</layout>
    <url>http://maven.aliyun.com/nexus/content/groups/public</url>
    <!-- 是否开启快照版构件下载 -->
    <snapshots>
        <enabled>false</enabled>
    </snapshots>
    <!-- 是否开启发布版构件下载-->
    <releases>
        <enabled>true</enabled>
    </releases>
</repository>
```

添加依赖后的 pom.xml 文件内容为：

```xml
<project xmlns="http://maven.apache.org/POM/4.0.0"
xmlns:xsi="http://www.w3.org/2001/XMLSchema-instance"
xsi:schemaLocation="http://maven.apache.org/POM/4.0.0https://maven.apache.org/xsd/maven-4.0.0.xsd">
    <modelVersion>4.0.0</modelVersion>
    <groupId>com.bigdata</groupId>
    <artifactId>hdfs</artifactId>
```

```xml
        <version>0.0.1-SNAPSHOT</version>
        <repositories>
        <!--设置远程仓库镜像为https://maven.aliyun.com/repository/public -->
            <repository>
                <id>alimaven</id>
                <name>aliyun maven</name>
                <layout>default</layout>
                <url>http://maven.aliyun.com/nexus/content/groups/public</url>
                <!-- 是否开启快照版构件下载 -->
                <snapshots>
                    <enabled>false</enabled>
                </snapshots>
                <!-- 是否开启发布版构件下载-->
                <releases>
                    <enabled>true</enabled>
                </releases>
            </repository>
        </repositories>
        <properties>
            <project.build.sourceEncoding>UTF-8</project.build.sourceEncoding>
        </properties>
        <dependencies>
            <dependency>
                <groupId>org.apache.hadoop</groupId>
                <artifactId>hadoop-client</artifactId>
                <version>3.1.2</version>
                <scope>provided</scope>
            </dependency>
            <dependency>
                <groupId>junit</groupId>
                <artifactId>junit</artifactId>
                <version>3.8.2</version>
                <scope>test</scope>
            </dependency>
        </dependencies>
</project>
```

单击工具栏上的保存按钮圖，系统会自动从远程仓库下载各种依赖包到本地仓库中，如图3-29所示。首次执行下载依赖包过程可能会持续较长时间，用户需要等待下载完毕。

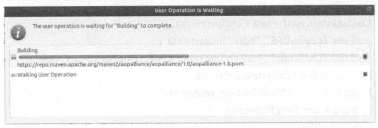

图3-29 下载依赖包到本地仓库

### 3．程序设计

配置好pom.xml文件并下载依赖包后，就可以创建Class进行业务处理了。

① 在 Eclipse 中选择窗口左侧"Package Explorer"中的工程，如 hdfs，展开文件夹，选择"src/main/java"文件夹并右键单击，在弹出菜单中选择"New→Class"选项，如图 3-30 所示。在弹出的"New Java Class"窗口中进行新建 Class 的设置，包括 Package、Name 设置，如图 3-31 所示。

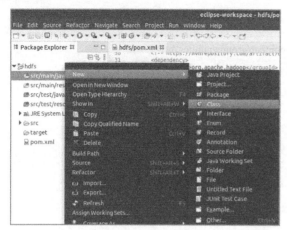

图 3-30　新建 Class

图 3-31　Class 设置

② 单击"Finish"按钮，完成 Class 的创建。然后，在代码编辑窗口进行程序设计。

例如，指定一个 HDFS 文件名，判断该文件是否存在。如果该文件存在，则输出该文件内容；如果该文件不存在，则创建该文件，并将/usr/local/hadoop/README.txt 文件内容复制到该文件中。代码为：

```
package com.bigdata.example;
import java.io.InputStreamReader;
import java.io.BufferedReader;
import java.io.FileInputStream;
import java.io.BufferedInputStream;
import java.io.InputStream;
import org.apache.hadoop.conf.Configuration;
import org.apache.hadoop.fs.FSDataInputStream;
import org.apache.hadoop.fs.FSDataOutputStream;
import org.apache.hadoop.fs.FileSystem;
import org.apache.hadoop.fs.Path;
import org.apache.hadoop.io.IOUtils;
public class HDFS {
    public static void main(String[] args) {
        try {
            Configuration conf = new Configuration();
            conf.set("fs.defaultFS", "hdfs://master:9000");
            conf.set("fs.hdfs.impl", "org.apache.hadoop.hdfs.DistributedFileSystem");
            FileSystem fs = FileSystem.get(conf);
            String fileName = "/hdfs/hadoop_readme.txt";
            Path file = new Path(fileName);
            if (fs.exists(file)) {
                FSDataInputStream in = fs.open(file);
                BufferedReader br = new BufferedReader(new InputStreamReader(in));
                String content = null;
```

```
                while ((content = br.readLine()) != null) {
                    System.out.println(content);
                }
                br.close();
                in.close();
                fs.close();
            } else {
                InputStream in = new BufferedInputStream(
                    new FileInputStream("/usr/local/hadoop/README.txt"));
                FSDataOutputStream out = fs.create(file);
                IOUtils.copyBytes(in, out, 4096, true);
                out.close();
                fs.close();
                System.out.print("create " + fileName + " successful!");
            }
        } catch (Exception e) {
            e.printStackTrace();
        }
    }
}
```

需要注意的是，如果程序中没有使用 Configuration 对象的 set 方法进行系统配置属性设置，可以将 Hadoop 平台的配置文件 core-site.xml 和 hdfs-site.xml 复制到工程下的/src/main/resources 文件夹中，程序运行时将根据配置文件进行设置。

③ 编写完程序代码后，单击运行按钮 ●，运行程序。

程序第 1 次运行结果如图 3-32 所示，显示创建文件成功；程序第 2 次运行时输出结果如图 3-33 所示，输出文件内容。

图 3-32　显示文件创建成功

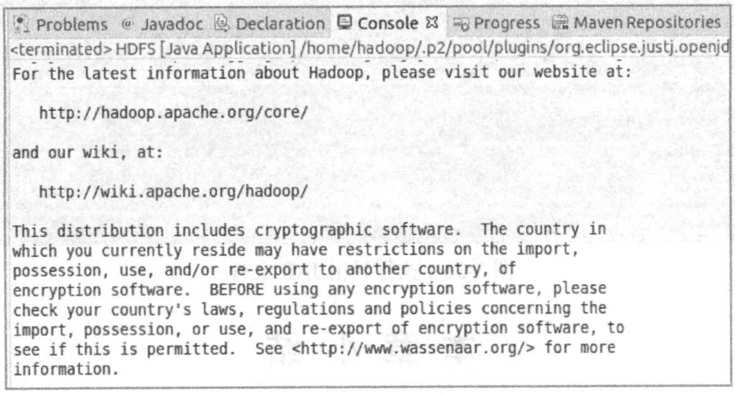

图 3-33　输出文件内容

### 4．工程打包、部署与运行

工程编写、调试完成后，通常需要以 JAR 文件形式导出，并部署到集群中以便执行。

为了在执行程序时动态指定文件路径和名称，可以在程序中通过参数获取文件路径和名称，只需将程序中的代码 String fileName = "/hdfs/hadoop_readme.txt"替换为下列代码即可：

```
String[] allArgs = (new GenericOptionsParser(conf, args)).getRemainingArgs();
String fileName=allArgs[0];
```

(1) 将工程导出为 JAR 文件

① 在 Eclipse 中选择工程 "hdfs" 并右键单击,在弹出菜单中选择 "Export" 选项,弹出导出类型选择界面,如图 3-34 所示。

② 选择 "Java→JAR file" 选项,打开导出设置界面。设置导出文件路径与名称后,如图 3-35 所示,单击 "Finish" 按钮,开始工程导出。

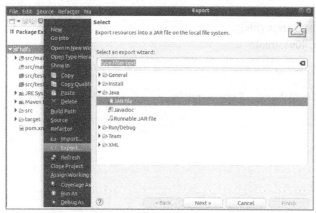

图 3-34　选择导出类型　　　　　　　　图 3-35　文件路径与名称设置

(2) 工程部署与执行

工程导出完成后,可以将导出的 JAR 文件部署到集群,并执行 JAR 文件。

所谓工程文件部署,是指将导出的 JAR 文件上传到集群的名称节点或者数据节点,以备执行。

执行 JAR 文件使用 hadoop jar 命令,语法为:

hadoop jar [jar文件位置] [ jar文件中的主类][参数]

例如,执行导出来的 hdfs.jar 文件,如图 3-36 所示。

图 3-36　执行导出 JAR 文件

# 本 章 小 结

本章首先介绍了 HDFS 的基本概念与工作机制,包括 HDFS 的架构、数据存储策略、数据读取策略、数据恢复机制、数据读写过程等,然后介绍了 HDFS Shell 常用命令,最后介绍了 HDFS Java 程序设计方法。通过本章的学习,读者可以掌握 HDFS 工作原理、HDFS 数据管理与开发。

# 思考题与习题

## 1. 简答题
（1）简述 HDFS 架构组件及其作用。
（2）简述 HDFS 数据存储策略。
（3）简述 HDFS 数据读取策略。
（4）简述 HDFS 数据恢复策略。

## 2. 选择题
（1）在 Hadoop2.x 中，数据块的默认大小为（　　）。
　　A．32MB　　　　　　B．64MB　　　　　　C．128MB　　　　　　D．256MB
（2）HDFS 是基于流数据模式访问和处理超大文件的需求而开发的，具有高容错性、高可靠性、高可扩展性、高吞吐率等特征，适合的读写任务是（　　）。
　　A．一次写入，少次读写　　　　　　B．多次写入，一次读写
　　C．一次写入，多次读取　　　　　　D．多次写入，多次读写
（3）关于 HDFS 的文件写入，正确的是（　　）。
　　A．支持多用户对同一文件的写操作　　B．用户可以在文件任意位置进行修改
　　C．默认将数据块复制成 3 份存放　　　D．复制的数据块默认都存在同一机架上
（4）hadoop dfs 中的 get 和 put 命令操作对象是（　　）。
　　A．本地文件　　　　　　　　　　　B．本地目录
　　C．集群文件与目录　　　　　　　　D．本地或集群上的文件或目录
（5）在 HDFS 文件系统中更改文件权限的命令是（　　）。
　　A．-chgrp　　　　　B．-chmod　　　　　C．-chown　　　　　D．-count
（6）关于 DataNode 节点描述不正确的是（　　）。
　　A．存储数据
　　B．存放集群运行过程中产生的日志信息
　　C．数据节点以心跳的方式向名称节点发送消息
　　D．保存数据块与本地文件之间的映射
（7）下面哪个程序负责 HDFS 数据存储？（　　）
　　A．NameNode　　　B．DataNode　　　C．ResourceManager　　D．NodeManager
（8）HDFS 中的数据块的副本数量在哪个配置文件中设置？（　　）
　　A．core-site.xml　　B．hdfs-site.xml　　C．yarn-site.xml　　D．mapreduce-site.xml
（9）客户端上传文件时，下列哪项不正确？（　　）
　　A．客户端通过 NameNode 确定要写入的 DataNode 地址和数据块
　　B．客户端将文件以数据块为单位，以管道方式依次传到 DataNode
　　C．客户端只上传数据到一个 DataNode，然后由 NameNode 负责数据块复制工作
　　D．当某个 DataNode 失败，客户端会继续传给其他 DataNode
（10）关于 NameNode 描述不正确的是（　　）。
　　A．存储集群元数据　　　　　　　　　　　　B．数据保存在内存中
　　C．保存文件、数据块、DataNode 之间的映射关系　　D．保存数据

## 3. 实训题

（1）利用 HDFS Shell 命令完成下列操作。
① 在 HDFS 根目录下创建 test 文件夹。
② 在 HDFS 中的/test 目录下创建 file.txt 文件。
③ 将 HDFS 中的/test/file.txt 文件改名为 file2.txt。
④ 将 file2.txt 文件复制到 HDFS 根目录下。
⑤ 在 Linux 本地创建 data.txt 文件并上传到 HDFS 根目录下。
⑥ 查看 HDFS 中的/test/file2.txt 文件大小。
⑦ 查看 HDFS 中的/test/file2.txt 文件的备份数。
⑧ 将 HDFS 中的/test/file2.txt 文件下载到本地/data 目录下。
⑨ 查看 HDFS 根目录及其所有子目录中的文件信息。
⑩ 删除 HDFS 中的/test 目录。
（2）创建一个 Maven 工程，将 3.3.2 节中介绍的 Java API 方法封装起来。
（3）利用 Java 程序设计完成下列任务。
① 在 HDFS 的/user/hadoop 目录下创建一个目录"exercise"。
② 将本地 Linux 系统上的 test.txt 文件复制到 exercise 目录下。
③ 查看 HDFS 中的 test.txt 内容，并在终端输出。
④ 将 HDFS 中的 test.txt 文件内容保存到另一个名为 new_test.txt 文件中。
⑤ 查看 HDFS 中 exercise 目录下各个文件的状态信息。

# 第 4 章 MapReduce

MapReduce 是 Hadoop 的核心组件，是 Google MapReduce 的开源实现，用于分布式计算。本章将介绍 MapReduce 分布式计算框架的理论知识和应用开发。
- MapReduce 概述：MapReduce 简介、计算模型、编程模型。
- MapReduce 架构：MapReduce V1 架构、基于 YARN 的 MapReduce V2 架构。
- MapReduce 编程组件：MapReduce 编程流程、编程组件、序列化与反序列化。
- MapReduce 程序设计实例：WordCount 程序开发详解。
- MapReduce 开发典型案例：数据去重、数据排序、计算平均值。
- MapReduce 网站日志分析应用：利用 MapReduce 进行网站日志分析。

## 4.1 MapReduce 概述

### 4.1.1 MapReduce 简介

MapReduce 最早是由 Google 公司提出的一种面向大规模数据处理的并行计算平台和软件架构。

MapReduce 是一个基于集群的高性能并行计算平台，允许用市场上普通的商用服务器构成一个包含数十、数百甚至数千个节点的分布式并行计算的集群。

MapReduce 是一个并行计算软件架构，能够在集群节点上自动完成计算任务的并行化处理，包括自动划分数据和计算任务、自动执行任务及收集计算结果，将数据分布存储、数据通信、容错处理等并行计算涉及的系统底层的复杂细节交由系统负责处理，大大减轻了软件开发人员的负担。

MapReduce 是一个并行程序设计模型。MapReduce 将复杂的、运行于大规模集群上的并行计算过程高度地抽象到了 map() 和 reduce() 两个函数。map() 函数负责把任务分解成多个子任务，reduce() 函数负责把结果汇总起来。

Hadoop MapReduce 是开源项目 Lucene（一种搜索引擎程序库）和 Nutch（搜索引擎）的创始人 Doug Cutting 基于 Google 公司 2004 年 12 月发表的 MapReduce 论文，对 Google MapReduce 的开源实现，目前已成为市场上最为成功、最广为接受和最易于使用的大数据并行处理技术之一。

Hadoop MapReduce 具有下列特性：

① 采用"分而治之，迭代汇总"的策略，把大规模数据集分发给一个主节点管理下的各个分节点并行完成处理，然后主节点通过整理各个从节点的操作结果得到最后结果。简单地说，MapReduce 就是"任务的分解与结果的汇总"，主要解决海量数据的并行计算问题。

② 采用"计算向数据靠拢"，而不是"数据向计算靠拢"的设计理念，因而减少了移动数据需要大量的网络传输开销，提高了系统的性能。

③ 采用主从（Master/Slave）架构，与 HDFS 类似，包括一个主节点和多个子节点。

④ Hadoop 框架是用 Java 实现的，但是 MapReduce 应用程序可以使用各种计算机编程语言

（如 Java、Python、C++、Ruby 等）编写。

HadoopMapReduce 具有下列优点：

① 易于编程。用户只需要编写 map()函数和 reduce()函数即可完成分布式程序设计。

② 易于扩展。基于分布式集群，可以根据业务需要动态增加集群节点，扩展集群规模，从而提高 MapReduce 的计算能力，能够处理 PB 级以上的海量数据的离线处理。

③ 容错性强。当某个节点宕机或任务中断时，MapReduce 会自动重启该节点或任务；如果某个节点损坏，MapReduce 会在其他节点启动一个一模一样的任务。

但是由于 MapReduce 工作机制的延迟性，无法满足毫秒级或秒级的应用要求，无法处理动态变化的数据，无法进行流式计算、实时计算。

此外，MapReduce 也不适合多个具有依赖关系的任务执行，因为一个任务的输出作为另一个任务的输入，会产生大量的磁盘 I/O 操作，降低系统性能。

### 4.1.2 MapReduce 计算模型

一个 MapReduce 作业通常会把输入的数据集（HDFS 文件）切分为若干独立的数据分片（InputSplit），由 Map 任务以完全并行的方式处理。MapReduce 框架会对 Map 任务的输出进行 Shuffle（合并、分区、排序、归并），然后将结果输入给 Reduce 任务，Reduce 处理后以分区（Partition）为单位输出到 HDFS，如图 4-1 所示。典型的作业输入、输出及中间结果都会被存储在 HDFS 文件系统中。整个框架负责任务的调度、监控和失败任务的重新执行。

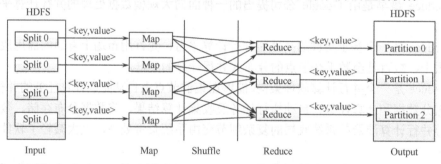

图 4-1 MapReduce 计算模型

MapReduce 在进行并行计算时可以细分为 5 阶段进行。

① 数据输入。MapReduce 对输入的 HDFS 文件进行分片，每个分片对应一个 Map 任务，输入分片存储的并非数据本身，而是一个记录分片存储位置、分片长度和分片起始位置的数组。

② Map 任务。将分片中的每条记录转换为一个键值对<key,value>，输入给 map()函数，映射到一组中间格式的键值对<key,value>集合。在 Map 阶段，执行 Map 任务的各个节点之间相互独立，不进行信息传递和交互。

③ Shuffle 任务。完成 Map 阶段输出数据的分区、排序、合并、归并等工作。

④ Reduce 任务。以一系列<key, list(value)>中间结果作为输入，执行用户定义的逻辑，与同一个键（key）关联的一组中间数值集规约为一个更小的数值集。

⑤ 结果输出。MapReduce 将处理结果按指定格式输出到 HDFS 存储。

### 4.1.3 MapReduce 编程模型

MapReduce 对数据集的处理整体可以分为 Map 和 Reduce 两大阶段：第一阶段是 Map 阶段，读取输入的数据，以键值对形式输出处理的中间结果，即"分而治之"；第二阶段是 Reduce 阶段，读取 Map 阶段输出的中间结果，根据键值进行分组合并，输出最终的结果，即"迭代汇总"。MapReduce 框架提供这两个阶段函数的实现，即 map()函数和 reduce()函数。MapReduce 程序设计就是用户根据业务需要，重写 map()函数和 reduce()函数，并为 map()函数和 reduce()函数开辟多个线程并发执行。

map()函数以键值对（key-value）作为输入，产生另外一系列键值对作为中间结果，输出到磁盘。MapReduce 框架会自动将这些中间数据按照键值进行聚合，且键值相同的数据被统一提交给 reduce()函数处理。

reduce()函数以键（key）及其对应的值（value）列表作为输入，因为 map()函数的输出被送到 Reduce 的过程中会根据 key 进行合并，把相同 key 的数据合并在一起，相同 key 对应的 value 在合并过程中放到一个集合列表中。所以，最终 map()函数的输出到达 Reduce 时就变成了一个 key 对应一个 value 列表。经过 reduce()函数的聚合运算后，得到最终的 key 及 value 值，并写入 HDFS 中。

例如，在词频统计应用中，map()函数接收的键值对为<k1,v1>，其中，k1 为字符偏移量，v1 为一行文本。map()函数输出为一系列<k2,v2>，其中 k2 为单词，v2 为数字 1。reduce()函数输入参数为<k2,list<v2>>，即每个单词与其所有的值的列表。reduce()函数的输出为<k2,v3>，即每个单词与出现总次数的键值对。如图 4-2 所示。

图 4-2 map()函数与 reduce()函数

## 4.2 MapReduce 架构

在 Hadoop 1.0 到 Hadoop 2.0 的发展过程中，MapReduce 架构也经历了 MapReduce V1 到 MapReduce V2 的版本变化。为了解决 MapReduce V1 资源共享、资源分配与回收方面的性能瓶颈，MapReduce V2 中引入了 YARN（Yet Another Resource Negotiator），它是一种新的 Hadoop 资源管理器，可以为上层应用提供统一的资源管理与调度。YARN 的引入，为集群在利用率、资源统一管理和数据共享等方面带来了巨大好处。

了解了 MapReduce V1 架构，再学习 MapReduce V2 架构就非常容易了。

### 4.2.1 MapReduce V1 架构

**1. MapReduce V1 架构组件**

与 HDFS 架构类似，MapReduce 也采用主从架构。在 MapReduce V1 架构中，主要由 Client、JobTracker、TaskTracker 和 Task 组成，如图 4-3 所示。

（1）Client

用户编写的 MapReduce 程序通过 Client 提交到 JobTracker。用户可通过 Client 提供的一些接口查看作业运行状态。

（2）JobTracker

JobTracker 负责资源监控和作业调度：监控所有 TaskTracker 与作业的健康状况，一旦发现失败，就将相应的任务转移到其他节点；跟踪任务的执行进度、资源使用量等信息，并将这些信息告诉任务调度器（TaskScheduler），而调度器会在资源出现空闲时，选择合适的任务去使用这些资源。

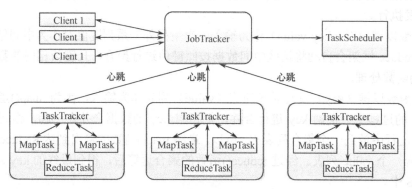

图 4-3 MapReduce V1 架构

（3）TaskTracker

TaskTracker 会周期性地通过"心跳"将本节点上资源的使用情况和任务的运行进度汇报给 JobTracker，同时接收 JobTracker 发送过来的命令并执行相应的操作（如启动新任务、杀死任务等）。如果 TaskTracker 发生失败，JobTracker 会重新把这些任务分配到其他的 TaskTracker 上运行；即使 TaskTracker 没有失败，也可以被 JobTracker 列入黑名单。TaskTracker 使用"slot"（任务槽）等量划分本节点上的资源量，slot 代表 CPU、内存等计算资源。一个 Task 获取到一个 slot 后才有机会运行，而 Hadoop 调度器的作用就是将各个 TaskTracker 上的空闲 slot 分配给 Task 使用。slot 分为 MapSlot 和 ReduceSlot 两种，分别供 MapTask 和 ReduceTask 使用。TaskTracker 通过 slot 数据（可配置参数）限制 Task 的并发数。

（4）Task

Task 分为 MapTask 和 ReduceTask 两种，均由 TaskTracker 启动。其中，MapTask 数量由分片数（split）决定，而 ReduceTask 数量由 Reduce 数量决定，默认为 1，可以由用户进行参数设置。

**2. MapReduce V1 架构工作流程**

MapReduce V1 架构的工作流程如图 4-4 所示。

① 一个 MapReduce 程序就是一个作业（Job），运行作业时，生成一个 JobClient。

② JobClient 向 JobTracker 请求一个新的作业 ID。

③ JobClient 将运行作业所需要的资源复制到 JobTracker 的文件系统中以作业 ID 命名的目录下。

④ JobClient 准备工作做好了之后，JobClient 向 JobTracker 提交任务。

⑤ JobTracker 接收到作业提交信息后，将其放入内部队列，交由作业调度器进行调度，并对其进行初始化，创建一个正在运行的作业对象，以便 JobTracker 跟踪作业的状态和进程。

⑥ 初始化完毕后，作业调度器会获取输入分片信息，每个分片创建一个 Map 任务。

⑦ 每个 TaskTracker 定期发送心跳给 JobTracker，告知自己的状态，并附带消息说明自己是否已准备好接受新任务。JobTracker 以此来分配任务，并使用心跳的返回值与 TaskTracker 通信。JobTracker 利用调度算法先选择一个作业然后选此作业的一个 Task 分配给 TaskTracker。

每个 TaskTracker 会有固定数量的 Map 和 Reduce 任务槽 slot，数量由 TaskTracker 的数量和内存大小来决定。JobTracker 会先将 TaskTracker 的所有 Map 任务槽填满，然后才填此 TaskTracker 的 Reduce 任务槽。

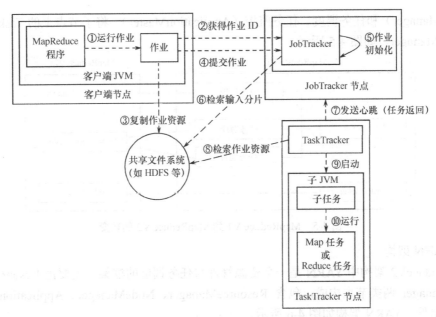

图 4-4　MapReduce V1 架构的工作流程

JobTracker 分配 Map 任务时会选取与输入分片最近的 TaskTracker，即数据 TaskTracker 优化。在分配 Reduce 任务用不着考虑 TaskTracker。

⑧ TaskTracker 分配到一个任务后，首先从 HDFS 中把作业的 JAR 文件及运行所需要的全部文件（DistributedCache 设置的）复制到 TaskTracker 本地。接下来 TaskTracker 为任务新建一个本地工作目录，并把 JAR 文件的内容解压到这个文件夹下。

⑨ TaskTracker 新建一个 TaskRunner 实例来运行该任务。

⑩ TaskRunner 启动一个新的 JVM 来运行每个任务。

### 3．MapReduce V1 架构缺点

MapReduce V1 架构存在下列问题。

① JobTracker 单点故障。由于 JobTracker 是 MapReduce 的集中处理节点，如果出现单点故障，集群将不能使用，集群的高可用性得不到保障。

② JobTracker 任务过重。JobTracker 节点完成了太多的任务，会造成过多的资源消耗，当任务非常多时，会造成很大的内存开销，潜在地增加了 JobTracker 节点死机的风险。

③ 资源利用率低。基于 slot 的资源分配，而 slot 是一种粗粒度的资源划分单位，通常一个任务不会用完分配的所有资源，但其他任务也无法使用这些空闲的资源。此外，MapSlot 与 ReduceSlot 之间无法共享，当所有 Map 任务完成之前，ReduceSlot 处于空闲状态，而所有 Map 任务完成后，MapSlot 处于空闲状态，从而导致资源利用不平衡。

④ 无法支持多种计算框架。MapReduce V1 只支持 MapReduce 计算，不支持内存计算框架、流式计算框架等。

为了从根本上解决 MapReduce V1 架构存在的性能瓶颈，Hadoop 2.x 中对 MapReduceV1 进行了重构，引入了 YARN 资源管理与调度组件，新的 MapReduce 架构称为 MapReduce V2 或 MapReduce on YARN。

## 4.2.2　MapReduce V2 架构

在 MapReduce V2 架构中，将 MapReduce V1 架构中主节点 JobTracker 功能拆分为资源管理

（ResourceManager）和任务调度、任务监控（ApplicationMaster），将子节点上的 TaskTracker 转变为 NodeManager，如图 4-5 所示。

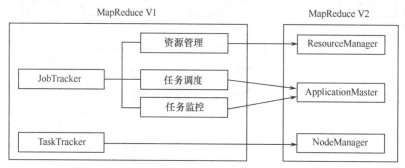

图 4-5　MapReduce V1 到 MapReduce V2 的转变

### 1. YARN 架构

MapReduceV2 架构中 YARN 是一个资源管理与任务调度的框架，主要由 ResourceManager 和 NodeManager 两类节点构成，包含 ResourceManager、NodeManager、ApplicationMaster 和 Container 组件。YARN 架构如图 4-6 所示。

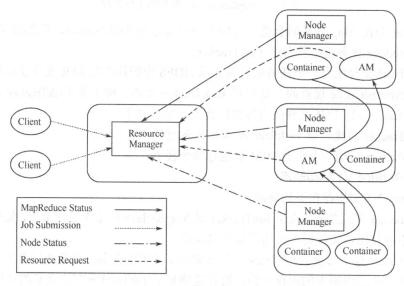

图 4-6　YARN 架构

#### （1）ResourceManager

ResourceManager 简称 RM，负责整个集群的资源管理与分配，是一个全局的资源管理系统。NodeManager 以心跳的方式向 ResourceManager 汇报资源使用情况。ResourceManager 只接收 NodeManager 的资源汇报信息，对具体的资源处理则交给 NodeManager 处理。ResourceManager 接收用户的请求，负责应用程序的调度管理，启动应用程序对应的 ApplicationMaster，并为每一个应用程序分配所需的资源。它主要由两个组件构成：调度器（Scheduler）和应用程序管理器（Applications Manager，ASM）。

调度器根据容量、队列等限制条件（如每个队列分配一定的资源，最多执行一定数量的作业等），将系统中的资源分配给各个正在运行的应用程序。调度器仅根据各个应用程序的资源需求进行资源分配，而资源分配单位用一个抽象概念"资源容器"（Resource Container，简称 Container）表示。Container 是一个动态资源分配单位，它将内存、CPU、磁盘、网络等资源封装在一起，从而限定每个任务使用的资源量。应用程序管理器主要负责管理整个系统中所有的应用程序，接收 Map/Reduce 作业的提交请求，为应用程序分配第一个 Container 来运行 ApplicationMaster，包括应用程序提交、与调度器协商资源以启动 ApplicationMaster、监控 ApplicationMaster 运行状态并在失败时重新启动它等。

（2）NodeManager

NodeManager 简称 NM，是每个节点上的资源和任务管理器，相当于所在机器的代理，负责该节点程序运行、资源管理和监控。集群中的每个节点都会运行一个 NodeManager 守护进程，定时向 ResourceManager 汇报本节点资源（如内存、CPU 等）的使用情况和 Container 的运行状态。当 ResourceManager 通信失败时，NodeManager 自动连接 ResourceManager 的备用节点。NodeManager 接收并处理来自 ApplicationMaster 的 Container 启动、停止等各种请求。

（3）ApplicationMaster

ApplicationMaster 简称 AM，每次提交一个应用程序便产生一个用于跟踪和管理这个程序的 ApplicationMaster，该程序运行在 ResourceManager 节点以外的机器上，负责向 ResourceManager 调度器申请资源，获取一个 Container，用于任务的运行和监控：将得到的任务进一步分配给内部的任务，即资源二次分配；与 ResourceManager 通信以启动或停止任务；监控所有任务的运行状态，并在任务运行失败时重新为任务申请资源和重启任务。

（4）Container

Container 是 YARN 中的资源抽象，它封装了某个节点上的多维度资源，如内存、CPU、磁盘、网络等，当 AM 向 RM 申请资源时，RM 为 AM 返回的资源便是用 Container 表示的。YARN 会为每个任务分配一个 Container，且该任务只能使用该 Container 中描述的资源。Container 是一个动态资源分配单位，根据应用程序的需求动态生成。目前，YARN 仅支持 CPU 和内存两种资源，且使用了轻量级资源隔离机制进行资源隔离。

## 2. MapReduce V2 工作流程

当用户向 YARN 提交一个 MapReduce 应用程序后，YARN 将分两个阶段运行该应用程序。第一阶段是启动 ApplicationMaster。第二阶段是由 ApplicationMaster 创建应用程序，为它申请资源，并监控它的整个运行过程，直到运行完成。

基于 YARN 的 MapReduce 作业执行流程如图 4-7 所示。

MapReduce 作业执行主要包括下列基本步骤。

（1）作业提交

client 调用 job.waitForCompletion 方法，向整个集群提交 MapReduce 作业（第 1 步）。新的作业 ID（应用程序 ID）由资源管理器分配（第 2 步）。作业的 client 核实作业的输出，计算输入的 split，将作业的资源（包括 JAR 文件、配置文件、split 信息）复制给 HDFS（第 3 步）。最后，通过调用资源管理器的 submitApplication()来提交作业（第 4 步）。

（2）作业初始化

当资源管理器收到 submitApplciation()请求时，就将该请求发给调度器（Scheduler），调度器分配 Container，然后资源管理器在该 Container 内启动 Application Master 进程，由节点管理器监控（第 5 步）。MapReduce 作业的 Application Master 是一个主类为 MRAppMaster 的 Java 应用，其通过创造一些 bookkeeping 对象来监控作业的进度，得到任务的进度和完成报告（第 6 步）。通过分布式文件系统得到由客户端计算好的输入 split（第 7 步），然后为每个输入 split 创建一个 Map 任务，根据 mapreduce.job.reduces 创建 Reduce 任务对象。

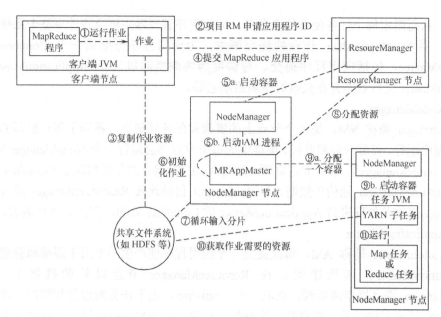

图 4-7 基于 YARN 的 MapReduce 作业执行流程

（3）任务分配

如果作业很小，Application Master 会选择在其自己的 JVM 中运行任务。如果不是小作业，那么 Application Master 向资源管理器请求 Container 来运行所有的 Map 任务和 Reduce 任务（第 8 步）。这些请求是通过心跳来传输的，包括每个 Map 任务的数据位置，比如存放输入 split 的主机名和机架。调度器利用这些信息来调度任务，尽量将任务分配给存储数据的节点，或者分配给和存放输入 split 的节点相同机架的节点。

（4）任务运行

当一个任务由资源管理器的调度器分配给一个 Container 后，Application Master 通过联系 Node Manager 节点来启动 Container（第 9 步）。任务由一个主类为 YarnChild 的 Java 应用执行。在运行任务之前，首先本地化任务需要的资源，比如作业配置、JAR 文件及分布式缓存的所有文件（第 10 步）。最后，运行 Map 或 Reduce 任务（第 11 步）。

YarnChild 运行在一个专用的 JVM 中，但是 YARN 不支持 JVM 重用。

（5）进度和状态更新

YARN 中的任务将其进度和状态返回给 Application Master，客户端每秒（通过 mapreduce.client.progressmonitor.pollinterval 设置）向 Application Master 请求进度更新，展示给用户。

（6）作业完成

除向 Application Master 请求作业进度外，客户端每 5 分钟都会通过调用 waitForCompletion() 来检查作业是否完成。时间间隔可以通过 mapreduce.client.completion.pollinterval 来设置。作业完成之后，Application Master 和 Container 会清理工作状态，OutputCommiter 的作业清理方法也会被调用。作业的信息会被作业历史服务器存储，以备之后用户核查。

## 4.3 MapReduce 编程组件

### 4.3.1 MapReduce 编程流程

由 4.1.2 节的介绍可知，MapReduce 程序包括数据输入、Map 处理、Shuffle 处理、Reduce

处理和结果输出等多个过程。MapReduce 框架屏蔽了复杂的分布式计算底层实现，为每个计算过程提供了编程组件，例如 InputFormat、InputSplit、RecordReader、Mapper、Partitioner、Combiner、Reducer、OutputFormat 等，如图 4-8 所示。

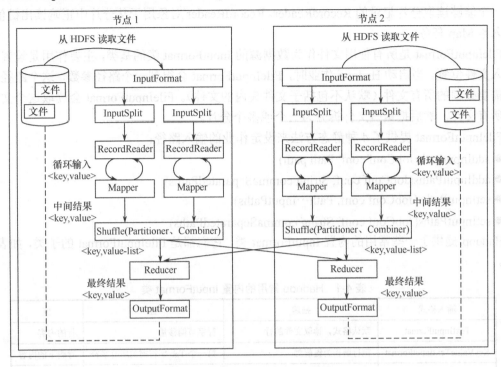

图 4-8　MapReduce 编程组件与流程

由于 Hadoop 已经为 InputFormat、Partitioner、Combiner 和 OutputFormat 提供了默认的实现，因此在大多数情况下，用户只需要编写 Mapper 和 Reducer 即可。编写好的 MapReduce 程序，需要打包上传到 Hadoop 大数据平台并提交给作业管理器加以执行。

MapReduce 程序的一般编写流程为：

① 编写 InputFormat（可选）；
② 编写 Mapper 类，重写 map()函数；
③ 编写 Partitioner（可选）；
④ 编写 Reducer 类，重新 reduce()函数；
⑤ 编写 OutputFormat（可选）；
⑥ 编写作业启动程序（函数）。

### 4.3.2　InputFormat

MapReduce 程序的输入文件一般是存在 Hadoop 集群上的 HDFS 文件，文件格式是任意的，可以是基于行的日志文件，也可以是二进制格式、多行输入记录或其他一些格式。这些文件通常都比较大，达到 GB、TB 数量级。

MapReduce 使用 InputFormat 类为作业设置输入文件，将文件分割为若干逻辑意义上的分片（InputSplit），并通过 RecordReader 对象，将 InputSplit 中的数据转化为<key,value>传输给 Mapper。

InputFormat 类提供了两个抽象方法。

① getSplit(JobContext context)：负责将一个大的数据文件在逻辑上折分成许多分片（InputSplit），每个分片对应一个 Map 任务。

② CreateRecordReader(InputSplit split,TaskAttemptContext context)：根据分片给定的方法，返回一个能够读取分片记录的 RecordReader。RecordReader 对象用于从分片中正确读出键值对，并输入给 Map 任务。

FileInputFormat 是所有使用文件作为数据源的 InputFormat 类的基类，主要作用是设置作业的输入文件位置。当启动 Hadoop 作业时，FileInputFormat 会得到一个路径参数，这个路径内包含了需要处理的所有文件（默认不包括子文件夹内的文件）。FileInputFormat 会读取这个文件夹中的所有文件，然后把这些文件拆分为一个或多个分片。

FileInputFormat 提供了 4 种静态方法来设定作业的输入路径：

- addInputPath(JobConf conf, Path path)
- addInputPaths(JobConf conf, String commaSeparatedPaths)
- setInputPaths(JobConf conf, Path…inputPaths)
- setInputPaths(JobConf conf, String commaSeparatedPaths)

Hadoop 提供了一些常用的内置 InputFormat 类，它们都是 FileInputFormat 的子类，如表 4-1 所示。

表 4-1 Hadoop 常用的内置 InputFormat 类

| 输入格式 | 描述 | 键 | 值 |
| --- | --- | --- | --- |
| TextInputFormat | 默认格式，读取文件的行 | 行字节偏移量 | 行的内容 |
| KeyValueTextInputFormat | 把行解析为键值对 | 第一个 Tab 字符前的所有字符 | 行剩下的内容 |
| NLineInputFormat | 将 N 行作为一个单位输入 | 行字节偏移量 | N 行内容 |
| SequenceFileInputFormat | Hadoop 定义的高性能二进制格式 | 用户自定义 | 用户自定义 |

默认的输入格式是 TextInputFormat。它把输入文件每一行作为单独的一个记录，但不做解析处理，这对那些没有被格式化的数据或是基于行的记录来说非常有用，如日志文件。KeyValueTextInputFormat 格式也是把输入文件每一行作为单独的一个记录，不同的是，TextInputFormat 把行字节偏移量作为键（key），整个文件行作为值（value），而 KeyValueTextInputFormat 则是通过搜索参数 mapreduce.input.keyvaluelinerecordreader.key.value.separator 指定的分隔符，默认为 Tab 字符（'\t'），把第一个 Tab 字符之前的内容作为键（key），而把行的其余部分作为值（value）。当将一个 MapReduce 作业的输出作为下一个作业的输入时，使用 KeyValueTextInputFormat 格式就非常适合，因为默认输出格式正是按 KeyValueTextInputFormat 格式输出数据的。SequenceFileInputFormat 格式可以读取 Hadoop 特定的二进制 Sequence 文件。Sequence 文件是块压缩的并提供对几种数据类型直接的序列化和反序列化操作，可以作为 MapReduce 作业的输出数据，并且用它做一个 MapReduce 作业到另一个作业的中间数据是非常高效的。

可以通过 JobConf 对象的 setInputFormatClass() 方法设定作业输入文件的输入格式。

### 4.3.3 InputSplit

源文件按特定的规则划分为一系列的 InputSplit，每个 InputSplit 有一个 Mapper 进行处理，Mapper 数量等于 InputSplit 数量，因此 MapReduce 数据处理的基本单位是 InputSplit。

对文件进行分片，并不是将文件切分开形成新的文件分片副本，而是形成一系列逻辑

InputSplit，InputSplit 中包含各个分片的数据信息，如文件块信息、起始位置、数据长度、所在节点列表等，因此，只需要根据 InputSplit 就可以找到分片的所有数据。

分片过程中最主要的任务就是确定参数 SplitSize，即分片数据大小，该值一旦确定，就将源文件按该值进行划分。如果文件小于该值，那么该文件就会成为一个单独的 InputSplit；如果文件大于该值，则按 SplitSize 进行划分后，剩下不足 SplitSize 的部分成为一个单独的 InputSplit。

SplitSize 参数值由 minSize、maxSize 和 blockSize 确定。

minSize：SplitSize 的最小值，由参数 mapreduce.input.fileinputformat.split.minsize 决定，可在 mapred-site.xml 文件中配置，默认为 0。

maxSize：SplitSize 的最大值，由参数 mapreduce.job.split.metainfo.maxsize 决定，可在 mapred-site.xml 文件中配置，默认为 10MB。

blockSize：HDFS 文件数据块大小，由参数 dbf.block.size 确定，可以在 hdfs-site.xml 中配置，默认为 128MB。

确定 SplitSize 参数值的规则为：

splitSize=max{minSize,min{maxSize,blockSize}}

可见，SplitSize 的大小一般在 minSize 和 blockSize 之间。通常情况下，InputSplit 大小等于数据块大小，这样一个 InputSplit 对应一个数据块，读取 InputSplit 时就不需要读取多个节点。如果一个 InputSplit 包含多个数据块，读取 InputSplit 时就要读取多个节点，这不仅会增加网络负担，还使得 Mapper 任务不能实现完全数据本地化。

### 4.3.4　RecordReader

MapReduce 程序执行时，InputFormat 接口调用 createRecordReader 方法为 InputSplit 创建 RecordReader 对象，并利用 getRecordReader 方法返回 InputSplit 的 RecordReader 对象。RecordReader 对象利用 RecordReader 接口提供的 createKey()和 createValue()方法，将 InputSplit 中的每行记录转换为键值对<key,value>，并将键值对传递给 Mapper 类的 map()函数处理。

RecordReader 实例是由输入格式决定的。对默认的 TextInputFormat 格式，RecordReader 实例是将 InputSplit 中每行数据都生成一条记录，表示成<key,value>。其中，key 是每个数据的记录在数据分片中的字节偏移量，数据类型是 LongWritable；value 是每行的内容，数据类型是 Text。对 KeyValueInputFormat 格式，RecordReader 实例是 KeyValueLineRecordReader，将 InputSplit 中每行数据特定分隔符之前的内容作为 key，而分隔符之后的内容作为 value，key 与 value 的数据类型都为 Text。对 SequenceFileInputFormat 格式，RecordReader 实例是 SequenceFileAsTextRecordReader。

### 4.3.5　Mapper

Hadoop MapReduce 框架为 InputFomat 类的每个 InputSplit 生成一个 Mapper 实例，每个 Mapper 实例只处理其对应的 InputSplit 数据，因此 Mapper 实例处理的数据是局部的。Mapper 实例接收键值对数据作为输入，执行 map()函数，输出一系列键值对，形成中间结果。每一个新的 Mapper 实例都会在单独的 Java 进程中被初始化，因此各个 Mapper 实例之间不能进行通信。

在 MapReduce 作业的主函数中，通过 job.setMapperClass(ＸＸＸ.Class)设置 Mapper 实例，例如 job.setMapperClass(wordCountMapper.Class)。MapReduce 框架会自动调用其 map()函数进行业务处理，通过 context.write()方法输出 0 个或多个键值对。map()函数输出键值对类型可以与输入键值对类型不同。

在 MapReduce 作业中，Map 任务数量由 InputSplit 数量决定，通常 InputSplit 大小为数据块大小,因此 Map 任务数量通常由输入文件数据块总数决定。Map 任务正常的并行规模是一个节点运行 10~100 个 Map 任务,对 CPU 消耗较小的 Map 任务可以设置 300 个左右。

图 4-9 显示了词频统计分析的 MapReduce 作业数据输入到 Mapper 输出的处理过程。

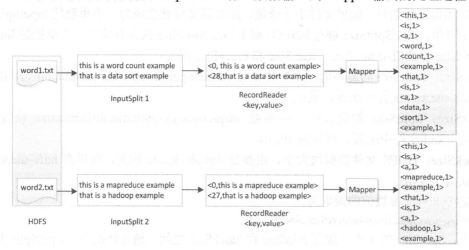

图 4-9　词频统计数据输入到 Mapper 输出的处理过程

## 4.3.6　Shuffle

map()函数输出的中间结果经过系统分区、排序、合并、归并等一系列处理后,成为最终 reduce()函数的输入数据,这个过程称为 Shuffle。Shuffle 对用户是透明的,采用可插拔的设计模式,方便用户应用自定义的逻辑替换内置的 Shuffle 逻辑,例如用户自定义分区、自定义分组、自定义排序等。

Shuffle 内部运行机制如图 4-10 所示,分为 Map 端 Shuffle 和 Reduce 端 Shuffle。

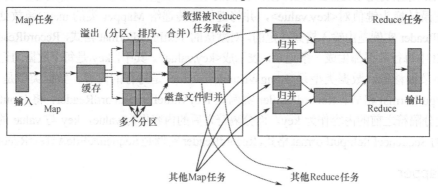

图 4-10　Shuffle 内部运行机制

(1) Map 端 Shuffle

Map 端 Shuffle 处理过程如图 4-11 所示。

图 4-11　Map 端 Shuffle 处理过程

map()函数产生的数据并不直接写入磁盘,而是先写入一个环形缓冲区(缓存),缓存大小由参数 mapreduce.task.io.sort.mb 设置（默认为 100MB），并且有一个写入阈值，阈值大小由参数 mapreduce.map.sort.spill.percent 设置（默认为 0.8，即 80%）。当写入缓存中的数据占比达到这一阈值时，map()函数会继续向剩下的缓存写入数据，但会在后台启动一个新线程，对前面 80%的缓存数据进行排序，然后写入本地磁盘文件（位置由参数 mapreduce.cluster.local.dir 设定），这一过程称为溢出操作（Spill），写入磁盘的文件称为溢出文件（Spill File）。如果剩下的 20%缓存已被写满，而前面的溢出操作还没有完成，Map 任务就会阻塞，直到溢出操作完成再继续向缓存中写入数据。

在启动溢出操作之前，首先需要把缓存中的数据进行分区（Partition），然后对每个分区的数据进行排序（Sort）和合并（Combine），之后再写入磁盘文件。每次溢出操作会生成一个新的磁盘文件，因此当每个 Map 任务完成时，磁盘中会生成多个溢出文件。每个溢出文件中的数据都是分区、有序的，但都是局部有序的。当某个 Map 任务完成溢出操作后，会对磁盘中该 Map 任务产生的所有临时溢出文件进行归并（Merge）、排序操作，生成最终的正式输出文件。归并是将所有溢出文件中的相同分区数据合并到一起，并对各个分区中的数据根据键（key）再进行一次排序，生成键值对列表（key, value-list）。在溢出文件归并的过程中，如果溢出文件数量大于某个值（默认为 3，由参数 min.num.spills.for.combine 设定），则会再次启动合并操作。两个键值对<a,1>和<a,1>，合并会得到<a,2>，归并会得到<a,{1,1}>。

（2）Reduce 端 Shuffle

Reduce 端 Shuffle 包括数据提取（数据复制）和归并两个过程，如图 4-12 所示。

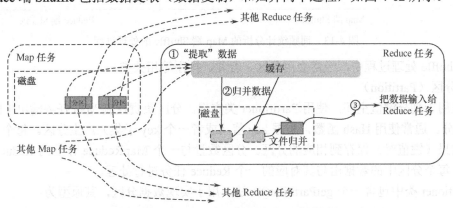

图 4-12　Reduce 端 Shuffle 处理过程

Reduce 任务的启动并不是所有的 Map 任务全部执行完成时。每个 Map 任务完成时间很难一致，所以当一个 Map 任务完成时，Reduce 任务就开始从 Map 任务输出中将自己要处理的数据提取回来。Reduce 提取数据的线程数通过参数 mapreduce.reduce.shuffle.parallelcopies 设定，默认值为 5。

Reduce 从不同 Map 端提取过来的数据不会直接写入磁盘，而是存储在 JVM 的内存中。当内存中数据达到一定阈值时，会对内存中的数据进行排序并写入本地磁盘，其处理方式和 Map 端的溢出操作类似，不同在于 Map 端的溢出操作进行的是简单的二次排序，Reduce 端由于内存中是多个已排好序的数据，因此采用的是归并排序。Reduce 内存溢出阈值由参数 mapreduce.reduce.merge.inmem.threshold 设定，默认值为 1000，当从 Map 端接收的文件数达到这一值时，进行归并排序。

如果 Reduce 端从 Map 端提取的数量量很大，可能需要对内存进行多次的归并排序后才能接收完所有的 Map 数据，此时会产生多个溢出文件。如果这些本地文件数量超过一定阈值（由参数 mapreduce.task.io.sort.factor 设定，默认为 10，也称归并因子），则需要对这些溢出文件进行归并排序，以减少文件数量。该操作并不是要求将所有溢出文件归并为一个文件，而是当归并后的文件数量少于或等于归并因子就停止了，此时可以将所有剩余文件一起进行归并排序，结果直接传给 Reduce 处理。

如果 Reduce 提取数据比较少，从内存归并排序得到的溢出文件并不多，可以将内存数据与溢出文件中的数据一起进行归并排序，直接传给 Reduce 处理。

图 4-13 显示了词频统计分析的 Map 端 Shuffle 的处理过程。

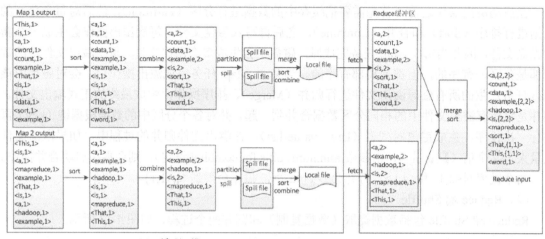

图 4-13　词频统计分析的 Map 端 Shuffle 的处理过程

在 Shuffle 处理过程中，经常会用分区、合并、排序等相关类。

① 分区（Partition）

分区用于划分键值空间，使用 Partitioner 类实现。分区负责控制 Map 任务输出中间结果的 key 的划分。通常使用 Hash 函数，根据 key 值（或者一个 key 子集）产生分区。每个 key 对应的中间结果（键值对）保存到相应的分区。分区数量与一个 MapReduce 作业的 Reduce 任务数量相同。每个分区中的数据由与其对应的一个 Reduce 任务进行处理。

Partitioner 类中包含一个 getPartition()方法，用于实现数据分区，其原型为：

● abstract int getPartition(KEY key, VALUE value, int numPartitions)

其中，key 和 value 为 Shuffle 传输的<key,value>，numPartitions 是分区总数。根据 getPartition()方法实现不同，Partitioner 类有 BinaryPartitioner、HashPartitioner、KeyFieldBasedPartitioner、RehashPartitioner、TotalOrderPartitioner 等多个实现类，其中默认使用的是基于 Hash 分区的 HashPartitioner。

用户可以自定义继承 Partitioner 类的分区类，实现 getPartition()方法，例如：

```
public class WordCountPartitioner extends Partitioner<Text,IntWritable>{
    @Override
    public int getPartition(Text key,IntWritable value,int numPartitions)
    {
        return (Integer.parseInt(key.toString())& Integer.MAX_VALUE)%numPartitions;
    }
}
```

然后在 MapReduce 作业的主函数中设置自定义的分区类名，例如：

job.setPartitionerClass(WordCountPartitioner.class)

需要注意的是，如果只有一个 Reduce 任务，不需要进行分区操作。

② 排序（Sort）

排序是 MapReduce 计算过程中的核心功能，默认是按照 key 进行排序的。如果 key 值是 Text 类型，则按字典顺序对字符串进行升序排序；如果 key 值是 IntWritable 类型，则按数字大小进行升序排序。

如果业务需要的排序方法与默认排序方法不同，例如按 key 降序排序，则需要用户自定义排序类来满足业务需求。自定义排序类需要继承 WritableComparator 类，重写 compare()方法，对接收到的 key 的类型，可以通过当前构造方法 super()来设置。

例如，定义一个按 key 降序排序的类。

```
public static class WordCountDesendingComparator extends WritableComparator{
    public int compare(WritableComparator a,WritableComparator b){
        return -super.compare(a,b);
    }
    public int compare(byte[] b1,int s1,int l1,byte[] b2,int s2,int l2) {
        return -super.compare(b1,s1,l1,b2,s2,l2);
    }
}
```

然后在 MapReduce 作业的主函数中设置自定义的排序类名，例如：

job.setSortComparatorClass(DesendingComparator.class)

可以使用 job.setNumReduceTask()设置 Reduce 任务数量，即分区数量。

③ 合并（Combiner）

MapReduce 框架使用 Map 任务将数据处理成一个个键值对，通过 Shuffle 处理，传递给 Reduce 任务处理并最终输出。在此过程中，如果有 10 亿个数据，Map 任务会生成 10 亿个键值对在网络间传输，所有数据都经过 Reduce 任务处理，不仅严重占用了网络带宽，还大大降低了程序的性能。为了解决该问题，MapReduce 框架提供了一个 combiner()函数。

combiner()函数与 reduce()函数的形式相同，是 reduce()函数的一个实现，不同之处在于它的输出类型是中间的键值对类型，这些中间值可以输入给 reduce()函数做最终的处理。combiner() 函数对 Map 端的输出先进行一个合并，减少在 Map 和 Reduce 节点之间的数据传输量，以提高网络 I/O 性能。combiner()函数并不适合所有的 MapReduce 应用，只适合具有结合律的操作，如求和、求最大值、求最小值等，但是，求平均值就不能进行合并运算了。因此，MapReduce 架构没有默认的实现，需要在 MapReduce 作业的主函数中通过 job.setCombinerClass(XXX.Class)设置。例如，job.setCombinerClass(WordCountReducer.class)。

### 4.3.7 Reducer

每个 Reduce 任务都会创建一个 Reducer 实例，Reducer 实例对具有相同键（key）的一组中间值进行规约。Reducer 实例的 reduce()函数接收一个键和关联到键的所有值的一个迭代器，迭代器会以一个未定义的顺序返回关联到同一个键的所有值。一个作业中 Reduce 任务数量通过 Job.setNumReduceTasks(n)设置。

在 MapReduce 作业的主函数中，通过 job.setReducerClass(XXX.Class)设置 Reducer 实例，例如，job.setReducerClass(wordCountReducer.Class)。MapReduce 框架会自动调用其 reduce()函数进行业务处理，通过 context.write()方法将输出键值对结果写入 HDFS 文件中。

### 4.3.8 OutputFormat

MapReduce 通过 InputFormat 类描述输入格式和规范，通过 OutputFormat 类描述输出格式和规范。MapReduce 框架依靠 OutputFormat 类完成输出规范检查，例如检查输出目录是否存在，并提供 RecordWriter 类的实现，将输出结果存储在文件系统中。

Hadoop 中提供的 OutputFormat 类的实现类为 FileOutputFormat，其直接子类包括 MapFileOutputFormat、SequenceFileOutputFormat 和 TextOutputFormat。其中，TextOutputFormat 是默认的输出格式，把每条记录写成文本行，以一个键值对<key\t value>（默认以 Tab 为键与值的分隔符）的方式写入一个文本文件中。这样，后面的 MapReduce 任务就可以通过 KeyValueTextInputFormat 类重新读取所需的输入数据了。SequenceFileOutputForma 输出格式更适合于在 MapReduce 作业间使用，它可以快速地序列化任意的数据类型到文件中，而对应 SequenceFileInputFormat 则会把文件反序列化为相同的类型并提交为下一个 Mapper 的输入数据，方式和前一个 Reducer 的生成方式一样。可以通过 JobConf 对象的 setOutputFormatClass() 方法设定作业输出文件的格式。

如果用户要基于 Hadoop 内置的输出格式和内置的 RecordWriter 进行定制，则需要重载 FileOutputFormat 类的 getRecordWriter() 方法以便获取新的 RecordWriter。

每一个 Reducer 会把结果输出到指定目录的一个单独的文件内，文件的命名一般是 part-r-nnnnn，其中，nnnnn 是关联到某个 Reduce 任务的分区 ID。

可以在 MapReduce 作业的主函数中，通过 JobConf 对象的 setOutputFormat() 方法设定作业输出文件的格式。可以通过 FileOutputFormat.setOutputPath() 来设置输出文件的目录。

图 4-14 描述了词频统计作业中 Reduce 数据处理及结果输出到 HDFS 的过程。

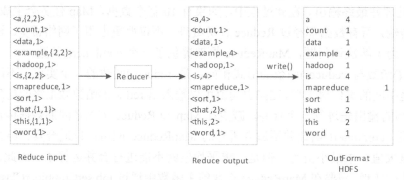

图 4-14　Reduce 数据处理及结果输出到 HDFS 的过程

### 4.3.9 序列化与反序列化

序列化是指将结构化对象转化为字节流，以便进行网络传输或者写入磁盘进行永久保存。反序列化则是序列化的逆过程，是指将从网络或者文件读取的字节流转化为结构化对象。

在 Hadoop 架构中，由于频繁进行进程间通信和 HDFS 文件读写，因此序列化与反序列化是必需的。

进程间通信：在 Hadoop 集群运行过程中，系统节点上进程间通信是通过远程过程调用（RPC）实现的。RPC 协议将消息序列化成二进制字节流后发送到远程节点，远程节点再将接收到的信息反序列化成原始消息，实现进程间的数据传输通信。

永久性存储：HDFS 是专门用来存储海量数据文件的，它是分布式存储的，将一个大文件切分成多个数据块，然后分发到不同的节点上进行永久性存储。此外，在 MapReduce 作业过程中，也频繁进行 HDFS 文件的读写。

虽然 Hadoop 是使用 Java 语言进行开发的，并且 Java 语言有专门的序列化技术，但是由于 Java 语言的序列化技术在处理海量数据时表现得并不优秀，以及为了减少与 Java 语言的耦合度，Hadoop 采用了一套专门应用于大数据的序列化框架 Writable。

在 Writable 序列化框架中，实现 Writable 接口的类都具有序列化与反序列化的功能。如果序列化类型需要进行同类型的比较，则需要实现 WritableComparable 接口。Hadoop 中 Writable 类的层次结构如图 4-15 所示。

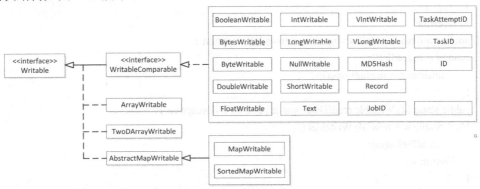

图 4-15 Writable 类的层次结构

Hadoop 中预定义的数据类型都实现了 WritableComparable 接口，包括面向所有基本数据类型的封装类。Java 数据类型与 Hadoop 序列数据类型的对应关系如表 4-2 所示。

表 4-2 Java 数据类型与 Hadoop 序列数据类型的对应关系

| Hadoop 数据类型 | Java 数据类型 |
| --- | --- |
| BooleanWritable | boolean |
| ByteWritable | byte |
| DoubleWritable | double |
| FloatWritable | float |
| IntWritable | int |
| LongWritable | long |
| Text | String |
| NullWritable | null |
| MapWritable | map |
| ArrayWritable | array |

如果自定义的序列化类型，不需要作为键值参与排序，那么就只需要实现 Writable 接口。如果需要作为键值参与排序，则需要实现 WritableComparable 接口，额外实现一个基于键值进行对象比较的方法 compareTo()。

Writable 接口的定义如下，需要实现 write()和 readFields()两个函数。
```
public interface Writable {
    void write(DataOutput var1) throws IOException;
    void readFields(DataInput var1) throws IOException;
}
```
例如，定义一个实现 Writable 接口的序列化/反序列化类：

```java
public class MyWritable implements Writable {
    private int counter;
    private long timestamp;
    MyWritable() { }
    @Override
    public void write(DataOutput out) throws IOException {
    out.writeInt(counter);
        out.writeLong(timestamp);
    }
    @Override
    public void readFields(DataInput in) throws IOException {
    counter = in.readInt();
        timestamp = in.readLong();
    }
    public static MyWritable read(DataInput in) throws IOException {
    MyWritable w = new MyWritable();
        w.readFields(in);
        return w;
    }
}
```

例如，定义一个实现 WritableComparable 接口的序列化/反序列化类：

```java
public class MyWritableComparable implements WritableComparable {
    private int counter;
    private long timestamp;
    @Override
    public void write(DataOutput out) throws IOException {
    out.writeInt(counter);
        out.writeLong(timestamp);
    }
    @Override
    public void readFields(DataInput in) throws IOException {
    counter = in.readInt();
        timestamp = in.readLong();
    }
    @Override
    public int compareTo(MyWritableComparable o) {
        int thisValue = this.value;
        int thatValue = o.value;
        return (thisValue < thatValue ? -1 : (thisValue==thatValue ? 0 : 1));
    }
    public int hashCode() {
        final int prime = 31;
        int result = 1;
        result = prime * result + counter;
        result = prime * result + (int) (timestamp ^ (timestamp >>> 32));
        return result
    }
}
```

## 4.4 WordCount 程序设计实例

本节将介绍 4.3 节中介绍的词频统计案例 MapReduce 程序设计的整个过程。词频统计是指统计一个或多个具有相同格式的文件中单词出现的频数，输出文件中每个单词及其频数占一行，以制表符（\t）分隔，并按照单词字母顺序排序。

### 4.4.1 准备输入文件

根据图 4-9 中的描述，首先在本地 Linux 系统宿主目录下创建两个文本文件，分别为 word1.txt 和 word2.txt，在每个文件中写入相应的内容，相邻两个单词之间使用空格分隔。

```
$ cd ~
$ gedit word1.txt
$ gedit word2.txt
```

然后在 HDFS 中创建/mapreduce/wordcount/input 目录，并将 word1.txt 和 word2.txt 文件上传到/mapreduce/wordcount/input 目录中，可以查看文件列表及文件内容。

```
$ hdfs dfs -mkdir -p /mapreduce/wordcount/input
$ hdfs dfs -put    word1.txt word2.txt /mapreduce/wordcount/input
$ hdfa dfs -ls/mapreduce/wordcount/input
$ hdfa dfs -cat /mapreduce/wordcount/input/word1.txt
$ hdfa dfs -ls /mapreduce/wordcount/input/word2.txt
```

执行过程如图 4-16 所示。

```
hadoop@Master:~$ gedit word1.txt
hadoop@Master:~$ gedit word2.txt
hadoop@Master:~$ hdfs dfs -mkdir -p /mapreduce/wordcount/input
hadoop@Master:~$ hdfs dfs -put word1.txt word2.txt /mapreduce/wordcount/input
hadoop@Master:~$ hdfs dfs -ls /mapreduce/wordcount/input
Found 2 items
-rw-r--r--   1 hadoop supergroup         57 2021-01-09 16:19 /mapreduce/wordcount/input/word1.txt
-rw-r--r--   1 hadoop supergroup         53 2021-01-09 16:19 /mapreduce/wordcount/input/word2.txt
hadoop@Master:~$ hdfs dfs -cat /mapreduce/wordcount/input/word1.txt
this is a word count example
that is a data sort example
hadoop@Master:~$ hdfs dfs -cat /mapreduce/wordcount/input/word2.txt
this is a mapreduce example
that is a hadoop example
```

图 4-16　准备输入文件

### 4.4.2 创建 Maven 工程

打开 Eclipse，创建 Maven 工程。

① 选择菜单“File→New→Project”，打开“New Project”窗口，选择“Maven”列表下的“Maven Project”选项。

② 单击“Next”按钮，进入“New Maven Project”界面，勾选“Create a simple project”和“Use default Workspace location”选项。

③ 单击“Next”按钮，进入图 4-17 所示的工程组与工程名称设置界面。

④ 单击“Finish”按钮，完成 Maven 工程的创建。

### 4.4.3 配置 Maven 工程

① 将 Hadoop 配置文件 core-site.xml、hdfs-site.xml（在/usr/local/hadoop/etc/hadoop 目录中）复制到工程 mapreduce 的/src/main/resources 目录中，如图 4-18 所示。

图 4-17　设置工程组与工程名称

图 4-18　复制 Hadoop 配置文件

② 编辑 pom.xml 文件，添加 Hadoop 依赖。

添加各种属性及 hadoop-client 等依赖。

```xml
<properties>
    <project.build.sourceEncoding>UTF-8</project.build.sourceEncoding>
    <hadoop.version>3.1.2</hadoop.version>
</properties>
<dependencies>
    <dependency>
        <groupId>org.apache.hadoop</groupId>
        <artifactId>hadoop-client</artifactId>
        <version>${hadoop.version}</version>
    </dependency>
    <dependency>
        <groupId>junit</groupId>
        <artifactId>junit</artifactId>
        <version>3.8.2</version>
        <scope>test</scope>
    </dependency>
</dependencies>
```

### 4.4.4　程序设计

（1）创建 WordCount 类

① 在 Eclipse 中选择窗口左侧"Package Explorer"中的"mapreduce"工程，展开文件夹，选择"src/main/java"文件夹并右键单击，在弹出菜单中选择"New→Class"。在弹出的"New Java Class"窗口中进行新建 Class 的设置，包括 Package、Name 设置，如图 4-19 所示。

② 单击"Finish"按钮，完成 Class 的创建。然后，在代码编辑窗口进行程序设计。例如，引入程序所需的各种类，如图 4-20 所示。

（2）编写 Mapper 内嵌类

编写一个继承 Mapper 类的内嵌类 WordCountMapper，重写 map() 函数。代码如下：

```java
public static class WordCountMapper extends Mapper<LongWritable, Text, Text, IntWritable> {
    //定义一个静态序列化常量one，类型为IntWritable，并将值初始化为1
    private final static IntWritable one = new IntWritable(1);
    //定义一个用于保存序列化key值的变量word，类型为Text
    private Text word = new Text();
```

```
/*
 * 重定义map()函数,其中第一个参数key为分片中每行文本的偏移量地址,
 * 第二个参数value为一行文本,第三个参数context为MapReduce框架的上下文环境,
 *并将map()函数输出保存到context中,然后由MapReduce框架进行后续处理。
 */
public void map(LongWritable key, Text value, Context context) throws IOException, InterruptedException
{
        //将参数传进来的Text类型文本转换为Java String数据类型
        String line=value.toString();
        //以空格为分隔符,拆分出一行所有单词,保存到迭代器中
        StringTokenizer token = new StringTokenizer(line);
        //循环遍历迭代器,以每个单词为key,常量one为值,构成键值对
        while (token.hasMoreTokens()) {
            word.set(token.nextToken());
        //将map()函数输出的键值对写入MapReduce上下文环境
            context.write(word, one);
        }
    }
}
```

图 4-19　创建 WordCount 类

图 4-20　代码编辑窗口

在 WordCountMapper 的定义中,Mapper<LongWritable, Text, Text, IntWritable>泛型的前两个参数为 map()函数的输入键值对的数据类型,其中 LongWritable 为输入 key 类型,Text 为输入值类型;后两个参数为 map()函数的输出中间键值对数据类型,其中 Text 为输入 key 类型,IntWritable 为输出值类型。

在对一行文本进行单词拆分时,需要根据输入文本文件的格式进行。StringTokenizer(String str)默认的分隔符是"空格""制表符('\t')""换行符('\n')""回车符('\r')",可以采用 StringTokenizer(String str, String delim)函数指定一个分隔符,进行文本拆分。

最后,使用 context.write()方法向 MapReduce 上下环境输出时,需要保证输出键值对类型与泛型最后两个参数指定的数据类型一致。

(3) 编写 Reducer 内嵌类

编写一个继承 Reducer 类的内嵌类 WordCountReducer,重写 reduce()函数。代码如下:

```
public static class MyReducer extends Reducer<Text, IntWritable, Text, IntWritable> {
    //定义一个变量result,类型为IntWritable,保存词频数量
```

```java
            private IntWritable result = new IntWritable();
            /*
            * 重定义reduce()函数，其中第一个参数key是map()函数处理后输出的中间结果键值对的
            *键，即单词。第二参数values是map()函数处理后输出与key对应的中间结果值的列表
            * 第三个参数context为MapReduce框架的上下文环境，将reduce()函数输出保存到
            *context中，最后由MapReduce框架写入HDFS。
            */
            public void reduce(Text key, Iterable<IntWritable> values, Context context)
                        throws IOException, InterruptedException {
                //定义一个计数器，用于累加运算，初始值为0
                int sum = 0;
                //循环遍历key对应的values列表中所有的值，然后进行累加计算
                for (IntWritable val : values) {
                //进行累加计算时，需要将IntWritable类型转换为int类型
                    sum += val.get();
                }
                result.set(sum);
                //将Reduce输出结果写入MapReduce框架的上下文环境context中
                context.write(key, result);
            }
        }
```

在 WordCountReducer 定义中，Reducer<Text, IntWritable, Text, IntWritable>泛型的前两个参数为 reduce()函数输入键值对数据类型，即 map()函数的输出中间结果的键值对的数据类型，因此这两个参数对应 Mapper<LongWritable, Text, Text, IntWritable>泛型的后两个参数，其中 Text 对应 key，即单词；IntWritable 对应值，即序列化值1的列表。Reducer<Text, IntWritable, Text, IntWritable>泛型的后两个参数为 reduce()函数的输出结果键值对数据类型，其中 Text 为输入 key 类型，即单词；IntWritable 为输出值类型，即词频数。

最后，使用 context.write()方法向 MapReduce 上下环境输出时，需要保证输出键值对类型与泛型最后两个参数指定的数据类型一致。

（4）编写 main()函数

定义好 Mapper 内嵌类和 Reducer 内嵌类后，就可以在主类的 main()函数中创建一个 MapReduce 作业，并进行作业的设置，包括输入文件目录、输出文件目录、主类、Mapper 类、Reducer 类、输出键值对类型等。在 MapReduce 编程模型中，业务处理就是以作业为单位进行的。

```java
        public static void main(String[] args) throws Exception{
            //创建Hadoop集群配置对象
            Configuration conf = new Configuration();
            //获取程序运行时输入的输入目录和输出目录，第一个参数为输入目录，第二参数为输出目录
            String[] pathArgs = new GenericOptionsParser(conf,args).getRemainingArgs();
            //检查参数个数，如果少于2个，则不满足程序运行要求
            if (pathArgs.length != 2)
            {
                System.err.println("Usage: WordCount <in><out>");
                System.exit(2);
            }
            Path path = new Path(pathArgs[1]);
            FileSystem fs = FileSystem.get(conf);
            //检查输出    目录是否已经存在，如果存在，删除该目录（程序执行时自动创建输出目录）
```

```
        if (fs.exists(path)) {
                fs.delete(path,true);
        }
        Job job = Job.getInstance(conf,"word count");    //根据系统配置信息，新建一个Job作业
        job.setJarByClass(WordCount.class);//设置主类
        job.setMapperClass(WordCountMapper.class); //设置自定义的Mapper类
        job.setReducerClass(WordCountReducer.class);//设置自定义的Reducer类
        job.setOutputKeyClass(Text.class);            //设置输出键值对键的类型
        job.setOutputValueClass(IntWritable.class);    //设置输出键值对值的类型
        FileInputFormat.addInputPath(job, new Path(pathArgs[0]));    //设置输入目录
        FileOutputFormat.setOutputPath(job, new Path(pathArgs[1]));  //设置输出目录
        System.exit(job.waitForCompletion(true) ? 0 : 1);//执行作业，等待作业执行完毕退出
}
```

main()函数是 MapReduce 作业执行的入口。在 main()函数中首先创建一个 Configuration 对象，用于保存所有的配置信息。该对象在创建时会读取工程下/src/main/resources 目录中的配置文件，如 core-site.xml、hdfs-site.xml 等，根据配置文件进行对象初始化。也可以通过配置对象的 set 方法进行属性设置。

需要注意的是，在 main()函数中设置的输出键值对类型必须与 Reducer<Text, IntWritable, Text, IntWritable>泛型的最后两个参数指定的数据类型一致。

如果 MapReduce 中包括用户自定义的排序、分区、合并等操作时，也需要在 main()函数中指定相应的排序类、分区类、合并类等，详见 4.4.5 节介绍。

（5）程序调试

代码编写完成后，可以进行程序测试。

① 选择 Eclipse 窗口左侧 "Package Explorer" 中的 "mapreduce" 工程，展开文件夹，选择 "src/main/java" 文件夹中的 WordCount.java 文件并右键单击，在弹出菜单中选择 "Run As→Run Configurations"，弹出 "Run Configurations" 对话框，如图 4-21 所示。

② 在 "Run Configurations" 对话框中，"Name" 框中选择 "WordCounts"，在 Arguments 选项卡中输入两个参数，第一个参数为输入文件路径信息/mapreduce/wordcount/input，第二个参数为输出文件路径信息/mapreduce/wordcount/output。然后单击 "Run" 按钮，执行程序。

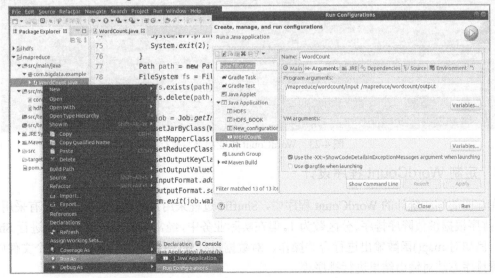

图 4-21 输入参数执行 WordCount 作业

③ 作业执行完成后，可以通过 Shell 命令查看执行结果，如图 4-22 所示。

```
$ hdfs dfs -ls /mapreduce/wordcount/output
$ hdfs dfs -cat /mapreduce/wordcount/output/part-r-00000
```

```
hadoop@Master:~$ hdfs dfs -ls /mapreduce/wordcount/output
Found 2 items
-rw-r--r--   1 hadoop supergroup          0 2021-01-09 16:29 /mapreduce/wordcount/output/_SUCCESS
-rw-r--r--   1 hadoop supergroup         83 2021-01-09 16:29 /mapreduce/wordcount/output/part-r-00000
hadoop@Master:~$ hdfs dfs -cat /mapreduce/wordcount/output/part-r-00000
a       4
count   1
data    1
example 4
hadoop  1
is      4
mapreduce       1
sort    1
that    2
this    2
word    1
```

图 4-22 WordCount 作业执行结果

由于默认情况下，MapReduce 只启动一个 Reduce 任务，即只有一个分区，分区 ID 为 00000，因此输出文件名为 part-r-00000。

### 4.4.5 工程打包、部署与运行

与 HDFS 开发类似，MapReduce 程序测试成功后，需要形成 JAR 文件导出，部署到 Hadoop 集群，然后进行执行。具体过程参见 3.4.4 节。

① 在 Eclipse 窗口中选择工程“mapreduce”并右键单击，在弹出菜单中选择“Export”，弹出导出类型选择界面。

② 选择“Java→JAR file”，打开导出设置界面。设置导出文件路径与名称后，单击“Finish”按钮，开始工程导出。

③ 工程导出完成后，可以将导出的 JAR 文件部署到集群，并执行 JAR 文件，如图 4-23 所示。

```
hadoop@Master:~$ ls
data     Documents  eclipse            examples.desktop  Music     Public     Videos     word2.txt
Desktop  Downloads  eclipse-workspace  hdfs.jar          Pictures  Templates  word1.txt  wordcount.jar
hadoop@Master:~$ hadoop jar wordcount.jar com.bigdata.example.WordCount /mapreduce/wordcount/input /mapr
educe/wordcount/output2
hadoop@Master:~$ hdfs dfs -ls /mapreduce/wordcount/output2
Found 2 items
-rw-r--r--   1 hadoop supergroup          0 2021-01-10 15:15 /mapreduce/wordcount/output2/_SUCCESS
-rw-r--r--   1 hadoop supergroup         83 2021-01-10 15:15 /mapreduce/wordcount/output2/part-r-00000
hadoop@Master:~$ hdfs dfs -cat /mapreduce/wordcount/output2/part-r-00000
a       4
count   1
data    1
example 4
hadoop  1
is      4
mapreduce       1
sort    1
that    2
this    2
word    1
```

图 4-23 WordCount 作业部署与执行

### 4.4.6 定制 WordCount 程序设计

在前面的词频统计的 WordCount 程序中，Shuffle 过程采用系统默认形式，即没有采用合并操作、排序根据键值降序排序、分区数为 1。但在实际业务中，经常需要根据业务需要进行 Shuffle 设置，例如对 map()函数输出进行合并操作、对数据进行分区存储（结果保存在多个文件中）、按设置排序方法对输出结果进行排序等。

例如，在词频统计的 MapReduce 作业中，要求进行数据合并操作，同时对数据分析结果按

单词的词频数进行倒序排序，把词频数小于 2 的结果保存到一个文件中，把词频数大于 2 的结果保存到另一个文件中。

（1）程序分析

① 合并：为了减少 Map 输出与 Reduce 输入之间数据的传输量，减轻网络负载，提高系统性能，可以在 WordCount 工作的 Shuffle 过程中进行数据合并操作。为此，需要在工程的 main() 函数中利用函数 job.setCombinerClass() 设置数据合并类。

② 分区：默认 MapReduce 任务只有一个分区，可以在工程主类中自定义一个继承 Partitioner 类的内嵌分区类，并重写 getPartition() 函数。然后，在 main() 函数中利用 job.setPartitionerClass() 函数设置分区类、利用 job.setNumReduceTasks() 函数设置分区数量（Reduce 任务数）。

③ 排序：在 MapReduce 任务中，默认按键值升序排序，如果需要按键值降序排序，或者按业务要求进行排序，需要用户在工程主类中自定义一个继承 WritableComparator 类的内嵌分区类，并重写 compare() 函数。然后，在 main() 函数中利用 job.setSortComparatorClass() 函数设置排序类。

在业务要求中，根据词频数量进行分区、排序，因此这实质是两个 MapReduce 作业。第一个作业就是按默认方式输出单词与其词频数，如 4.4 节中的介绍；第二个作业是读取第一个作业结果作为输入，然后对 map() 函数输出的中间键值对进行键与值的交换，即原来的键（key）为单词、值（value）为词频数，交换后键（key）为词频数、值（value）为单词，例如，原来的<example,4>转换为<4,example>。键值交换可以通过内置 InverseMapper.Class 类实现，在 main() 函数中使用 setMapperClass(InverseMapper.Class) 设置。键值交换后，基于新的键值对进行分区、排序。

（2）程序设计

① 选择 Eclipse 窗口左侧 "Package Explorer" 中的 "mapreduce" 工程，展开文件夹，选择 "src/main/java" 文件夹，创建一个名为 WordCount2 的类。

②在程序头部，引入各种内置类。代码如下：

```
package com.bigdata.example;
import java.io.IOException;
import java.util.StringTokenizer;
import org.apache.hadoop.conf.Configuration;
import org.apache.hadoop.fs.FileSystem;
import org.apache.hadoop.fs.Path;
import org.apache.hadoop.io.IntWritable;
import org.apache.hadoop.io.LongWritable;
import org.apache.hadoop.io.Text;
import org.apache.hadoop.io.WritableComparable;
import org.apache.hadoop.mapreduce.Job;
import org.apache.hadoop.mapreduce.Mapper;
import org.apache.hadoop.mapreduce.Partitioner;
import org.apache.hadoop.mapreduce.Reducer;
import org.apache.hadoop.mapreduce.lib.input.FileInputFormat;
import org.apache.hadoop.mapreduce.lib.input.SequenceFileInputFormat;
import org.apache.hadoop.mapreduce.lib.input.TextInputFormat;
import org.apache.hadoop.mapreduce.lib.map.InverseMapper;
import org.apache.hadoop.mapreduce.lib.output.FileOutputFormat;
import org.apache.hadoop.mapreduce.lib.output.SequenceFileOutputFormat;
```

```java
import org.apache.hadoop.mapreduce.lib.output.TextOutputFormat;
import org.apache.hadoop.util.GenericOptionsParser;
```

需要注意的是，在 Hadoop 3.x 中，FileInputFormat、FileOutputFormat 等类在 org.apache.hadoop.mapreduce.lib.***包中，而非 Hadoop 2.x 中的 org.apache.hadoop.mapred.***。

③ 在 WordCount 类中编写一个内嵌类 WordCountMapper。代码与 4.4.4 节中 WordCountMapper 类相同。

④ 在 WordCount 类中编写一个内嵌类 WordCountReducer。代码与 4.4.4 节中 WordCountReducer 类相同。

⑤ 在 WordCount 类中编写一个内嵌类 WordCountPartitioner，代码如下：

```java
public static class WordCountPartitioner extends Partitioner<IntWritable,Text>{
    public int getPartition(IntWritable key,Text value, int numPartitions) {
        int n = Integer.parseInt(key.toString());
        if(n<=2){
            return 0;
        }
        return 1;
    }
}
```

Partitioner<IntWritable,Text>泛型的两个参数对应 map()函数输出的中间结果键值对的类型，第一个参数为 key 数据类型，第二个参数为 value 数据类型。在本例中，经过 InverseMapper.Class 类转换后的键为词频数，类型为 IntWritable；值的类型为单词，类型为 Text。

⑥ 在 WordCount 类中编写一个内嵌类 IntWritableDecreasingComparator，用于按键值降序排序。代码如下：

```java
private static class IntWritableDecreasingComparator extends IntWritable.Comparator {
    public int compare(WritableComparable a, WritableComparable b) {
        return -super.compare(a, b);//降序，只需将父类排序结果取负即可
    }
    public int compare(byte[] b1, int s1, int l1, byte[] b2, int s2, int l2) {
        return -super.compare(b1, s1, l1, b2, s2, l2);
    }
}
```

在本例中，自定义排序类继承 IntWritable.Comparator，这是因为用于排序的键的数据类型为 IntWritable。

⑦ 编写 main()函数，进行作业创建与设置。代码如下：

```java
public static void main(String[] args) throws Exception{
    //创建Hadoop集群配置对象，自动读取配置文件
    Configuration conf = new Configuration();
    //获取程序运行时输入的输入目录和输出目录，第一个参数为输入目录，第二参数为输出目录
    String[] pathArgs = new GenericOptionsParser(conf,args).getRemainingArgs();
    //检查参数个数，如果少于2个，则不满足程序运行要求
    if (pathArgs.length != 2)
    {
        System.err.println("Usage: WordCount <in><out>");
        System.exit(2);
    }
    //检查输出目录是否已经存在，如果存在，删除该目录（程序执行时自动创建输出目录）
```

```
        Path path = new Path(pathArgs[1]);
        FileSystem fs = FileSystem.get(conf);
        if (fs.exists(path)) {
            fs.delete(path,true);
        }
        //设置保存第一个作业输出结果的临时目录
        Path tempDir=new Path("/mapreduce/wordcount/temp");
        if (fs.exists(tempDir)) {
            fs.delete(tempDir,true);
        }
        try {
            //第一个MapReduce作业
            Job job = Job.getInstance(conf,"word count");//根据系统配置信息，新建第一个Job作业
            job.setJarByClass(WordCount2.class);//设置主类
            job.setMapperClass(WordCountMapper.class);    //设置自定义的Mapper类
            job.setReducerClass(WordCountReducer.class); //设置自定义的Reducer类
            job.setOutputKeyClass(Text.class);//设置输出键值对的键的类型
            job.setOutputValueClass(IntWritable.class);//设置输出键值对的值的类型
            job.setCombinerClass(WordCountReducer.class);//设置Shuffle中的合并类
            FileInputFormat.addInputPath(job, new Path(pathArgs[0]));//设置输入目录
            FileOutputFormat.setOutputPath(job, tempDir);//设置输入目录（临时目录）
            job.setInputFormatClass(TextInputFormat.class);      //设置输入文件格式
            job.setOutputFormatClass(SequenceFileOutputFormat.class); //设置输出文件格式
            if(job.waitForCompletion(true)) {//执行第一个作业，如果成功，继续第二个作业
                Job sortJob=Job.getInstance(conf,"word sort");   //创建第2个Job作业
                sortJob.setJarByClass(WordCount2.class);//设置主类
                sortJob.setMapperClass(InverseMapper.class);//设置键值交换类
                sortJob.setInputFormatClass(SequenceFileInputFormat.class);   //设置输入文件格式
                sortJob.setOutputFormatClass(TextOutputFormat.class);    //设置输出文件格式
                sortJob.setOutputKeyClass(IntWritable.class);//设置输出键值对的键的类型
                sortJob.setOutputValueClass(Text.class);//设置输出键值对的值的类型
                FileInputFormat.addInputPath(sortJob,tempDir); //设置输入目录（临时目录）
                FileOutputFormat.setOutputPath(sortJob,new Path(pathArgs[1])); //设置输出目录
                sortJob.setSortComparatorClass(IntWritableDecreasingComparator.class);//设置排序类
                sortJob.setPartitionerClass(WordCountPartitioner.class);//设置分区类
                sortJob.setNumReduceTasks(2);//设置分区数量
                if(sortJob.waitForCompletion(true) {  //执行第2个作业
                    System.out.print("sort successful!");
                }
            }
        }catch(Exception ex) {
            ex.printStackTrace();
        }finally {
            fs.delete(tempDir,true);  //两个作业执行完成后，删除临时目录
        }
    }
}
```

⑧ 程序调试。参考4.4.4节程序调试方法，在"Run Configurations"对话框中，"Name"框中选择"WordCount2"，在 Arguments 选项卡中输入两个参数，第一个参数为输入文件路径

信息 /mapreduce/wordcount/input，第二个参数为输出文件路径信息 /mapreduce/wordcount/output2。然后单击"Run"按钮，执行程序。作业执行完成后，可以通过 Shell 命令查看执行结果，如图 4-24 所示。

```
$ hdfs dfs -ls /mapreduce/wordcount/output2
$ hdfs dfs -cat /mapreduce/wordcount/output/part-r-00000
$ hdfs dfs -cat /mapreduce/wordcount/output/part-r-00001
```

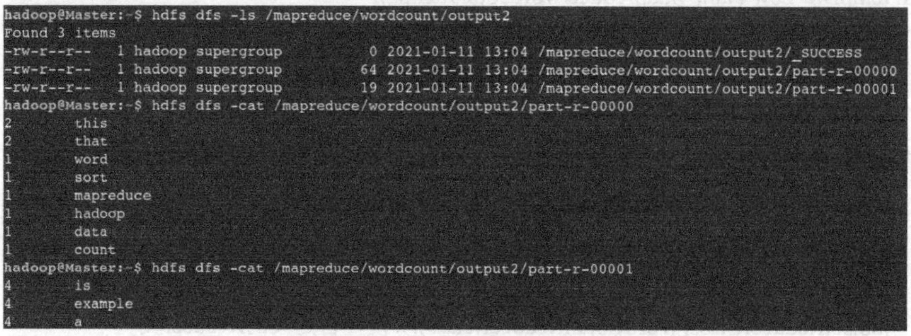

图 4-24  WordCount2 作业执行结果

## 4.5 MapReduce 开发典型案例

掌握了 MapReduce 程序设计的基本思想与流程后，本节将介绍 3 个 MapReduce 程序设计案例，包括数据去重、数据排序、计算平均值，这些案例也是 MapReduce 的典型应用。

### 4.5.1 数据去重

**1. 任务描述**

数据去重是利用并行技术对大数据集数据进行有意义的筛选。在大数据文件中包含了大量的记录，每条记录记载了某事物的一些属性，需要根据某几个属性的组合，去除相同的重复组合，并统计其中某属性的统计值。

本案例计算在网站日志中每天访问网站的 IP，要求时间与 IP 相同的数据保留一个。图 4-25 描述了日志文件中的数据格式和预期的输出结果。

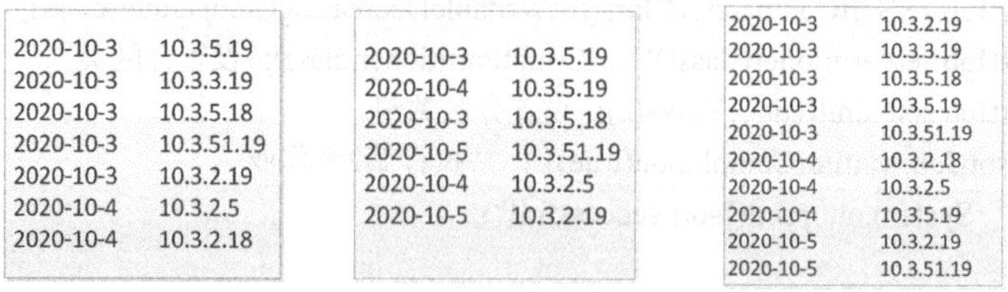

| log1.txt | | log2.txt | | 预期输出结果 | |
| --- | --- | --- | --- | --- | --- |
| 2020-10-3 | 10.3.5.19 | 2020-10-3 | 10.3.5.19 | 2020-10-3 | 10.3.2.19 |
| 2020-10-3 | 10.3.3.19 | 2020-10-4 | 10.3.5.19 | 2020-10-3 | 10.3.3.19 |
| 2020-10-3 | 10.3.5.18 | 2020-10-4 | 10.3.5.18 | 2020-10-3 | 10.3.5.18 |
| 2020-10-3 | 10.3.51.19 | 2020-10-5 | 10.3.51.19 | 2020-10-3 | 10.3.5.19 |
| 2020-10-3 | 10.3.2.19 | 2020-10-4 | 10.3.2.5 | 2020-10-3 | 10.3.51.19 |
| 2020-10-4 | 10.3.2.5 | 2020-10-5 | 10.3.2.19 | 2020-10-4 | 10.3.2.18 |
| 2020-10-4 | 10.3.2.18 | | | 2020-10-4 | 10.3.2.5 |
| | | | | 2020-10-4 | 10.3.5.19 |
| | | | | 2020-10-5 | 10.3.2.19 |
| | | | | 2020-10-5 | 10.3.51.19 |

图 4-25  去重作业输入数据和预期的输出结果

**2. 设计思路**

① map() 函数将每条记录中日期与 IP 地址属性组合作为键（key），将空字符串作为值（value），生成的键值对<key,value>作为中间值输出。

② reduce() 函数把输入的中间结果的键值对的键（key）作为输出结果的键（key），value 仍然取空字符串，作为输出结果的值（value）。

因为具有相同键(key)的中间结果都被送到同一个 Reducer，而 Reducer 只输出一个键(key)，从而实现了去重的目的。

## 3. 程序设计

创建一个名为 Distinct 的类，编写 Mapper 类、Reducer 类及 main()函数。代码如下：

```java
package com.bigdata.example;
import java.io.IOException;
import java.util.StringTokenizer;
import org.apache.hadoop.conf.Configuration;
import org.apache.hadoop.fs.FileSystem;
import org.apache.hadoop.fs.Path;
import org.apache.hadoop.io.LongWritable;
import org.apache.hadoop.io.Text;
import org.apache.hadoop.mapreduce.Job;
import org.apache.hadoop.mapreduce.Mapper;
import org.apache.hadoop.mapreduce.Reducer;
import org.apache.hadoop.mapreduce.lib.input.FileInputFormat;
import org.apache.hadoop.mapreduce.lib.output.FileOutputFormat;
import org.apache.hadoop.util.GenericOptionsParser;
public class Distinct {
    public static class DistinctMapper extends Mapper<LongWritable, Text, Text, Text> {
        private Text line = new Text();
        public void map(LongWritable key, Text value, Context context)
            throws IOException, InterruptedException {
            line=value;
            context.write(line, new Text(""));
        }
    }
    public static class DistinctReducer extends Reducer<Text, Text, Text, Text> {
        public void reduce(Text key, Iterable<Text> values, Context context)
            throws IOException, InterruptedException {
            context.write(key, new Text(""));
        }
    }
    public static void main(String[] args) throws Exception {
        Configuration conf = new Configuration();
        String[] pathArgs = new GenericOptionsParser(conf, args).getRemainingArgs();
        if (pathArgs.length != 2) {
            System.err.println("Usage: wordcount <in><out>");
            System.exit(2);
        }
        Path path = new Path(pathArgs[1]);
        FileSystem fs = FileSystem.get(conf);
        if (fs.exists(path)) {
            fs.delete(path, true);
        }
        Job job = Job.getInstance(conf,"distinct");
        job.setJarByClass(Distinct.class);
        job.setMapperClass(DistinctMapper.class);
        job.setReducerClass(DistinctReducer.class);
        job.setOutputKeyClass(Text.class);
        job.setOutputValueClass(Text.class);
```

```
            FileInputFormat.addInputPath(job, new Path(pathArgs[0]));
            FileOutputFormat.setOutputPath(job, new Path(pathArgs[1]));
            System.exit(job.waitForCompletion(true) ? 0 : 1);
        }
    }
```

#### 4．程序执行

（1）数据准备

按图 4-25 的描述，在 Linux 本地创建 log1.txt 和 log2.txt 两个文件，并输入相应内容。然后，在 HDFS 上创建输入目录/mapreduce/distinct/input。最后，将 log1.txt 和 log2.txt 文件上传到 HDFS 的/mapreduce/distinct/input 目录中。如图 4-26 所示。

```
hadoop@Master:~$ gedit log1.txt
hadoop@Master:~$ gedit log2.txt
hadoop@Master:~$ hdfs dfs -mkdir -p /mapreduce/distinct/input
hadoop@Master:~$ hdfs dfs -put log1.txt log2.txt /mapreduce/distinct/input
hadoop@Master:~$ hdfs dfs -ls /mapreduce/distinct/input
Found 2 items
-rw-r--r--   1 hadoop supergroup        140 2021-01-12 15:15 /mapreduce/distinct/input/log1.txt
-rw-r--r--   1 hadoop supergroup        120 2021-01-12 15:15 /mapreduce/distinct/input/log2.txt
```

图 4-26　去重作业输入数据准备

（2）导出、部署、执行程序

程序调试成功后，以 JAR 文件形式导出，部署到 Hadoop 集群中。

使用下列 hadoop jar 命令执行 JAR 文件：

```
$ hadoop jar distinct.jar com.bigdata.example.Distinct /mapreduce/distinct/input /mapredut/distinct/output
```

执行成功后，可以查看执行结果，如图 4-27 所示。

```
hadoop@Master:~$ hadoop jar distinct.jar com.bigdata.example.Distinct /mapreduce/distinct/input
/mapredut/distinct/output
hadoop@Master:~$ hdfs dfs -ls /mapreduce/distinct/output
Found 2 items
-rw-r--r--   1 hadoop supergroup          0 2021-01-12 15:33 /mapreduce/distinct/output/_SUCCESS
-rw-r--r--   1 hadoop supergroup        211 2021-01-12 15:33 /mapreduce/distinct/output/part-r-00000
hadoop@Master:~$ hdfs dfs -cat /mapreduce/distinct/output/part-r-00000
2020-10-3    10.3.2.19
2020-10-3    10.3.3.19
2020-10-3    10.3.5.18
2020-10-3    10.3.5.19
2020-10-3    10.3.51.19
2020-10-4    10.3.2.18
2020-10-4    10.3.2.5
2020-10-4    10.3.5.19
2020-10-5    10.3.2.19
2020-10-5    10.3.51.19
```

图 4-27　去重作业执行及结果

### 4.5.2　数据排序

#### 1．任务描述

数据排序是许多实际任务执行时要完成的第一项工作，比如学生成绩评比、数据建立索引等。数据排序和数据去重类似，都是先对原始数据进行初步处理，为进一步的数据操作打好基础。

在本案例中，输入文件是包含一系列数字的多个文件，数字之间使用空格分隔。要求在输出文件中每行有两个数字，第 1 个数字代表原始数据在原始数据集中的位次，第 2 个数字代表原始数据。

图 4-28 描述了包含输入数据格式及 3 种预期的输出结果。

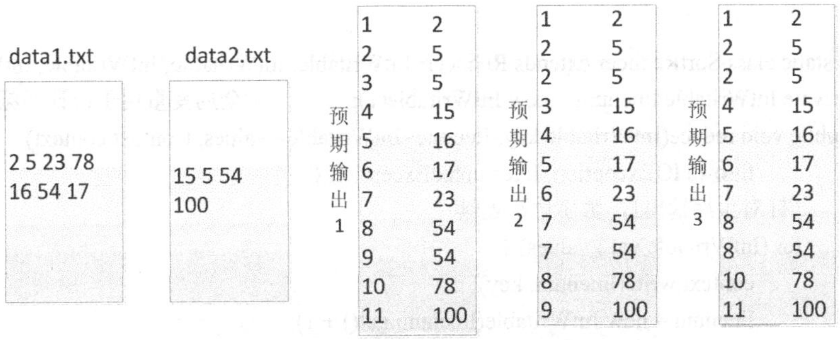

图 4-28  排序作业输入数据及预期输出结果

### 2. 设计思路

由于 MapReduce 程序在 Shuffle 阶段具有按键值降序排序的功能，因此数据排序操作过程中，将需要排序的数据作为 map()函数输出的键值对的键（key），而键值对的值可以是任意的。reduce()函数接收到<key,value-list>之后，将输入的键（key）作为输出的值（value），而输出的键（key）就是该键（key）在输入数据中的位次，可以使用一个全局变量表示。需要注意的是，在数据排序中没有合并操作，也就是在 MapReduce 过程中不使用 Combiner。

### 3. 程序设计

创建一个名为 Sort 的类，编写 Mapper 类、Reducer 类及 main()函数。代码如下：

```java
package com.bigdata.example;
import java.io.IOException;
import java.util.StringTokenizer;
import org.apache.hadoop.conf.Configuration;
import org.apache.hadoop.fs.FileSystem;
import org.apache.hadoop.fs.Path;
import org.apache.hadoop.io.IntWritable;
import org.apache.hadoop.io.LongWritable;
import org.apache.hadoop.io.Text;
import org.apache.hadoop.mapreduce.Job;
import org.apache.hadoop.mapreduce.Mapper;
import org.apache.hadoop.mapreduce.Reducer;
import org.apache.hadoop.mapreduce.lib.input.FileInputFormat;
import org.apache.hadoop.mapreduce.lib.output.FileOutputFormat;
import org.apache.hadoop.util.GenericOptionsParser;
public class Sort {
    public static class SortMapper extends Mapper<LongWritable, Text, IntWritable, IntWritable> {
        private final static IntWritable one = new IntWritable(1);
        private IntWritable data = new IntWritable();
        public void map(LongWritable key, Text value, Context context) throws IOException, InterruptedException {
            StringTokenizer token = new StringTokenizer(value.toString()); //空格为分隔符进行拆分
            while (token.hasMoreTokens()) {
                int num=Integer.parseInt(token.nextToken()); //将数字由String类型转换为int类型
                data.set(num);          //将int类型的数字转换为IntWritable类型，赋给变量data
                context.write(data, one); //输出<数字,1>的键值对
            }
        }
    }
```

```java
public static class SortReducer extends Reducer<IntWritable, IntWritable, IntWritable, IntWritable> {
    private IntWritable linenum = new IntWritable(1);        //全局变量用于统计词频数
    public void reduce(IntWritable key, Iterable<IntWritable> values, Context context)
            throws IOException, InterruptedException {
        //针对预期输出1，数字位次连续
        for (IntWritable val : values) {
            context.write(linenum, key);
            linenum = new IntWritable(linenum.get() + 1);
        }
        /*
        *针对预期输出2，数字位次连续，相同数字位次相同
        * for (IntWritable val : values) {
        *     context.write(linenum, key);
        * }
        * linenum = new IntWritable(linenum.get() + 1);
        */


        /*
        *针对预期输出3，数字位次不连续，相同数字位次相同
        * int count=0;
        * for (IntWritable val : values) {
        *     context.write(linenum, key);
        *     count++;
        * }
        * linenum = new IntWritable(linenum.get() + count);
        */
    }
}
public static void main(String[] args) throws Exception {
    Configuration conf = new Configuration();
    String[] pathArgs = new GenericOptionsParser(conf,args).getRemainingArgs();
    if (pathArgs.length != 2) {
        System.err.println("Usage: sort<in><out>");
        System.exit(2);
    }
    Path path = new Path(pathArgs[1]);
    FileSystem fs = FileSystem.get(conf);
    if (fs.exists(path)) {
        fs.delete(path, true);
    }
    Job job = Job.getInstance(conf," Sort" );
    job.setJarByClass(Sort.class);
    job.setMapperClass(SortMapper.class);
    job.setReducerClass(SortReducer.class);
    job.setOutputKeyClass(IntWritable.class);
    job.setOutputValueClass(IntWritable.class);
    FileInputFormat.addInputPath(job, new Path(pathArgs[0]));
    FileOutputFormat.setOutputPath(job, new Path(pathArgs[1]));
```

```
        System.exit(job.waitForCompletion(true) ? 0 : 1);
    }
}
```

### 4．执行结果

（1）数据准备

按图 4-28 的描述，在 Linux 本地创建 data1.txt 和 data2.txt 两个文件，并输入相应内容。然后，在 HDFS 上创建输入目录/mapreduce/sort/input。最后，将 data1.txt 和 data2.txt 文件上传到 HDFS 的/mapreduce/sort/input 目录中。如图 4-29 所示。

```
hadoop@Master:~$ gedit data1.txt
hadoop@Master:~$ gedit data2.txt
hadoop@Master:~$ hdfs dfs -mkdir -p /mapreduce/sort/input
hadoop@Master:~$ hdfs dfs -put data1.txt data2.txt /mapreduce/sort/input
hadoop@Master:~$ hdfs dfs -ls /mapreduce/sort/input
Found 2 items
-rw-r--r--   1 hadoop supergroup         19 2021-01-12 22:20 /mapreduce/sort/input/data1.txt
-rw-r--r--   1 hadoop supergroup         12 2021-01-12 22:20 /mapreduce/sort/input/data2.txt
```

图 4-29　排序作业输入数据准备

（2）导出、部署、执行程序

程序调试成功后，以 JAR 文件形式导出，部署到 Hadoop 集群中，然后使用 hadoop jar 命令执行该文件。

```
$ hadoop jar sort.jar com.bigdata.example.Sort /mapreduce/sort/input /mapredut/sort/output
```

执行成功后，可以查看执行结果，如图 4-30 所示。

```
hadoop@Master:~$ hdfs dfs -ls /mapreduce/sort
Found 4 items
drwxr-xr-x   - hadoop supergroup          0 2021-01-12 22:20 /mapreduce/sort/input
drwxr-xr-x   - hadoop supergroup          0 2021-01-13 15:13 /mapreduce/sort/output1
drwxr-xr-x   - hadoop supergroup          0 2021-01-12 22:58 /mapreduce/sort/output2
drwxr-xr-x   - hadoop supergroup          0 2021-01-12 22:59 /mapreduce/sort/output3
hadoop@Master:~$ hdfs dfs -cat /mapreduce/sort/output1/part-r-00000
1       2
2       5
3       5
4       15
5       16
6       17
7       23
8       54
9       54
10      78
11      100
```

```
hadoop@Master:~$ hdfs dfs -cat /mapreduce/sort/output2/part-r-00000
1       2
2       5
2       5
3       15
4       16
5       17
6       23
7       54
7       54
8       78
9       100
```

```
hadoop@Master:~$ hdfs dfs -cat /mapreduce/sort/output3/part-r-00000
1       2
2       5
2       5
4       15
5       16
6       17
7       23
8       54
8       54
10      78
11      100
```

图 4-30　数据排序作业执行及结果

## 4.5.3　计算平均值

### 1．任务描述

求平均值是 MapReduce 比较常见的应用，例如求学生多门课程的平均成绩、求不同类型商品的平均点击次数等。本例通过求不同类型商品平均点击次数介绍 MapReduce 求平均值的算法。

输入文件中每行数据包括商品类别编号和点击次数，以制表符（\t）分隔。输出文件中每行包括商品类型及平均点击次数。如图 4-31 所示。

### 2．设计思路

基本思路是 Map 端读取数据，在数据输入到 Reduce 之前先经过 Shuffle，将 map()函数输出的 key 值相同的所有 value 值形成一个集合 value-list，然后将该集合输入到 Reduce 端，Reduce 端汇总并且统计记录数，然后作商即可。

| goods1.txt | | goods2.txt | | 预期输出结果 | |
|---|---|---|---|---|---|
| 52127 | 5 | 52132 | 30 | 52006 | 462 |
| 52120 | 93 | 52132 | 45 | 52009 | 2615 |
| 52092 | 93 | 52132 | 24 | 52024 | 347 |
| 52132 | 38 | 52009 | 2615 | 52090 | 11 |
| 52006 | 462 | 52132 | 25 | 52092 | 93 |
| 52109 | 28 | 52090 | 13 | 52109 | 35 |
| 52109 | 43 | 52132 | 6 | 52120 | 93 |
| 52132 | 0 | 52136 | 0 | 52127 | 5 |
| 52132 | 34 | 52090 | 10 | 52132 | 23 |
| 52132 | 9 | 52024 | 347 | 52136 | 0 |

图 4-31 求平均值作业的输入数据及预期输出结果

### 3. 程序设计

创建一个名为 Score 的类，编写 Mapper 类、Reducer 类及 main()函数。代码如下：

```java
package com.bigdata.example;

import java.io.IOException;
import org.apache.hadoop.conf.Configuration;
import org.apache.hadoop.fs.FileSystem;
import org.apache.hadoop.fs.Path;
import org.apache.hadoop.io.IntWritable;
import org.apache.hadoop.io.LongWritable;
import org.apache.hadoop.io.Text;
import org.apache.hadoop.mapreduce.Job;
import org.apache.hadoop.mapreduce.Mapper;
import org.apache.hadoop.mapreduce.Reducer;
import org.apache.hadoop.mapreduce.lib.input.FileInputFormat;
import org.apache.hadoop.mapreduce.lib.output.FileOutputFormat;
import org.apache.hadoop.util.GenericOptionsParser;
public class Score {
    public static class ScoreMapper extends Mapper<LongWritable, Text, Text, IntWritable> {
        private Text id = new Text();
        private IntWritable num = new IntWritable();
        public void map(LongWritable key, Text value, Context context)
                throws IOException, InterruptedException {
            String line = value.toString();
            System.out.println(line);
            String[] arr = line.split("\t");
            id.set(arr[0]);
            num.set(Integer.parseInt(arr[1]));
            context.write(id, num);
        }
    }
    public static class ScoreReducer extends Reducer<Text, IntWritable, Text, IntWritable> {
        public void reduce(Text key, Iterable<IntWritable> values, Context context)
                throws IOException, InterruptedException {
            int sum = 0;
            int count = 0;
            for (IntWritable val : values) {
```

```java
                sum += val.get();
                count++;
            }
            int average = (int) sum / count;// 计算平均成绩
            context.write(key, new IntWritable(average));
        }
    }
    public static void main(String[] args) throws Exception {
        Configuration conf = new Configuration();
        String[] pathArgs = new GenericOptionsParser(conf, args).getRemainingArgs();
        if (pathArgs.length != 2) {
            System.err.println("Usage: Score <in><out>");
            System.exit(2);
        }
        Path path = new Path(pathArgs[1]);
        FileSystem fs = FileSystem.get(conf);
        if (fs.exists(path)) {
            fs.delete(path, true);
        }
        Job job = Job.getInstance(conf, "Score");
        job.setJarByClass(Score.class);
        job.setMapperClass(ScoreMapper.class);
        job.setReducerClass(ScoreReducer.class);
        job.setOutputKeyClass(Text.class);
        job.setOutputValueClass(IntWritable.class);
        FileInputFormat.addInputPath(job, new Path(pathArgs[0]));
        FileOutputFormat.setOutputPath(job, new Path(pathArgs[1]));
        System.exit(job.waitForCompletion(true) ? 0 : 1);
    }
}
```

### 4．执行结果

（1）数据准备

按图 4-31 的描述，在 Linux 本地创建 goods1.txt 和 goods2.txt 两个文件，并输入相应内容。然后，在 HDFS 上创建输入目录/mapreduce/score/input。最后，将 goods1.txt 和 goods2.txt 文件上传到 HDFS 的/mapreduce/score/input 目录中。如图 4-32 所示。

```
hadoop@Master:~$ gedit goods1.txt
hadoop@Master:~$ gedit goods2.txt
hadoop@Master:~$ hdfs dfs -mkdir -p /mapreduce/score/input
hadoop@Master:~$ hdfs dfs -put goods1.txt goods2.txt /mapreduce/score/input
hadoop@Master:~$ hdfs dfs -ls /mapreduce/score/input
Found 2 items
-rw-r--r--   1 hadoop supergroup        118 2021-01-13 16:04 /mapreduce/score/input/goods1.txt
-rw-r--r--   1 hadoop supergroup        121 2021-01-13 16:04 /mapreduce/score/input/goods2.txt
```

图 4-32 求平均值作业输入数据准备

（2）导出、部署、执行程序

程序调试成功后，以 JAR 文件形式导出，部署到 Hadoop 集群中，然后使用 hadoop jar 命令执行该文件。

$ hadoop jar score.jar com.bigdata.example.Score /mapreduce/score/input /mapredut/score/output

执行成功后，可以查看执行结果，如图 4-33 所示。

```
hadoop@Master:~$ hdfs dfs -ls /mapreduce/score/output
Found 2 items
-rw-r--r--   1 hadoop supergroup          0 2021-01-13 21:26 /mapreduce/score/output/_SUCCESS
-rw-r--r--   1 hadoop supergroup         92 2021-01-13 21:26 /mapreduce/score/output/part-r-00000
hadoop@Master:~$ hdfs dfs -cat /mapreduce/score/output/part-r-00000
52006   462
52009   2615
52024   347
52090   11
52092   93
52109   35
52120   93
52127   5
52132   23
52136   0
```

图 4-33  求平均值作业执行及结果

## 4.6  网站浏览量统计分析

本节将介绍如何利用 MapReduce 统计分析每日访问网站 IP 数量，即每日的页面浏览量。

程序设计的基本思路是，map()函数从接收的输入值（value）中分离出日期和 IP，然后以日期为键（key）、IP 为值（value），输出中间结果键值对<日期，IP>，经过 Shuffle 的合并操作后，同一个日期（key）及其对应的 IP 列表（list-values）传输给同一个 reduce()函数，reduce()函数进行值的合并，最后输出日期（key）和对应 IP 总和（value）。

### 1．准备数据

创建/mapreduce/logs/input 目录，并将日志文件上传到该目录中，如图 4-34 所示。

```
hadoop@Master:~$ hdfs dfs -mkdir -p /mapreduce/logs/input
hadoop@Master:~$ hdfs dfs -put log-20190501 /mapreduce/logs/input
hadoop@Master:~$ hdfs dfs -ls /mapreduce/logs/input
Found 1 items
-rw-r--r--   1 hadoop supergroup     956192 2021-01-14 17:31 /mapreduce/logs/input/log-20190501
```

图 4-34  网站日志分析数据准备

### 2．程序设计

创建一个 weblog 项目，然后创建一个名为 PV 的类。代码如下：

```java
package com.bigdata.example;
import java.io.IOException;
import java.text.ParseException;
import java.text.SimpleDateFormat;
import java.util.Date;
import java.util.Locale;
import org.apache.hadoop.conf.Configuration;
import org.apache.hadoop.fs.FileSystem;
import org.apache.hadoop.fs.Path;
import org.apache.hadoop.io.IntWritable;
import org.apache.hadoop.io.LongWritable;
import org.apache.hadoop.io.Text;
import org.apache.hadoop.mapreduce.Job;
import org.apache.hadoop.mapreduce.Mapper;
import org.apache.hadoop.mapreduce.Reducer;
import org.apache.hadoop.mapreduce.lib.input.FileInputFormat;
import org.apache.hadoop.mapreduce.lib.output.FileOutputFormat;
import org.apache.hadoop.util.GenericOptionsParser;
public class PV {
    public static class PVMapper extends Mapper<LongWritable, Text, Text, Text> {
```

```java
        private Text dt = new Text();         //定义一个变量表示日期
        private Text ip = new Text();         //定义一个变量表示IP地址
        public void map(LongWritable key, Text value, Context context) throws IOException,
InterruptedException {
            String line = value.toString();
            String[] arr = line.split(" ");
            String dateString1 = arr[3].substring(1);      //获取日期字符串
            SimpleDateFormat df1 = new SimpleDateFormat("dd/MMM/yyyy:HH:mm:ss", Locale.US);
            Date date = null;
            try {
                date = df1.parse(dateString1);    //将获取的日期字符串转换为日期
            } catch (ParseException e) {
                e.printStackTrace();
            }
            SimpleDateFormat df2 = new SimpleDateFormat("yyyyMMdd", Locale.US);
            String dateString2 = df2.format(date);    //将日期转换为字符串
            dt.set(dateString2);
            ip.set(arr[0]);
            context.write(dt, ip);
        }
    }
    public static class PVReducer extends Reducer<Text, Text, Text, IntWritable> {
        private IntWritable result = new IntWritable();
        public void reduce(Text key, Iterable<Text> values, Context context) throws IOException,
InterruptedException {
            if (key.toString() == null || key.toString().equals(""))    //如果日期字符串为空,返回
                return;
            int count = 0;
            for (Text val : values) {
                count++;
            }
            result.set(count);
            context.write(key, result);
        }
    }
    public static void main(String[] args) throws Exception {
        Configuration conf = new Configuration();
        String[] otherArgs = (new GenericOptionsParser(conf, args)).getRemainingArgs();
        if (otherArgs.length != 2) {
            System.err.println("Usage: pv <in><out>");
            System.exit(2);
        }
        FileSystem fileSystem = FileSystem.get(conf);
        Path path = new Path(otherArgs[1]);
        FileSystem fs = FileSystem.get(conf);
        if (fs.exists(path)) {
            fs.delete(path, true);
        }
        Job job = Job.getInstance(conf, "PV");
```

```
            job.setJarByClass(PV.class);
            job.setMapperClass(PVMapper.class);          //设置Map处理类
            job.setReducerClass(PVReducer.class);         //设置Reduce处理类
            job.setMapOutputKeyClass(Text.class);         //设置Map输出键的类型
            job.setMapOutputValueClass(Text.class);       //设置Map输出值的类型
            job.setOutputKeyClass(Text.class);            //设置Reduce输出键的类型
            job.setOutputValueClass(IntWritable.class);   //设置Reduce输出值的类型
            FileInputFormat.addInputPath(job, new Path(otherArgs[0]));   //设置输入目录
            FileOutputFormat.setOutputPath(job, new Path(otherArgs[1])); //设置输出目录
            System.exit(job.waitForCompletion(true) ? 0 : 1);
        }
    }
```

### 3. 程序执行

程序调试成功后，以 JAR 文件形式导出，然后利用 hadoop jar 命令执行该文件。

```
$ hadoop jar pv.jar com.bigdata.example.PV /mapreduce/logs/input /mapreduce/logs/output
```

执行成功后，可以查看执行结果，如图 4-35 所示。

```
hadoop@Master:~$ hdfs dfs -ls /mapreduce/logs/output
Found 2 items
-rw-r--r--   1 hadoop supergroup          0 2021-01-15 20:48 /mapreduce/logs/output/_SUCCESS
-rw-r--r--   1 hadoop supergroup        390 2021-01-15 20:48 /mapreduce/logs/output/part-r-00000
hadoop@Master:~$ hdfs dfs -cat /mapreduce/logs/output/part-r-00000
20190401        316
20190402        199
20190403        102
20190404        208
20190405        52
20190406        139
```

图 4-35 每日网站浏览量统计结果

# 本 章 小 结

本章系统介绍了 MapReduce 分布式计算框架，包括 MapReduce 编程模型、MapReduce 架构、运行机制等理论知识，以及 MapReduce 编程模型、开发组件、程序设计思想、程序设计案例等应用开发知识。通过 WordCount 程序设计详解、典型 MapReduce 案例（数据去重、数据排序、计算平均值）和 MapReduce 网站浏览量统计分析的学习，读者可以掌握 MapReduce 应用开发的思想与方法。

# 思考题与习题

### 1. 简答题

（1）简述基于 YARN 的 MapReduce 架构及工作机制。
（2）简述 MapReduce 编程模型。
（3）简述 MapReduce 数据处理流程。
（4）简述 MapReduce 常用的编程组件。
（5）简述 Hadoop 数据序列化与反序列化的必要性。

### 2. 选择题

（1）关于 MapReduce 的描述不正确的是（　　）。
　　A. MapReduce 是 Hadoop 的核心组件　　　　B. MapReduce 是一个并行计算框架

C．MapReduce 适合流式数据处理　　　D．MapReduce 适合批量数据处理

（2）下面过程生成中间键值对的是（　　）。

A．Reducer　　B．Mapper　　C．Combiner　　D．Partitioner

（3）在 MapReduce 中，Map 任务数量取决于下列哪个因素？（　　）

A．数据块大小　　B．输入数据　　C．输出数据　　D．数据分片数量

（4）在 Hadoop 中，下面哪个是默认的 InputFormat 类型，它将每行内容作为新值，而将字节偏移量作为 key 值？（　　）

A．FileInputFormat　　　　　　　B．KeyValueTextInputFormat

C．TextInputFormat　　　　　　　D．SequenceFileInputFormat

（5）下列关于 MapReduce 说法不正确的是（　　）。

A．MapReduce 是一种计算框架

B．MapReduce 隐藏了并行计算的细节，方便使用

C．MapReduce 程序只能用 Java 语言编写

D．MapReduce 来源于 Google 的学术论文

（6）有关 MapReduce 的输入、输出，下列说法错误的是（　　）。

A．链接多个 MapReduce 作业时，序列文件是首选格式

B．FileInputFormat 中实现 getSplits()将输入数据分片，分片数目和大小任意定义

C．想完全禁止输出，可以使用 NullOutputFormat

D．每个 Reduce 需将它的输出写入自己的文件中，输出无须分片

（7）MapReduce 框架提供了一种序列化键值对的方法，支持这种序列化的类能够在 Map 和 Reduce 过程中充当键或值，以下说法错误的是（　　）。

A．实现 Writable 接口的类是值

B．实现 WritableComparable 接口的类可以是值或键

C．Hadoop 的基本类型 Text 并不实现 WritableComparable 接口

D．键和值的数据类型可以超出 Hadoop 自身支持的基本类型

（8）关于 MapReduce 中的键值对，下面说法正确的是（　　）。

A．Key 类必须实现 Writable　　　　B．Key 类必须实现 WritableComparable

C．Value 类必须实现 WritableComparable　　D．Value 类必须继承 WritableComparable

（9）在 Hadoop 的分区阶段，默认的 Partitioner 是（　　）。

A．BinaryPartitioner　　　　　　　B．KeyFieldBasedPartitioner

C．RehashPartitioner　　　　　　　D．HashPartitioner

（10）关于 MapReduce 工作流程描述不正确的是（　　）。

A．不同的 Map 任务之间不会进行通信

B．不同的 Reduce 任务之间不会发生任何信息交换

C．Map 数据是全局性的，Reduce 数据是局部性的

D．只有当 Map 任务全部结束后，Reduce 任务才能开始

3．实训题

（1）创建一个输入文件，保存商品 id 及用户点击次数信息。其中，第一字段为商品 id，第二字段为用户点击次数，两列之间使用制表符（\t）分隔，如图 4-36 中输入文件。请编写一个 MapReduce 程序，将商品按点击次数从低到高排序，结果如图 4-36 中的输出文件。

（2）编写一个输入文件，保存学生学号及课程成绩信息。其中，第一字段为学生学号，第

二到第六字段为学生不同科目的成绩，字段之间使用制表符（\t）分隔，如图 4-37 所示。请编写一个 MapReduce 程序，输出学生学号及平均成绩。

输入文件

| | |
|---|---|
| 1010037 | 100 |
| 1010102 | 100 |
| 1010152 | 97 |
| 1010178 | 96 |
| 1010280 | 104 |
| 1010320 | 103 |
| 1010510 | 104 |
| 1010603 | 96 |
| 1010637 | 97 |

输出文件

| | |
|---|---|
| 96 | 1010603 |
| 96 | 1010178 |
| 97 | 1010637 |
| 97 | 1010152 |
| 100 | 1010102 |
| 100 | 1010037 |
| 103 | 1010320 |
| 104 | 1010510 |
| 104 | 1010280 |

图 4-36 按点击次数排序

输入文件

| | | | | |
|---|---|---|---|---|
| 18001050101 | 76 | 81 | 83 | 81 | 91 |
| 18001050102 | 75 | 79 | 69 | 90 | 86 |
| 18001050104 | 72 | 85 | 60 | 77 | 87 |
| 18001050105 | 76 | 76 | 78 | 83 | 88 |
| 18001050106 | 70 | 76 | 40 | 73 | 84 |
| 18001050107 | 80 | 84 | 65 | 83 | 91 |
| 18001050108 | 85 | 92 | 60 | 78 | 91 |
| 18001050109 | 80 | 78 | 41 | 76 | 90 |

图 4-37 求学生平均成绩

# 第 5 章 分布式数据库 HBase

HBase 是一个运行在 Hadoop 集群上的列式存储的分布式数据库，提供大规模数据的实时随机访问。本章将介绍分布式数据库 HBase 的理论体系与应用开发。
- HBase 概述：Hbase 简介、特性、适用场景。
- HBase 数据模型：HBase 数据模型基本概念、概念视图、物理视图。
- HBase 体系结构：HBase 架构组成、运行机制。
- HBase 安装与配置：HBase 运行模式、安装准备、伪分布式安装与配置。
- HBase Shell：Hbase Shell 简介、HBase 命令组成、DDL 命令组、DML 命令组。
- HBase 程序设计：Hbase Java API 简介、表管理程序设计、数据操作程序设计。
- HBase 与 MapReduce 融合：HBase 与 MapReduce 融合概述、HBaseMapReduce Java API、Hbase MapReduce 程序设计。
- HBase 学生成绩分析：综合利用 HBase 与 MapReduce 程序设计分析学生平均成绩。

## 5.1 HBase 概述

### 5.1.1 HBase 简介

HBase 是一个开源的、分布式 NoSQL 数据库系统，是 Google 的 Bigtable 的开源实现，是 Apache 基金会的顶级项目。HBase 使用 Java 语言实现，利用 Hadoop 分布式文件系统 HDFS 作为其文件存储系统，支持使用 MapReduce 分布式批量处理 HBase 中存储的海量数据，利用 Zookeeper 协同管理数据，实现稳定服务和失败恢复。

作为一个典型的分布式数据库系统，HBase 具有高可靠性、高性能、可伸缩等分布式架构特性。虽然 HBase 底层采用了 HDFS 的存储，但与 HDFS 支持顺序存取不同，HBase 采用列式存储结构，支持数据的随机存取功能。因此，HBase 是一个可以进行随机存取和检索数据的存储平台。

### 5.1.2 HBase 特性

与传统的关系型数据库相比，HBase 具有下列特征。

① 容量巨大：一个 HBase 表可以达到数十亿行、数百万列，可以在横向和纵向两个维度插入数据，具有很大的弹性。

② 数据类型简单：所有半结构化数据和非结构化数据在 HBase 中都被序列化为字节数组，用户可以根据需要解析为不同的数据类型。

③ 列式存储：HBase 中的数据是基于列族存储的，同一列族中的数据集中存储在几个文件中。

④ 稀疏表：HBase 表中的空单元格不占用存储空间。

⑤ 数据多版本：HBase 中每个单元格中的数据可以有多个版本，时间戳就是单元值的版本号。

⑥ 数据操作简单：在 HBase 中只存在数据插入、查询、删除、清空等操作。更新数据的方法是生成一个新版本的数据。数据删除本质上是插入带有删除标记的数据。

### 5.1.3 HBase 适用场景

HBase 并不适合所有应用场景。具有下列某些特点的应用，采用 HBase 数据库会具有更高的性能。

① 数据模式是动态的或者可变的，且支持半结构化和非结构化的数据。

② 数据库中的很多列都包含很多空字段，HBase 中的空字段不会像在关系数据库中占用空间。

③ 需要很高的吞吐量，瞬间写入量很大。

④ 数据有很多版本需要维护，HBase 利用时间戳来区分不同版本的数据。

⑤ 具有高可扩展性，能动态地扩展整个存储系统。

在实际应用中，有很多公司使用 HBase，如 Facebook 公司的 Social Inbox 系统，使用 HBase 作为消息服务的基础存储设施，每月可处理几千亿条的消息；Yahoo 公司使用 HBase 存储检查近似重复的指纹信息的文档，它的集群中分别运行着 Hadoop 和 HBase，表中存了上百万行数据；Adobe 公司使用 Hadoop+HBase 的生产集群，将数据直接持续地存储在 HBase 中，并将 HBase 作为数据源进行 MapReduce 的作业处理；Apache 公司使用 HBase 来维护 Wiki 的相关信息。

## 5.2 HBase 数据模型

### 5.2.1 HBase 基本概念

在 HBase 中，数据存储在由行与列组成的表中，这与关系数据库（RDBMS）很相似，但两者是不同的。在关系数据库中，通过行与列二维确定一个单元值，而在 HBase 中是通过行键、列（列族:列族限定符）和版本唯一确定一个单元值。HBase 表结构如图 5-1 所示。

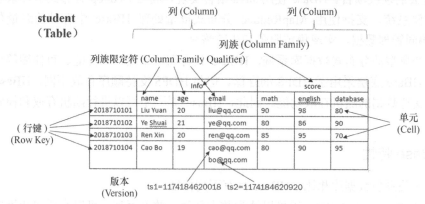

图 5-1 HBase 表结构

（1）命名空间（Namespace）

HBase 中命名空间是表的逻辑分组，类似于关系数据库系统中的数据库，任何表都属于特定的命名空间。在 HBase 中有两个预定义命名空间：hbase 和 default。hbase 是系统命名空间，所有 HBase 内部表都属于 hbase 命名空间；其他所有没有显示指定命名空间的表，都自动归属于 default 命名空间。

命名空间与表的归属表示形式为<namespace>:<table>，例如 default:student。

（2）表（Table）

HBase 采用表来组织数据。表名在模式定义时预先声明，例如表 student。

(3）行键（Row Key）

每个 HBase 表都由若干行组成，每个行由行键来唯一标识，类似于关系数据库中的主键。HBase 中访问表中数据有 3 种方法：第一种方法是通过单个行键进行访问；第二种方法是通过一个行键的区间进行访问；第三种方法是进行全表扫描。

行键可以是任意字符串，最大长度是 64KB。在 HBase 内部，行键保存为字节数组。在存储时，数据按行键的字典顺序排序存储。因此，在设计行键时，要充分考虑排序存储这个特性，将经常一起读取的行放到一起。例如，当行键为网站域名时，应使用倒序法存储：org.apache.www、org.apache.hbase、org.apache.hive，这样与 apache 相关的网页在一个 HBase 表中的存储位置就是临近的，而不会因为域名首单词（www、hbase、hive）差异太大而分散在不同的位置。

（4）列（Column）

在 HBase 中，列由列族（Column Family）和列族限定符（Column Family Qualifier）构成，中间使用冒号（:）分隔，列族名称作为列族成员的前缀，例如，info:name、info:age、info:email 都是列族 info 的成员。列族需要在创建表模式时指定，列族限定符可以在数据操作时动态指定。

（5）列族（Column Family）

一个 HBase 表由列族组成，列族是基本的访问控制单元和存储单元。每个列族可以包含任意个列族限定符。列族是表模式的一部分，必须在使用表之前定义。在 HBase 中，数据存储是以列族为单位进行的，同一个列族成员集中存储在一起，这就是 HBase 支持基于列的随机访问的原因。例如，在 student 表中包括 info 和 score 两个列族。在一个 HBase 表中，列族数量不要超过两个，过多列族将影响系统性能。

（6）列族限定符（Column Family Qualifier）

列族里的数据通过列族限定符来定位。列族限定符不用预先定义，不需要在不同行之间保持一致。可以在应用中动态为列族添加列族限定符。列族限定符都以列族作为前缀。例如，student 表中列族 score 包括 math、english 和 database 三个列族限定符。

（7）单元格（Cell）

在 HBase 表中，通过行键、列族和列族限定符确定一个"单元格"，单元格中存储的数据没有数据类型，都是未经解释的字节数组 byte[]。每个单元格中可以保存一个数据的多个版本。

（8）版本（Version）

在 HBase 表中，通过<rowkey,column,version>来确定一个单元值。一个单元格中可以存在数据的多个版本，通过一个长整型（Long Integer）数据标识。默认采用时间戳来作为标识单元值的版本。单元格版本可以由 HBase 在数据写入时自动赋值，也可以由用户显式赋值。如果应用程序要避免数据版本冲突，就必须自己生成具有唯一性的版本号。在每个单元格中，数据按照版本号倒序排序，即最新版本的数据排在最前面。例如，student 表中行键为 2018710104 的学生存在两个 email，分别对应版本 ts1 和 ts2。为了查询该学生的 email，需要通过行键（2018710104）、列族（info）、列族限定符（email）和版本（时间戳 ts1 或 ts2）定位到某个具体的单元值。

为了避免数据存在过多版本造成数据存储和索引的负担，HBase 提供了两种数据版本回收方式：一是设置保存数据的版本数量，二是设置单元值的有效期。

（9）主键与索引

在 HBase 表中，行键就是行的唯一标识，因此行键就是主键，例如 2018710101、2018710102。在 HBase 中，唯一确定一个值的索引由行键、列族、列族限定符、版本组合而成。例如，在 student 表中，<"2018710104","info","email",ts1 >确定了单元值 cao@qq.com，而<"2018710104","info","email",ts2 >确定了单元值 bo@qq.com。

### 5.2.2 概念视图

在 HBase 的概念视图中，一个表可以是一个稀疏、多维的映射关系。

例如，表 5-1 是一个用于存储网页的 HBase 表的概念视图。表中有一个行键和两个列族。其中，行键是一个反向的域名 URL。列族 contents 包含一个名为 html 的列族限定符；列族 anchor 包含 cssnsi.com 和 my.look.ca 两个列族限定符。在 HBase 表中，一个新建的列族限定符可以随时添加到已存在的列族中。

表 5-1　HBase 表概念视图

| 行键 | 时间戳 | 列族 contents | 列族 anchor |
|---|---|---|---|
| "com.cnn.www" | t9 | | anchor:cnnsi.com = "CNN" |
| | t8 | | anchor:my.look.ca = "CNN.com" |
| | t6 | contents:html = "<html>…" | |
| | t5 | contents:html = "<html>…" | |
| | t3 | contents:html = "<html>…" | |

在同一个单元格中可以有多个数据，这些数据具有不同的时间戳。从表的概念视图可以看出，每个行都包含相同的列族，但不需要每个列族包含相同的列族限定符，也不需要每个单元格都有数据。因此，HBase 表是一个稀疏表。

### 5.2.3 物理视图

HBase 表采用基于列族的存储方式，因此表 5-1 所示的概念视图在存储时会分成两部分，按照列族 anchor 和列族 contents 分别存放，如表 5-2、表 5-3 所示，属于同一个列族的数据保存在一起。与每个列族一起存放的还包括行键和时间戳。

表 5-2　列族 anchor 物理视图

| 行键 | 时间戳 | 列族 anchor |
|---|---|---|
| "com.cnn.www" | t9 | anchor:cnnsi.com = "CNN" |
| "com.cnn.www" | t8 | anchor:my.look.ca = "CNN.com" |

表 5-3　列族 contents 物理视图

| 行键 | 时间戳 | 列族 contents |
|---|---|---|
| "com.cnn.www" | t6 | contents:html = "<html>…" |
| "com.cnn.www" | t5 | contents:html = "<html>…" |
| "com.cnn.www" | t3 | contents:html = "<html>…" |

概念视图中的空单元格实际上是没有进行物理存储的，因此对返回时间戳为 t8 的 contents:html 的值的请求，结果为空。同样，一个返回时间戳为 t9 的 anchor:my.look.ca 的值的请求，结果也为空。

如果没有指定时间戳，那么会返回特定单元格的最新值。对有多个版本的单元格，优先返回最新版本的值，因为单元值是按照版本递减顺序存储的，所以对返回行键为 com.cnn.www 的数据并且没有指定时间戳的请求，返回结果如图 5-2 所示。

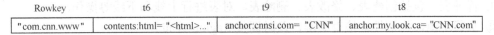

图 5-2  查询返回结果

## 5.3  HBase 体系结构

HBase 采用主从（Master/Slave）服务器架构，由一个 HMaster 服务器和多个 HRegionServer 构成。其中，HMaster 服务器负责管理所有的 HRegionServer，HRegionServer 负责存储许多 Region，每一个 Region 是 HBase 表的一个分块。HBase 中的所有服务器都通过 ZooKeeper 来进行协调，并处理各个服务器在运行期间可能遇到的错误。HBase 体系结构如图 5-3 所示。

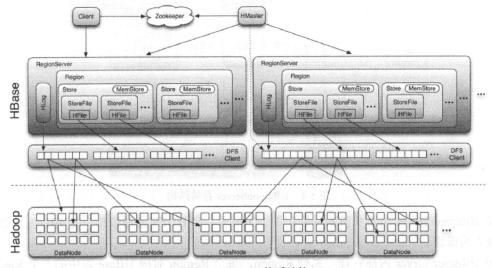

图 5-3  HBase 体系结构

（1）Client

Client 使用 HBase RPC 机制与 Hmaster 服务器和 HRegionServer 进行通信。

① 对管理类的操作，Client 与 HMaster 服务器进行 RPC。

② 对数据读写类的操作，Client 与 HRegionServer 进行 RPC。

③ Client 缓存 Region 位置，以便提高对 HBase 的访问效率。

（2）Zookeeper

Zookeeper 为 HBase 集群提供协同管理服务，具有下列功能。

① 通过选举机制保证任何时候 HBase 集群中只有一个 HMaster 服务器，避免了 HMaster 服务器的单点问题。

② 每个 HRegionServer 上线时会在 Zookeeper 中注册为一个临时节点，Zookeeper 实时监控各个 HRegionServer 的健康状况。当发现某个 HRegionServer 宕机时，能及时通知 Hmaster 服务器进行相应处理。

③ 存储 HBase 的目录表 hbase:meta，该表包含系统中所有 Region 的列表。

（3）HMaster

在 HBase 中可以启动多个 HMaster 服务器，通过 Zookeeper 的选举机制保证系统中总有一个 HMaster 服务器处于运行状态，其他 HMaster 服务器处于备用状态。HMaster 服务器的具体功能包括：
① 管理表，包括创建表、修改表、删除表，对表进行上线和下线等操作。
② 负载均衡，周期性地检查集群，及时调整 Region 分布。
③ 在 Region 分裂后，负责新 Region 的分布。
④ 在 HRegionServer 停机后，负责失效 HRegionServer 上 Region 的迁移工作。

（4）HRegionServer

HBase 数据库中所有数据都保存在 HDFS 中，用户通过一系列 HRegionServer 获取这些数据。一台机器上只能运行一个 HRegionServer，每个 HRegionServer 可以维护多个 Region，每个区段的 Region 只能被一个 HRegionServer 维护。HRegionServer 存储结构如图 5-4 所示。

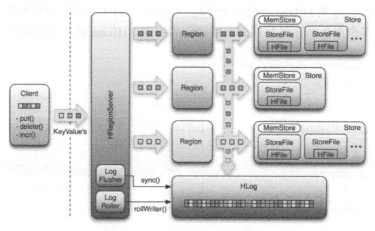

图 5-4　HRegionServer 存储结构

① HRegionServer 主要负责响应用户的 I/O 请求，向 HDFS 读写数据，是 HBase 体系结构中最核心的模块。

② HRegionServer 内部管理一系列 Region，每个 Region 对应 HBase 表中的一个 Region。每个 Region 由多个 Store 组成，每个 Store 对应表中一个列族的存储，可以看出每个列族就是一个集中的存储单元。

③ HRegionServer 还内部管理一个 HLog，HLog 存储数据日志，当 HRegionServer 发生故障时，利用 HLog 进行故障恢复。

（5）Region

HBase 表中的所有数据都按照行键的字典顺序排列存储，在行的方向上分隔为多个 Region。Region 是 HBase 中分布式存储和负载均衡的最小单元，即不同的 Region 可以分别在不同的 HRegionServer 上，但同一个 Region 是不会拆分到多个 HRegionServer 上的。

在表初建时只有一个 Region，随着数据不断插入，Region 不断增大，当增大到指定阈值（由参数 hbase.hregion.max.filesize 确定，默认 256MB）时，HBase 会使用中间的行键将表水平分裂成两个新的 Region。随着表中数据的不断增多，会逐渐分裂成若干个 Region，即一个表的数据被保存到多个 Region 中。

每个 Region 保存一个表中某段连续的数据。每个 Region 都有一个唯一的 RegionID。Region 在 hbase:meat 中以键值对形式存储，键为"表名，开始行键，RegionID"，值为"info:regioninfo,info:server,info:serverstartcode"。定位一个 Region 就是通过该键值对进行的。

当对表进行读写操作时，首先读取 Zookeeper 中的 hbase:meta 表，根据行键确定要读取的 Region 所在 HRegionServer 的位置，然后通过 HRegionServer 进行数据读写操作。

### （6）Store

Store 是 HBase 存储的核心，由 MemStore 和 StoreFile 组成，MemStore 为内存中的缓存，StoreFile 对应磁盘上的文件（Hfile）。每个 Region 包含若干个 Store，每个 Store 对应相应表中的一个列族，即表中有几个列族，Region 就包含几个 Store。

用户向表中写入数据时，首先写入表对应 Region 所在 HRegionServer 的 HLog 中，然后写入相应 Store 的 MemStore，在一定条件下，MemStoreFlush 进程被触发，将 MemStore 内容写入一个 StoreFile 文件（底层实现为 HFile），同时在 HLog 中标记数据已经写入文件。当 StoreFile 文件数量增长到一定阈值（hbase.hstore.compaction.min，最小合并文件数，默认为 3；hbase.hstore.compaction.max，最大合并文件数，默认为 10）时，会触发 Compact 合并操作，将多个 StoreFile 合并成一个 StoreFile。在合并过程中，会进行版本合并和数据删除。多次执行 Compact 合并操作，形成的 StoreFile 会越来越大。当单个 StoreFile 大小超过一定阈值后（hbase.hregion.max.filesize），会触发 Split 拆分操作，同时把当前的 Region 拆分成两个 Region，原来 Region 会下线，新拆分出的两个 Region 会被 Hmaster 服务器分配到相应的 HRegionServer 上，使原来一个 Region 的压力得以分流到两个 Region 上。向 HBase 表中写入数据的过程如图 5-5 所示。

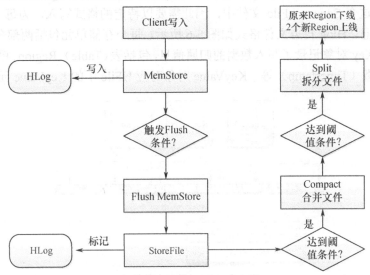

图 5-5　向 HBase 表中写入数据的过程

需要注意的是，MemStore 刷新时，最小刷新单元是每个 Region 中所有的 MemStore，而不是单个 MemStore，因此一个表中的列族数量不要超过两个。下列任何情况下都可以触发 MemStore 刷新。

① 当一个 MemStore 达到参数 hbase.hregion.memstore.flush.size 设定值时，一个 Region 中的所有 Store 的 MemStore 都将被刷新到磁盘。

② 当 MemStore 的总体使用率达到 hbase.regionserver.global.memstore.upperLimit 设定值时，来自不同 Region 的 MemStore 都将被刷新到磁盘，以减少 HRegionServer 中 MemStore 的总体使用量。刷新顺序是按 Region 中 MemStore 使用量降序进行的。直到 MemStore 总体使用率等于或低于参数 hbase.regionserver.global.memstore.lowerLimit 设定值为止。

③ 当指定 HRegionServer 中 WAL 日志条目数达到参数 hbase.regionserver.max.logs 设定值时，不同 Region 中的 MemStore 将被刷新到磁盘，以减少 WAL 中的日志条目数。此时，MemStore 刷新顺序基于时间进行。首先刷新具有最旧 MemStore 的 Region，直到 WAL 数量降至参数

hbase.regionserver.max.logs 设定值以下。

用户查询数据时，首先从相应 Region 的 Store 的 MemStore 中寻找数据，如果找不到，再到 StoreFile 中寻找，如果找到数据，先将数据缓存到 MemStore 中，然后返回给用户。

（7）HLog

在分布式系统环境中，无法避免系统出错或者宕机，一旦 HRegionServer 意外退出，MemStore 中的内存数据就会丢失，为此在 HRegionServer 引入了 HLog。

HBase 为每个 HRegionServer 配置了一个 HLog 文件，它是一种预写式日志（Write Ahead Log）文件。每次用户操作写入 MemStore 的同时，会写一份数据到 HLog 文件，HLog 文件定期会滚动出新文件，并删除旧的文件（已持久化到 StoreFile 中的数据）。当 HRegionServer 意外终止后，HMaster 会通过 Zookeeper 感知，HMaster 首先处理遗留的 HLog 文件，将不同 Region 的 HLog 数据拆分，分别放到相应 Region 目录下，然后将失效的 Region 重新分配，领取到这些 Region 的 HRegionServer 在加载 Region 的过程中，会发现有历史 HLog 需要处理，因此会重现 HLog 中的数据到 MemStore 中，然后刷新到 StoreFile，完成数据恢复。

HLog 文件是一个普通的 Hadoop SequenceFile 文件，无论哪个 Region 数据写入，都要直接以追加形式写入这个 SequenceFile 文件中，因此需要以特定的格式写入，为每一个操作记录添加相应的归属信息。Hlog 日志文件格式如图 5-6 所示，每个存储单元包括两部分：HLogKey 和 KeyValue。HLogKey 对象记录了写入数据的归属信息，包括表（Table）、Region、序列号（Sequence Number）、时间戳（Timestamp）等。KeyValue 对象对应 HFile 中的 KeyValue 元数据。

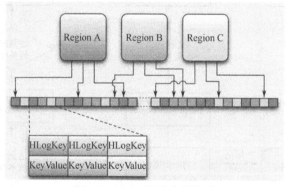

图 5-6　HLog 日志文件格式

## 5.4　HBase 安装与配置

### 5.4.1　HBase 运行模式

HBase 有 3 种运行模式，分别为独立模式（Standalone）、伪分布式模式（Pseudo-Distributed）和完全分布式模式（Fully Distributed）。

（1）独立模式

独立模式只在一个节点上配置 HBase，不使用 HDFS，而是直接使用 Linux 本地文件系统。在同一个 JVM 实例中运行所有 HBase 守护进程——HMaster、HRegionServer 和 Zookeeper。

（2）伪分布式模式

伪分布式模式与独立模式类似，同样只运行在一个节点上，既可以使用本地文件系统，也可以使用 HDFS。与独立模式所有 HBase 守护进程运行在一个 JVM 实例不同，HBase 的每个守护进程都作为一个独立进程运行。

（3）完全分布式模式

完全分布式模式是在分布式 Hadoop 集群环境下进行的配置。需要在多个节点上进行 HBase 配置，使用 HDFS 作为底层的存储结构。此外，还需要进行 SSH、节点时间同步等配置。

本节将介绍 HBase 2.2.6 伪分布式安装与配置，完全分布式安装与配置详见 10.2 节。

### 5.4.2　HBase 安装准备

（1）Zookeeper 选择

HBase 使用 Zookeeper 作为协同管理组件，可以使用独立安装的 Zookeeper，也可以使用 HBase 自带的 Zookeeper。在 HBase 配置时，进行 Zookeeper 的相应设置即可。

（2）JDK 版本选择

HBase 使用 Java 开发，运行时需要 JDK 环境，需要在配置文件设置 JAVA_HOME 环境变量。不同版本的 HBase 与 JDK 的对应关系如表 5-4 所示。

表 5-4　HBase 与 JDK 的对应关系

| Java 版本 | HBase 1.4+ | Hbase 2.2+ | Hbase 2.3+ |
|---|---|---|---|
| JDK7 | ✓ | ✗ | ✗ |
| JDK8 | ✓ | ✓ | ✓ |
| JDK11 | ✗ | ✗ | ! |

注：✓，完全支持；✗，不完全支持；!，未经测试。

（3）Hadoop 版本选择

如果以伪分布式或完全分布式运行 HBase，就需要选择合适的 Hadoop 版本。不同版本的 HBase 与 Hadoop 的兼容关系如表 5-5 所示。

表 5-5　HBase 与 Hadoop 的兼容关系

| Hadoop 版本 | Hbase 1.4.x | Hbase 1.6.x | Hbase 1.7.x | Hbase 2.2.x | Hbase 2.3.x |
|---|---|---|---|---|---|
| Hadoop 2.7.0 | ✗ | ✗ | ✗ | ✗ | ✗ |
| Hadoop 2.7.1+ | ✓ | ✗ | ✗ | ✗ | ✗ |
| Hadoop 2.8.[0-2] | ✗ | ✗ | ✗ | ✗ | ✗ |
| Hadoop 2.8.[3-4] | ! | ✗ | ✗ | ✗ | ✗ |
| Hadoop 2.8.5+ | ! | ✓ | ✓ | ✓ | ✗ |
| Hadoop 2.9.[0-1] | ✗ | ✗ | ✗ | ✗ | ✗ |
| Hadoop 2.9.2+ | ! | ✓ | ✓ | ✓ | ✗ |
| Hadoop 2.10.x | ! | ✓ | ✓ | ! | ✗ |
| Hadoop 3.1.0 | ✗ | ✗ | ✗ | ✗ | ✗ |
| Hadoop 3.1.1+ | ✗ | ✗ | ✗ | ✓ | ✓ |
| Hadoop 3.2.x | ✗ | ✗ | ✗ | ✓ | ✓ |

注：✓，完全支持；✗，不完全支持；!，未经测试。

（4）HBase 配置文件

HBase 使用与 Hadoop 相同的配置系统，所有配置文件都位于 HBase 安装目录的 conf 子目录中，需要保持集群中所有节点的同步。在手动安装与配置 HBase 之前，需要了解 HBase 配置文件的作用。

① backup-masters：默认情况下不存在。列出所有 HMaster 进程备份的机器名，每行记录一台机器名或 IP 地址。

② hadoop-metrics2-hbase.properties：用于连接 HBase 的 Hadoop Metrics2 框架。

③ hbase-env.sh 和 hbase-cmd.sh：设置 HBase 工作环境，包括 Java 相关环境变量设置。

④ hbase-policy.xml：RPC 服务器使用的默认策略配置文件，对客户端请求进行授权决策，仅在启用 HBase 安全性时使用。

⑤ hbase-site.xml：指定覆盖 HBase 的默认配置选项。HBase 默认配置在 lib/hbase-common.2.2.6/hbase-default.xml 中，也可以在 HBase Web UI 的 HBase 配置选项卡中查看集群的整体有效配置。

⑥ log4j.properties：通过 log4j 进行 HBase 日志记录的配置文件。修改该文件中的参数，可以改变 HBase 的日志级别。改动后，HBase 需要重新启动，以使配置生效。

⑦ regionservers：包含在 HBase 集群中运行的所有 HRegionServer 主机列表，每行一个主机名或 IP 地址，HBase 运行脚本会依次访问每一行来启动列表中的 HRegionServer。

### 5.4.3 HBase 伪分布式集群安装与配置

本节将介绍在 Hadoop 3.1.2 伪分布式环境下进行 Hbase 2.2.6 伪分布式安装与配置。

#### 1. HBase 安装

（1）软件下载

从 HBase 官网（http://hbase.apache.org/downloads.html）下载 hbase-2.2.6-bin.tar.gz，并上传到本地/Downloads 目录中。

（2）使用 tar 命令解压安装包 hbase-2.2.6-bin.tar.gz 至路径 /usr/local

```
$ cd   ~/Download        #进入源文件hbase-2.2.6-bin.tar.gz所在目录
$ sudo   tar  -zxvf   hbase-2.2.6-bin.tar.gz  -C  /usr/local
```

（3）将解压的文件名 hbase-2.2.6 重命名为 hbase

```
$ sudo  mv   /usr/local/hbase-2.2.6   /usr/local/hbase
```

（4）将目录/usr/local/hbas 目录所有者改为用户 hadoop

```
$ sudo chown -R hadoop /usr/local/hbase
```

（5）配置环境变量，并使其生效

① 使用 gedit 命令打开用户的配置文件.bashrc。

```
$ gedit   ~/.bashrc
```

② 在文件中加入下列内容：

```
export HBASE_HOME=/usr/local/hbase
export PATH=$PATH:$HBASE_HOME/bin
```

③ 使环境变量生效。

```
$ source   ~/.bashrc
```

#### 2．HBase 伪分布式配置

HBase 已经安装成功，就可以进行 HBase 伪分布式模式的配置。

（1）配置 hbase-env.sh 文件

在 hbase-env.sh 文件中配置环境标量 JAVA_HOME 和 HBASE_MANAGES_ZK。

使用 gedit 命令编辑 hbase-env.sh 文件：

```
$ gedit /usr/local/hbase/conf/hbase-env.sh
```

在文件中写入下列内容：

```
export JAVA_HOME=/usr/lib/jvm/jdk1.8.0_231
export HBASE_CLASSPATH=/usr/local/hadoop/etc/hadoop
export HBASE_MANAGES_ZK=true
```

注意：HBASE_MANAGES_ZK=true 指定采用 HBase 自带的 Zookeeper。

（2）配置 hbase-site.xml 文件

修改 hbase.rootdir，指定 HBase 数据在 HDFS 上的存储路径；将属性 hbase.cluter.distributed 设置为 true，设置集群处于伪分布式模式。

用 gedit 命令编辑 hbase-site.xml 文件：

`$ gedit /usr/local/hbase/cont/hbase-site.xml`

修改后<configuration>和</configuration>标记之间内容为：

```xml
<configuration>
    <property>
        <name>hbase.rootdir</name>
        <value>hdfs://master:9000/hbase</value>
    </property>
    <property>
        <name>hbase.cluster.distributed</name>
        <value>true</value>
    </property>
    <property>
        <name>hbase.unsafe.stream.capability.enforce</name>
        <value>false</value>
    </property>
</configuration>
```

### 3. HBase 启动与关闭

由于伪分布式和完合分布式 HBase 是运行在 Hadoop 集群之上的，因此启动 HBase 之前，需要先启动 Hadoop 集群。HBase 启动后，可以通过 Shell、Web 方式验证 HBase 配置结果。

（1）启动 Hadoop

```
$ start-dfs.sh
$ start-yarn.sh
```

（2）启动 HBase

`$ start-hbase.sh`

（3）查看 HBase 进程

可以通过运行 jps 命令查看 HBase 启动后启动的相关 Java 进程，如图 5-7 所示。

图 5-7 HBase 相关进程

（4）打开 HBase 命令行窗口

输入 hbase shell 命令后，可以进入 HBase 命令行窗口，如图 5-8 所示。

```
hadoop@Master:~$ hbase shell
HBase Shell
Use "help" to get list of supported commands.
Use "exit" to quit this interactive shell.
For Reference, please visit: http://hbase.apache.org/2.0/book.html#shell
Version 2.2.6, r88c9a386176e2c2b5fd9915d0e9d3ce17d0e456e, Tue Sep 15 17:36:14 CST 2020
Took 0.0104 seconds
hbase(main):001:0>
```

图 5-8　HBase 命令行窗口

（5）通过浏览器查看 HBase 运行状态

打开浏览器，输入网址 http://master:16010，可以查看 HBase 运行状态信息，如图 5-9 所示。

图 5-9　查看 HBase 运行状态

（6）关闭 HBase

$ stop-hbase.sh

## 5.5　HBase Shell

### 5.5.1　HBase Shell 简介

作为典型的 NoSQL 数据库，HBase 不支持 SQL 语句操作。可以通过 HBase Shell、Native Java API、Thrift Gateway、REST Gateway、Pig、Hive 等接口访问 HBase。

HBase Shell 是 HBase 提供的命令行工具，用户可以通过命令行模式操作 HBase，可以进行命名空间管理、对表进行 DDL 与 DML 操作、HBase 权限管理等。通过 hbase shell 命令打开 HBase Shell 窗口后，执行 help 命令，可以查看 HBase 支持的 Shell 命令，如图 5-10 所示。

HBase Shell 命令根据功能的不同分为多个组，本节将主要介绍 General、DDL、DML 组中的一些常用 Shell 命令。

可以执行 help "command" 查看命令的帮助信息。

```
hbase(main):004:0> help
COMMAND GROUPS:
  Group name: general
  Commands: processlist, status, table_help, version, whoami
  Group name: ddl
  Commands: alter, alter_async, alter_status, clone_table_schema, create, describe, disable, disable_all, drop, drop_all, enable,
  enable_all, exists, get_table, is_disabled, is_enabled, list, list_regions, locate_region, show_filters
  Group name: namespace
  Commands: alter_namespace, create_namespace, describe_namespace, drop_namespace, list_namespace, list_namespace_tables
  Group name: dml
  Commands: append, count, delete, deleteall, get, get_counter, get_splits, incr, put, scan, truncate, truncate_preserve
  Group name: tools
  Commands: assign, balance_switch, balancer, balancer_enabled, catalogjanitor_enabled, catalogjanitor_run, catalogjanitor_switch,
  cleaner_chore_enabled, cleaner_chore_run, cleaner_chore_switch, clear_block_cache, clear_compaction_queues, clear_deadservers,
  close_region, compact, compact_rs, compaction_state, compaction_switch, decommission_regionservers, flush, hbck_chore_run,
  is_in_maintenance_mode, list_deadservers, list_decommissioned_regionservers, major_compact, merge_region, move, normalize,
  normalizer_enabled, normalizer_switch, recommission_regionserver, regioninfo, rit, split, splitormerge_enabled,
  splitormerge_switch, stop_master, stop_regionserver, trace, unassign, wal_roll, zk_dump
  Group name: replication
  Commands: add_peer, append_peer_exclude_namespaces, append_peer_exclude_tableCFs, append_peer_namespaces, append_peer_tableCFs,
  disable_peer, disable_table_replication, enable_peer, enable_table_replication, get_peer_config, list_peer_configs, list_peers,
  list_replicated_tables, remove_peer, remove_peer_exclude_namespaces, remove_peer_exclude_tableCFs, remove_peer_namespaces,
  remove_peer_tableCFs, set_peer_bandwidth, set_peer_exclude_namespaces, set_peer_exclude_tableCFs, set_peer_namespaces,
  set_peer_replicate_all, set_peer_serial, set_peer_tableCFs, show_peer_tableCFs, update_peer_config
  Group name: snapshots
  restore_snapshot, snapshot
  Group name: configuration
  Commands: update_all_config, update_config
  Group name: quotas
  Commands: disable_exceed_throttle_quota, disable_rpc_throttle, enable_exceed_throttle_quota, enable_rpc_throttle,
  list_quota_snapshots, list_quota_table_sizes, list_quotas, list_snapshot_sizes, set_quota
  Group name: security
  Commands: grant, list_security_capabilities, revoke, user_permission
  Group name: procedures
  Commands: list_locks, list_procedures
```

图 5-10　HBase 支持的 Shell 命令

### 5.5.2　General 命令组

General 命令组中的 Shell 命令如表 5-6 所示。

表 5-6　General 命令组中的 Shell 命令

| Shell 命令 | 描述 | 样例 |
| --- | --- | --- |
| processlist | 显示 HRegionServer 中所有的任务列表 | processlist 'all' |
| status | 提供有关系统状态的详细信息，如集群中存在的服务器数量、活动服务器计数、平均负载值等 | status 'simple'<br>status 'replication' |
| version | 显示当前使用的 HBase 版本信息 | Version |
| table_help | 提供与表引用相关的 HBase Shell 命令及其语法信息 | table_help |
| whoami | 从 HBase 集群返回当前的 HBase 用户信息 | Whoami |

例如，查看 status 命令帮助信息，并利用 status 查看 HBase 状态信息。

hbase> help 'status'
hbase> status 'summary'

执行结果如图 5-11 所示。

```
hbase(main):001:0> help 'status'
Show cluster status. Can be 'summary', 'simple', 'detailed', or 'replication'. The
default is 'summary'. Examples:

  hbase> status
  hbase> status 'simple'
  hbase> status 'summary'
  hbase> status 'detailed'
  hbase> status 'replication'
  hbase> status 'replication', 'source'
  hbase> status 'replication', 'sink'
hbase(main):002:0> status 'summary'
1 active master, 0 backup masters, 1 servers, 0 dead, 3.0000 average load
Took 0.8680 seconds
```

图 5-11　status 命令帮助信息与执行结果

### 5.5.3　DDL 命令组

HBase 中与表管理相关的 Shell 命令放在 DDL 命令组中，提供了表的创建、删除及模式修改等命令。DDL 命令组中常用的 Shell 命令如表 5-7 所示。

表 5-7 DDL 命令组中常用的 Shell 命令

| Shell 命令 | 描述 | 样例 |
|---|---|---|
| alter | 修改列族模式 | alter 'tablename',family1[,…] |
| create | 创建表 | create 'tablename', 'family1' [,…] |
| describe | 显示表相关的详细信息 | describe 'tablename' |
| disable | 禁用指定的表 | diable 'tablename' |
| disable_all | 禁用所有匹配给定条件的表 | disable_all 'regex' |
| drop | 删除被禁用的表 | drop 'tablename' |
| drop_all | 删除所有匹配给定条件且处于禁用状态的表 | drop_all 'regex' |
| enable | 启用指定的表 | enable 'tablename' |
| enable_all | 启用所有匹配给定条件的表 | enable_all 'regex' |
| exists | 测试表是否存在 | exists 'tablename' |
| get_table | 返回一个表的引用 | get_table 'tablename' |
| is_enabled | 验证指定的表是否被启用 | is_enabled 'tablename' |
| is_disabled | 验证指定的表是否被禁用 | is_disabled 'tablename' |
| list | 列出 HBase 中存在的所有表 | list |
| show_filters | 显示 HBase 中所有的过滤器 | show_filters |

（1）create

create 命令创建表时需要指定命名空间、表名和列族，并使用单引号或双引号引起来。如果没有指定命名空间，默认为 default 命名空间。列族包括列族名称、版本数量、有效期等属性，如表 5-8 所示。

表 5-8 列族属性

| 名称 | 描述 | 取值 |
|---|---|---|
| NEW_VERSION_BEHAVIOR | 允许删除指定版本号的数据 | true/false，默认为 false |
| NAME | 列族名称 | 可打印的字符串 |
| BLOOMFILTER | 提高随机读取数据的性能 | NONE/ROWCOL/ROW(默认) |
| VERSIONS | 单元格数据版本的最大数量 | 默认为 1 |
| IN_MEMORY | 列族在缓存中是否拥有更高的优先级 | true/false，默认为 false |
| KEEP_DELETED_CELLS | 是否将标记为删除的数据从 HBase 中移除 | true/false，默认为 false |
| DATA_BLOCK_ENCODING | 数据块编码 | NONE(默认) |
| TTL | 数据有效时长，单位为 s | 默认 FOREVER |
| COMPRESSION | 压缩编码 | NONE(默认)/LZO/SNAPPY/GZIP |
| MIN_VERSIONS | 单元格数据版本的最小数量 | 默认为 0 |
| BLOCKCACHE | 是否需将数据块放入缓存 | true/false，默认为 true |
| BLOCKSIZE | 数据块大小。数据块越小，索引越大 | 默认 65536 字节 |
| REPLICATION_SCOPE | 数据复制范围 | 默认为 0，复制功能关闭 |

create 命令的基本形式为：

create '[namespace:]tablename', {NAME => 'FamName', VERSIONS => n …}

例如：
hbase> create 'ns1:t1', {NAME => 'f1', VERSIONS => 5}
hbase> create 't1', {NAME => 'f1'}, {NAME => 'f2'}, {NAME => 'f3'}
hbase> create 't1', 'f1', 'f2', 'f3'
hbase> create 't1', {NAME => 'f1', VERSIONS => 1, TTL => 2592000, BLOCKCACHE => true}

创建图 5-1 所示的 student 表，包含 info 和 score 两个列族，其中 score 列族中的单元格版本最大数量为 5。

hbase> create 'student','info',{NAME=>'score',VERSIONS=>5}

注意：HBase Shell 命令中的字母区分大小写。除数字外，其他属性值都需要用引号括起来。

（2）describe

表创建后，可以使用 describe 命令查看表的模式，describe 可以简写为 desc。

例如，查看 student 表的模式，如图 5-12 所示。

hbase> desc 'student'

```
hbase(main):004:0> desc 'student'
Table student is ENABLED
student
COLUMN FAMILIES DESCRIPTION
{NAME => 'info', VERSIONS => '1', EVICT_BLOCKS_ON_CLOSE => 'false', NEW_VERSION_BEHAVIOR => 'fals
e', KEEP_DELETED_CELLS => 'FALSE', CACHE_DATA_ON_WRITE => 'false', DATA_BLOCK_ENCODING => 'NONE'
, TTL => 'FOREVER', MIN_VERSIONS => '0', REPLICATION_SCOPE => '0', BLOOMFILTER => 'ROW', CACHE_IND
EX_ON_WRITE => 'false', IN_MEMORY => 'false', CACHE_BLOOMS_ON_WRITE => 'false', PREFETCH_BLOCKS_O
N_OPEN => 'false', COMPRESSION => 'NONE', BLOCKCACHE => 'true', BLOCKSIZE => '65536'}

{NAME => 'score', VERSIONS => '5', EVICT_BLOCKS_ON_CLOSE => 'false', NEW_VERSION_BEHAVIOR => 'fal
se', KEEP_DELETED_CELLS => 'FALSE', CACHE_DATA_ON_WRITE => 'false', DATA_BLOCK_ENCODING => 'NONE'
, TTL => 'FOREVER', MIN_VERSIONS => '0', REPLICATION_SCOPE => '0', BLOOMFILTER => 'ROW', CACHE_IN
DEX_ON_WRITE => 'false', IN_MEMORY => 'false', CACHE_BLOOMS_ON_WRITE => 'false', PREFETCH_BLOCKS_
ON_OPEN => 'false', COMPRESSION => 'NONE', BLOCKCACHE => 'true', BLOCKSIZE => '65536'}

2 row(s)

QUOTAS
0 row(s)
Took 0.3320 seconds
```

图 5-12　查看 student 表模式

（3）alter

表创建后，可以使用 alter 命令为表增加、删除或修改列族。

alter 命令的基本形式为：

alter '[namespace:]tablename', {NAME => 'FamName', VERSIONS => n …}

例如：
hbase> alter 't1', NAME => 'f1', VERSIONS => 5
hbase> alter 't1', 'f1', {NAME => 'f2', IN_MEMORY => true}, {NAME => 'f3', VERSIONS => 5}
hbase> alter 'ns1:t1', NAME => 'f1', METHOD => 'delete'

① 如果只修改一个列族，可以在 alter 语句中直接通过属性名（NAME）指定要修改的列族，并进行列族属性的设置。例如，将 student 表的 info 列族的单元时间版本数修改为 3。

hbase> alter 'student',NAME=>'info',VERSIONS=>3

② 如果要修改多个列族，则每个列族信息需要用花括号{}括起来，列族之间用逗号分开。例如，将 info 列族和 score 列族的单元格最大版本数都修改为 4。

hbase> alter 'student',{NAME=>'info',VERSIONS=>4},{NAME=>'score',VERSIONS=>4}

③ 如果指定修改的列族不存在，则向表中插入一个新列族。例如，修改列族 address 的 TTL 为 36000s。

hbase> alter 'student',{NAME=>'address',TTL=>36000}

④ 如果要删除一个列族，需要使用 METHOD=>'delete'短语。例如，删除列族 address。

```
hbase> alter 'student',{NAME=>'address',METHOD=>'delete'}
```

（4）list

可以使用 list 命令查看当前 HBase 中有哪些表，可以使用正则表达式进行表过滤。

例如：

```
hbase> list
hbase> list 'stud.*'
hbase> list 'ns:stud.*'
hbase> list 'ns:.*'
```

（5）drop

使用 drop 命令删除表之前，需要先使用 disable 命令禁用表。例如，删除 student 表：

```
hbase> disable 'student'
hbase> drop 'student'
```

### 5.5.4 DML 命令组

HBase 中操作表中数据的 Shell 命令放在 DML 命令组中，可以对表中数据进行插入、删除、修改、查询等。DML 命令组中常用的 Shell 命令如表 5-9 所示。

表 5-9 DML 命令组中常用的 Shell 命令

| 名称 | 描述 | 样例 |
| --- | --- | --- |
| put | 向指定单元格中插入数据 | put 'tablename','rowkey','column','value' |
| get | 获取指定条件的数据 | get 'tablename','rowkey',{COLUMN=>[column1…]} |
| scan | 查看表中所有数据 | scan 'tablename',{COLUMNS=>column,…, LIMIT=> num} |
| delete | 删除单元格 | delete 'tablename', 'rowkey', 'column', timestamp |
| deleteall | 删除指定的行 | deleteall 'tablename','rowkey' |
| truncate | 清空表 | truncate 'tablename' |
| count | 统计表中的行数 | count 'tablename' |

注意：样例中 column 是由列族和列族限定符构成的，即 family:qualifier。

（1）put

表创建后，可以使用 put 命令向表中插入数据。需要注意的是，在 HBase 中每次只能为一个单元格插入数据，通过行键、列（列族:列族限定符）定位到一个单元格，版本号可以为系统自动生成的时间戳，也可以由用户指定一个长整型（Long）值。

put 命令的基本形式为：

```
put '[namespace:]tablename','rowkey','family:qulifier',value[,timestamp]
```

例如：

```
hbase> put 'ns1:t1', 'r1', 'c1', 'value'
hbase> put 't1', 'r1', 'c1', 'value'
hbase> put 't1', 'r1', 'c1', 'value', ts1
```

向 student 表中插入两行数据，每次插入一个单元格的值。

```
hbase> put 'student','2018710101','info:name','LiuYuan'
hbase> put 'student','2018710101','info:age',20,1611644461753
hbase> put 'student','2018710101','score:math',90
hbase> put 'student','2018710101','score:math',80
hbase> put 'student','2018710102','info:name','YeShuai'
```

（2）get

数据插入 HBase 表之后，可以使用 get 命令进行单行数据查询。可以返回指定行键对应的一行数据、一个或多个列族的数据、一个或多个单元格值、特定版本单元格值等。

get 命令的基本形式为：

get '[namespace:]tablename','rowkey',{COLUMN=>'column'…}

例如：

hbase> get 'ns1:t1', 'r1'
hbase> get 't1', 'r1'
hbase> get 't1', 'r1', 'c1'
hbase> get 't1', 'r1', 'c1', 'c2'
hbase> get 't1', 'r1', {COLUMN => 'c1'}
hbase> get 't1', 'r1', {COLUMN => ['c1', 'c2', 'c3']}
hbase> get 't1', 'r1', {COLUMN => 'c1', TIMERANGE => [ts1, ts2], VERSIONS => 4}
hbase> get 't1', 'r1', {COLUMN => 'c1', TIMESTAMP => ts1, VERSIONS => 4}

查询 student 表中学号为 2018710101 的学生的信息：

hbase> get 'student','2018710101'
hbase> get 'student','2018710101','info','score'
hbase> get 'student','2018710101','info:name'
hbase> get 'student','2018710101',{COLUMN=>['info:name'],TIMESTAMP=>1611644461752}
hbase> get 'student','2018710101',{COLUMN=>'score:math',VERSIONS=>2}

执行结果如图 5-13 所示。

```
hbase(main):013:0> get 'student','2018710101'
COLUMN                    CELL
 info:age                 timestamp=1611644461753, value=20
 info:name                timestamp=1611922565737, value=LiuYuan
 score:math               timestamp=1611922595831, value=80
1 row(s)
Took 0.0648 seconds
hbase(main):014:0> get 'student','2018710101','info','score'
COLUMN                    CELL
 info:age                 timestamp=1611644461753, value=20
 info:name                timestamp=1611922565737, value=LiuYuan
 score:math               timestamp=1611922595831, value=80
1 row(s)
Took 0.0520 seconds
hbase(main):015:0> get 'student','2018710101','info:name'
COLUMN                    CELL
 info:name                timestamp=1611922565737, value=LiuYuan
1 row(s)
Took 0.0108 seconds
hbase(main):017:0> get 'student','2018710101',{COLUMN=>['info:name'],TIMESTAMP=>1611922565737}
COLUMN                    CELL
 info:name                timestamp=1611922565737, value=LiuYuan
1 row(s)
Took 0.0106 seconds
hbase(main):018:0> get 'student','2018710101',{COLUMN=>'score:math',VERSIONS=>2}
COLUMN                    CELL
 score:math               timestamp=1611922595831, value=80
 score:math               timestamp=1611922587336, value=90
1 row(s)
Took 0.0337 seconds
```

图 5-13　get 命令执行结果

默认情况下，查询只会返回单元格中最近的一个版本的数据，如果需要返回多个版本数据，需要使用 VERSIONS 参数设置。

（3）scan

使用 get 命令只能返回指定行键对应行中的数据，如果要返回多行数据，可以使用 scan 命令。

scan 命令的基本形式为：

scan '[namespace:]tablename',{COLUMNS=>column…}

在 scan 命令中，可以通过 TIMERANGE、FILTER、LIMIT、STARTROW、STOPROW、TIMESTAMP、COLUMNS、VERSIONS 等参数指定返回的数据。其中，COLUMNS 参数指定要返回的列，LIMIT 参数指定返回的行数，VERSIONS 参数指定返回的单元格版本数量。

例如：

hbase> scan 'hbase:meta'
hbase> scan 'hbase:meta', {COLUMNS => 'info:regioninfo'}
hbase> scan 't1', {COLUMNS => ['c1', 'c2'], LIMIT => 10, STARTROW => 'xyz'}
hbase> scan 't1', {COLUMNS => 'c1', TIMERANGE => [1303668804000, 1303668904000]}

利用 scan 查询 student 表中学生的信息，并返回数据两个最新版本。

hbase> scan 'student',{COLUMNS=>'score',VERSIONS=>2,LIMIT=>2}
hbase> scan 'student',{COLUMNS=>'score:math',VERSIONS=>2}
hbase> scan 'student',{COLUMNS=>['info','score'],VERSIONS=>2}

执行结果如图 5-14 所示。

```
hbase(main):019:0> scan 'student',{COLUMNS=>'score',VERSIONS=>2,LIMIT=>2}
ROW                     COLUMN+CELL
 2018710101             column=score:math, timestamp=1611922595831, value=80
 2018710101             column=score:math, timestamp=1611922587336, value=90
1 row(s)
Took 0.1200 seconds
hbase(main):020:0> scan 'student',{COLUMNS=>'score:math',VERSIONS=>2}
ROW                     COLUMN+CELL
 2018710101             column=score:math, timestamp=1611922595831, value=80
 2018710101             column=score:math, timestamp=1611922587336, value=90
1 row(s)
Took 0.0139 seconds
hbase(main):021:0> scan 'student',{COLUMNS=>['info','score'],VERSIONS=>2}
ROW                     COLUMN+CELL
 2018710101             column=info:age, timestamp=1611644461753, value=20
 2018710101             column=info:name, timestamp=1611922565737, value=LiuYuan
 2018710101             column=score:math, timestamp=1611922595831, value=80
 2018710101             column=score:math, timestamp=1611922587336, value=90
 2018710102             column=info:name, timestamp=1611922606248, value=YeShuai
2 row(s)
Took 0.0709 seconds
```

图 5-14　scan 命令执行结果

（4）delete 与 deleteall

可以使用 delete 命令删除指定单元格中的数据，使用 deleteall 命令删除指定行键对应的所有数据。

delete 与 deleteall 命令的基本形式为：

delete '[namespace:]tablename','rowkey','column'[,'timestamp']
delete '[namespace:]tablename','rowkey'

例如，删除 student 表中学号为 2018710101 的学生的信息：

hbase> delete 'student','2018710101','score:math',1611646169016
hbase> delete 'student','2018710101','score:math'
hbase> delete 'student','2018710101','score'
hbase> deleteall 'student','2018710101'

（5）truncate

可以使用 truncate 命令清空整个表。truncate 命令执行过程实质是顺序执行 disable 命令、drop 命令，以及执行 create 命令重新创建该表。

例如，清空 student 表的过程如图 5-15 所示。

```
hbase(main):024:0> truncate 'student'
Truncating 'student' table (it may take a while):
Disabling table...
Truncating table...
Took 3.6940 seconds
```

图 5-15 清空 student 表的过程

（6）count

可以使用 count 命令统计表中数据的行数。参数 INTERVAL 设置多少行显示一次对应的行键，默认为 1000。参数 CACHE 设置每次去取的缓存区大小，默认为 10，调整该参数可提高查询速度。

count 命令的基本形式为：

count '[namespace:]tablename'[,INTERVAL=>n,CACHE=>m]

例如，统计 student 表中数据的行数：

hbase> count 'student',INTERVAL=>10,CACHE=>10

### 5.5.5 查询过滤器

使用 get 命令或 scan 命令查询数据时，可以使用过滤器来设置数据查询的范围。HBase 中的过滤器类似 SQL 语句中的 where 条件。HBase 内置了多种类型的过滤器，执行 show_filters 命令可以查看当前 HBase 提供的所有内置过滤器。

在 get 和 scan 命令中使用过滤器的语法格式为：

get '[namespace:]tablename', 'rowkey', FILTER => "filtername(operator, 'compatator')"
scan '[namespace:]tablename',FILTER => "filtername(operator, 'comparator')"

其中，operator 为比较运算符，常用的比较运算符如表 5-10 所示。comparator 为比较器，常用的比较器如表 5-11 所示。

表 5-10 常用的比较运算符

| 比较运算符 | 说明 |
| --- | --- |
| = | 等于 |
| > | 大于 |
| >= | 大于或等于 |
| < | 小于 |
| <= | 小于或等于 |
| != | 不等于 |

表 5-11 常用的比较器

| 比较器 | 说明 |
| --- | --- |
| BinaryComparator | 按字节索引顺序比较完整字节数组 |
| BinaryPrefixComparator | 按字节索引顺序比较字节数组前缀 |
| BitComparator | 按位进行异或、与、并操作 |
| NullComparator | 判断给定的是否为空值 |
| RegexStringComparator | 匹配正则表达式，仅支持等或不等操作 |
| SubstringComparator | 判断提供的子字符串是否在单元格中 |

根据功能不同，HBase 过滤器分为行过滤器、列族过滤器、值过滤器、参考列过滤器和专用过滤器等多种类型。

（1）行过滤器

行过滤器基于行键来过滤数据。RowFilter 是通用的行过滤器，可以配合比较器和比较运算符，实现行键字符串的比较和过滤。例如，匹配行键中大于 1001 的数据，可使用 binary 比较器。匹配包含 10001 子字符串的行键，可使用 substring 比较器，substring 比较器不支持大于或小于运算符。

```
hbase(main):028:0> scan 'student',FILTER=>"RowFilter(=,'substring:10001')"
hbase(main):029:0> scan 'student',FILTER=>"RowFilter(>,'binary:10001')"
```

执行结果如图 5-16 所示。

```
hbase(main):028:0> scan 'student',FILTER=>"RowFilter(=,'substring:10001')"
ROW                    COLUMN+CELL
 10001                 column=info:name, timestamp=1611923090006, value=zhangsan
1 row(s)
Took 0.0928 seconds
hbase(main):029:0> scan 'student',FILTER=>"RowFilter(>,'binary:10001')"
ROW                    COLUMN+CELL
 10002                 column=info:name, timestamp=1611923105044, value=lisi
 10003                 column=info:name, timestamp=1611923118099, value=wangwu
2 row(s)
Took 0.0443 seconds
```

图 5-16  RowFilter 行过滤器执行结果

针对行键进行匹配的行过滤器还有 PrefixFilter、KeyOnlyFilter、FirstKeyOnlyFilter 和 InclusiveStopFilter，其具体含义和使用示例如表 5-12 所示。

表 5-12  行过滤器

| 行过滤器名称 | 说明 | 示例 |
| --- | --- | --- |
| PrefixFilter | 行键前缀过滤器，匹配行键前缀 | scan 'student', FILTER => "PrefixFilter('10001')" |
| KeyOnlyFilter | 返回行键、列族和列族限定符，不返回值 | scan 'student', FILTER => "KeyOnlyFilter()" |
| FirstKeyOnlyFilter | 只扫描显示相同键的第一个单元格，其键值对会显示出来 | scan 'student', FILTER => "FirstKeyOnlyFilter()" |
| InclusiveStopFilter | 替代 ENDROW 返回终止条件行 | scan 'student', { STARTROW => '10001', FILTER => "InclusiveStopFilter('binary:10002')" } |

示例执行结果如图 5-17 所示。

```
hbase(main):036:0> scan 'student', FILTER => "PrefixFilter('10001')"
ROW                    COLUMN+CELL
 10001                 column=info:name, timestamp=1611923090006, value=zhangsan
1 row(s)
Took 0.0162 seconds
hbase(main):037:0> scan 'student', FILTER => "KeyOnlyFilter()"
ROW                    COLUMN+CELL
 10001                 column=info:name, timestamp=1611923090006, value=
 10002                 column=info:name, timestamp=1611923105044, value=
 10003                 column=info:name, timestamp=1611923118099, value=
3 row(s)
Took 0.0248 seconds
hbase(main):038:0> scan 'student', FILTER => "FirstKeyOnlyFilter()"
ROW                    COLUMN+CELL
 10001                 column=info:name, timestamp=1611923090006, value=zhangsan
 10002                 column=info:name, timestamp=1611923105044, value=lisi
 10003                 column=info:name, timestamp=1611923118099, value=wangwu
3 row(s)
Took 0.0314 seconds
hbase(main):039:0> scan 'student', { STARTROW => '10001',FILTER => "InclusiveStopFilter('binary:10002')" }
ROW                    COLUMN+CELL
 10001                 column=info:name, timestamp=1611923090006, value=zhangsan
 10002                 column=info:name, timestamp=1611923105044, value=lisi
 10003                 column=info:name, timestamp=1611923118099, value=wangwu
3 row(s)
Took 0.0232 seconds
```

图 5-17  行过滤器示例执行结果

（2）列过滤器

列过滤器通过比较列族或列族限定符的名称返回结果。FamilyFilter 是列族过滤器，QualifierFilter、ColumnPrefixFilter、MultipleColumnPrefixFilter、ColumnRangeFilter 是针对列族限定符进行匹配的过滤器。常用的列过滤器如表 5-13 所示。

表 5-13 常用的列过滤器

| 列过滤器名称 | 说明 | 示例 |
|---|---|---|
| FamilyFilter | 对列族进行匹配 | scan 'student', FILTER=> "FamilyFilter(= , 'substring:info')" |
| QualifierFilter | 列族限定符过滤器，只显示对应限定符的数据 | scan 'student', FILTER => "QualifierFilter(=,'substring:math')" |
| ColumnPrefixFilter | 对列族限定符的前缀进行过滤 | scan 'student', FILTER => "ColumnPrefixFilter('math')" |
| MultipleColumnPrefixFilter | 可以指定多个列族限定符前缀进行过滤 | scan 'student', FILTER => "MultipleColumnPrefixFilter('name','math')" |
| ColumnRangeFilter | 过滤列族限定符的范围 | scan 'student', FILTER => "ColumnRangeFilter('math',true,'name',false)" |

为了演示列过滤器的应用效果，先向 student 表中插入下列数据：

```
hbase(main):015:0> put 'student',10001,'score:english',80
hbase(main):016:0> put 'student',10002,'score:english',90
hbase(main):017:0> put 'student',10002,'score:math',85
hbase(main):018:0> put 'student',10003,'score:math',75
hbase(main):019:0> put 'student',10003,'score:english',95
```

列过滤器示例执行结果如图 5-18 所示。

```
hbase(main):052:0> scan 'student', FILTER=>"FamilyFilter(= , 'substring:info')"
ROW                     COLUMN+CELL
 10001                  column=info:name, timestamp=1611923090006, value=zhangsan
 10002                  column=info:name, timestamp=1611923105044, value=lisi
 10003                  column=info:name, timestamp=1611923118099, value=wangwu
3 row(s)
Took 0.0337 seconds
hbase(main):053:0> scan 'student', FILTER => "QualifierFilter(=,'substring:math')"
ROW                     COLUMN+CELL
 10002                  column=score:math, timestamp=1611923521195, value=85
 10003                  column=score:math, timestamp=1611923530549, value=75
2 row(s)
Took 0.0090 seconds
hbase(main):054:0> scan 'student', FILTER => "ColumnPrefixFilter('math')"
ROW                     COLUMN+CELL
 10002                  column=score:math, timestamp=1611923521195, value=85
 10003                  column=score:math, timestamp=1611923530549, value=75
2 row(s)
Took 0.0072 seconds
hbase(main):055:0> scan 'student', FILTER => "MultipleColumnPrefixFilter('name','math')"
ROW                     COLUMN+CELL
 10001                  column=info:name, timestamp=1611923090006, value=zhangsan
 10002                  column=info:name, timestamp=1611923105044, value=lisi
 10002                  column=score:math, timestamp=1611923521195, value=85
 10003                  column=info:name, timestamp=1611923118099, value=wangwu
 10003                  column=score:math, timestamp=1611923530549, value=75
3 row(s)
Took 0.0514 seconds
hbase(main):056:0> scan 'student', FILTER => "ColumnRangeFilter('math',true,'name',false)"
ROW                     COLUMN+CELL
 10002                  column=score:math, timestamp=1611923521195, value=85
 10003                  column=score:math, timestamp=1611923530549, value=75
2 row(s)
Took 0.0240 seconds
```

图 5-18 列过滤器示例执行结果

（3）值过滤器

值过滤器是基于单元格的值来过滤数据的。常用的值过滤器如表 5-14 所示。

表 5-14　常用的值过滤器

| 值过滤器名称 | 说明 | 示例 |
|---|---|---|
| ValueFilter | 值过滤器，找到符合值条件的键值对 | scan 'student', FILTER => "ValueFilter(=,'substring:lisi')" |
| ColumnValueFilter | 在指定的列族和列族限定符中进行比较，只返回匹配的单元 | scan 'student', FILTER => "ColumnValueFilter ('score', 'math', =, 'binary:85')" |
| SingleColumnValueFilter | 在指定的列族和列族限定符中进行比较，返回匹配单元格所在的行 | scan 'student', FILTER => "SingleColumnValueFilter ('score', 'math', =, 'binary:85')" |
| SingleColumnValueExcludeFilter | 排除匹配成功的值 | scan 'student', FILTER => "SingleColumnValueExcludeFilter('score', 'math', =, 'binary:85')" |

值过滤器示例执行结果如图 5-19 所示。

```
hbase(main):067:0> scan 'student', FILTER => "ValueFilter(=,'substring:lisi')"
ROW                     COLUMN+CELL
 10002                  column=info:name, timestamp=1611923105044, value=lisi
1 row(s)
Took 0.0256 seconds
hbase(main):068:0> scan 'student', FILTER => " ColumnValueFilter ( 'score', 'math', =, 'binary:85' )"
ROW                     COLUMN+CELL
 10002                  column=score:math, timestamp=1611923521195, value=85
1 row(s)
Took 0.0161 seconds
hbase(main):069:0> scan 'student', FILTER => " SingleColumnValueFilter ( 'score', 'math', =, 'binary:85' )"
ROW                     COLUMN+CELL
 10001                  column=info:name, timestamp=1611923090006, value=zhangsan
 10001                  column=score:english, timestamp=1611923503132, value=80
 10002                  column=info:name, timestamp=1611923105044, value=lisi
 10002                  column=score:english, timestamp=1611923512867, value=90
 10002                  column=score:math, timestamp=1611923521195, value=85
2 row(s)
Took 0.0156 seconds
hbase(main):070:0> scan 'student', FILTER => "SingleColumnValueExcludeFilter('score', 'math', =, 'binary:85')"
ROW                     COLUMN+CELL
 10001                  column=info:name, timestamp=1611923090006, value=zhangsan
 10001                  column=score:english, timestamp=1611923503132, value=80
 10002                  column=info:name, timestamp=1611923105044, value=lisi
 10002                  column=score:english, timestamp=1611923512867, value=90
2 row(s)
Took 0.0167 seconds
```

图 5-19　值过滤器示例执行结果

（4）其他过滤器

还有一些其他的过滤器，其过滤方式和示例如表 5-15 所示。

表 5-15　其他过滤器

| 过滤器名称 | 说明 | 示例 |
|---|---|---|
| ColumnCountGetFilter | 限制每个逻辑行返回键值对的个数，在 get 方法中使用 | get 'student', '10001', FILTER => "ColumnCountGetFilter(3)" |
| TimestampsFilter | 时间戳过滤，支持等值，可以设置多个时间戳 | scan 'student', FILTER => "TimestampsFilter(1611923503132,1611923503932)" |
| InclusiveStopFilter | 设置停止行 | scan 'student', { STARTROW => '10001', ENDROW => '10003', FILTER => "InclusiveStopFilter('10003')" } |
| PageFilter | 对显示结果按行进行分页显示 | scan 'student', { STARTROW => '10001', ENDROW => '10003', FILTER => "PageFilter(2)" } |
| ColumnPaginationFilter | 对一行的所有列分页，只返回 [offset,offset+limit] 范围内的列 | scan 'student', { STARTROW => '10001', ENDROW => '10003', FILTER => "ColumnPaginationFilter(2,1)" } |

示例执行结果如图 5-20 所示。

```
hbase(main):001:0> get 'student', '10001', FILTER => "ColumnCountGetFilter(3)"
COLUMN                          CELL
 info:name                      timestamp=1611923090006, value=zhangsan
 score:english                  timestamp=1611923503132, value=80
1 row(s)
Took 0.8828 seconds
hbase(main):002:0> scan 'student', FILTER => "TimestampsFilter(1611923503132,1611923503932)"
ROW                             COLUMN+CELL
 10001                          column=score:english, timestamp=1611923503132, value=80
1 row(s)
Took 0.0602 seconds
hbase(main):003:0> scan 'student', { STARTROW => '10001', ENDROW => '10003', FILTER => "InclusiveStopFilter('10003')" }
ROW                             COLUMN+CELL
 10001                          column=info:name, timestamp=1611923090006, value=zhangsan
 10001                          column=score:english, timestamp=1611923503132, value=80
 10002                          column=info:name, timestamp=1611923105044, value=lisi
 10002                          column=score:english, timestamp=1611923512867, value=90
 10002                          column=score:math, timestamp=1611923521195, value=85
2 row(s)
Took 0.0574 seconds
hbase(main):004:0> scan 'student', { STARTROW => '10001', ENDROW => '10003', FILTER => "PageFilter(2)" }
ROW                             COLUMN+CELL
 10001                          column=info:name, timestamp=1611923090006, value=zhangsan
 10001                          column=score:english, timestamp=1611923503132, value=80
 10002                          column=info:name, timestamp=1611923105044, value=lisi
 10002                          column=score:english, timestamp=1611923512867, value=90
 10002                          column=score:math, timestamp=1611923521195, value=85
2 row(s)
Took 0.0663 seconds
hbase(main):005:0> scan 'student', { STARTROW => '10001', ENDROW => '10003', FILTER => "ColumnPaginationFilter(2,1)" }
ROW                             COLUMN+CELL
 10001                          column=score:english, timestamp=1611923503132, value=80
 10002                          column=score:english, timestamp=1611923512867, value=90
 10002                          column=score:math, timestamp=1611923521195, value=85
2 row(s)
Took 0.0518 seconds
```

图 5-20　其他过滤器示例执行结果

HBase Shell 命令都有相应的 Java API，这些 Java API 实质就是 HBase Shell 命令的封装，因此掌握了 HBase Shell 命令的使用，将更容易理解和掌握 Java API 的程序设计。

## 5.6　HBase 程序设计

### 5.6.1　HBase Java API 简介

HBase 提供了丰富的 Java API 进行 HBase 对象管理与数据操作，例如表的创建与修改、数据的插入与查询等。由于 HBase 是采用 Java 语言开发的，因此原生的 Java API 是 HBase 最常规和高效的访问接口，适合 Hadoop MapReduce 作业并行批处理 HBase 数据。常用的 HBase Java API 如表 5-16 所示。

表 5-16　常用的 HBase Java API

| HBase Java API | 功能描述 |
| --- | --- |
| HBaseConfiguration | 管理 HBase 的配置信息 |
| Connection | 建立与 HBase 的连接 |
| Admin | 管理 HBase |
| TableDescriptorBuilder | 表描述生成器 |
| TableDescriptor | 表描述 |
| ColumnFamilyDescriptorBuilder | 列族描述生成器 |
| ColumnFamilyDescriptor | 列族描述 |
| Table | 表实例，与 HBase 进行通信 |
| TableName | 表名 |
| Put | 要插入表中数据对象 |
| Get | 要查询的单行数据对象 |
| Delete | 要删除的数据对象 |
| Result | 保存返回的一行结果数据 |
| Scan | 要查询的多行数据对象 |
| ResultScanner | 保存查询返回的多行结果数据 |

（1）HBaseConfiguration

HBaseConfiguration 类的作用是根据配置文件内容创建配置对象，获取 HBase 配置信息。

对 HBase 的任何操作都需要首先创建 HBaseConfiguration 类的实例，即通过调用 HBaseConfiguration 类的 create()方法创建一个 Configuration 对象。用户可以调用 Configuration 对象的 set()方法重写 HBase 的配置信息。例如：

```
Configuration conf = HBaseConfiguration.create();
conf.set("hbase.rootdir","hdfs://master:9000/hbase");
```

（2）Connection

Connection 接口表示与 HBase 的连接，它可以让 HBase 客户端连接到 ZooKeeper 和个别的 HRegionServer。可以以 HBase 配置对象为参数，调用 ConnectFactory 类的 createConnection()方法创建 Connection 接口。操作完成后，需要调用 close()方法关闭连接以释放资源。例如：

```
Connection conn=ConnectFactory.createConnection(conf);
conn.close();
```

（3）Admin

对 HBase 的管理是通过 Admin 实例实现的。Admin 接口提供了众多的表管理的方法，可以用于创建表、删除表、修改表等。表 5-17 列出了 Admin 接口常用的方法。

表 5-17　Admin 接口常用的方法

| 方法 | 返回值 | 描述 |
| --- | --- | --- |
| addColumnFamily(TableName tableName, ColumnFamilyDescriptor columnFamily) | void | 向表中添加一个列族 |
| deleteColumnFamily(TableName tableName, byte[] columnFamily) | void | 从表中删除一个存在的列族 |
| modifyColumnFamily(TableName tableName, ColumnFamilyDescriptor columnFamily) | void | 修改表中存在的一个列族 |
| createTable(TableDescriptor desc) | void | 创建一个新表 |
| deleteTable(TableName tableName) | void | 删除一个表 |
| disableTable(TableName tableName) | void | 禁用表 |
| enableTable(TableName tableName) | void | 激活表 |
| getDescriptor(TableName tableName) | TableDescriptor | 获取一个表的描述 |
| listTableDescriptors() | List<TableDescriptor> | 列出所有表的描述 |
| listTableNames() | TableName[] | 列出所有表的名称 |
| modifyTable(TableDescriptor desc) | void | 修改一个存在的表 |
| tableExists(TableName tableName) | boolean | 检查表是否存在 |

Admin 实例的创建是通过调用 Connection 接口的 getAdmin()方法实现的。操作完成后，需要调用 close()方法关闭 Admin 实例以释放资源。例如：

```
Admin admin=conn.getAdmin();
admin.close();
```

（4）TableDescriptorBuilder

在 HBase2.x 中引入了表描述生成器 TableDescriptorBuilder 类，设置表中所有列族描述、表的类型、表是否只读、MemStore 最大容量、Region 什么时候应该分裂等，构建表描述

TableDescriptor 实例。表 5-18 列出了 TableDescriptorBuilder 类中常用的方法。

表 5-18 TableDescriptorBuilder 类中常用的方法

| 方法 | 返回值 | 描述 |
|---|---|---|
| build() | TableDescriptor | 构建表描述实例 |
| newBuilder(TableName name) | TableDescriptorBuilder | 创建表描述生成器 |
| setColumnFamily(ColumnFamilyDescriptor family) | TableDescriptorBuilder | 添加一个列族 |
| setColumnFamilies(Collection<ColumnFamilyDescriptor> families) | TableDescriptorBuilder | 添加多个列族 |
| modifyColumnFamily(ColumnFamilyDescriptor family) | TableDescriptorBuilder | 修改一个列族 |
| removeColumnFamily(byte[] name) | TableDescriptorBuilder | 删除一个列族 |
| setMaxFileSize(long maxFileSize) | TableDescriptorBuilder | 设置 Region 分裂阈值 |
| setMemStoreFlushSize(long memstoreFlushSize) | TableDescriptorBuilder | 设置缓存刷新阈值 |
| setReadOnly(boolean readOnly) | TableDescriptorBuilder | 设置表是否只读 |

（5）TableDescriptor

TableDescriptor 接口包含了表的详细信息，包括所有列族描述、表的类型、表是否只读、MemStore 最大容量、Region 什么时候应该分裂等。表 5-19 列出了 TableDescriptor 接口中常用的方法。

表 5-19 TableDescriptor 接口中常用的方法

| 方法 | 返回值 | 描述 |
|---|---|---|
| getColumnFamilies() | ColumnFamilyDescriptor[] | 返回表的所有列族描述的集合 |
| getColumnFamily(byte[] name) | ColumnFamilyDescriptor | 返回表指定名称的列族的描述 |
| getColumnFamilyNames() | Set<byte[]> | 返回表的所有列族名称的集合 |
| getMaxFileSize() | long | 返回设置 Region 分裂阈值 |
| getMemStoreFlushSize() | long | 返回缓存刷新阈值 |
| getTableName() | TableName | 返回表名 |
| isReadOnly() | boolean | 判断表是否只读 |

创建表时，先创建 TableDescriptorBuilder 实例，进行表属性、列族、MemStore 阈值等设置，然后构建 TableDescriptor 对象，最后以 TableDescriptor 对象为参数创建表。

（6）ColumnFamilyDescriptorBuilder

在 HBase2.x 中引入了列族描述生成器 ColumnFamilyDescriptorBuilder 类，设置列族的属性，如数据最大版本数量、数据有效时长等，构建列族描述 ColumnFamilyDescriptor 实例。表 5-20 列出了 ColumnFamilyDescriptorBuilder 类中常用的方法。

表 5-20 ColumnFamilyDescriptorBuilder 类中常用的方法

| 方法 | 返回值 | 描述 |
|---|---|---|
| build() | ColumnFamilyDescriptor | 构建列族描述实例 |
| newBuilder(byte[] name) | ColumnFamilyDescriptorBuilder | 创建列族描述生成器 |
| of(byte[] name) | ColumnFamilyDescriptor | 构建列族描述实例 |
| setBlocksize(int value) | ColumnFamilyDescriptorBuilder | 设置列族数据块大小 |

续表

| 方法 | 返回值 | 描述 |
|---|---|---|
| setInMemory(boolean value) | ColumnFamilyDescriptorBuilder | 设置列族在内存中的优先级 |
| setMaxVersions(int value) | ColumnFamilyDescriptorBuilder | 设置数据最大版本数量 |
| setMinVersions(int value) | ColumnFamilyDescriptorBuilder | 设置数据最小版本数量 |
| setTimeToLive(int value) | ColumnFamilyDescriptorBuilder | 设置数据有效时长 |

（7）ColumnFamilyDescriptor

ColumnFamilyDescriptor 接口包含列族的详细信息，包括版本数量、压缩设置等。ColumnFamilyDescriptor 实例由 ColumnFamilyDescriptorBuilder 类构建，用于创建表或向表中添加列族。表 5-21 列出了 ColumnFamilyDescriptor 接口中常用的方法。

表 5-21 ColumnFamilyDescripto 接口中常用的方法

| 方法 | 返回值 | 描述 |
|---|---|---|
| getBlocksize() | int | 获取列族数据块大小 |
| getMaxVersions() | int | 获取列族数据版本最大数量 |
| getMinVersions() | int | 获取列族数据版本最小数量 |
| getName() | byte[] | 获取列族名称 |
| getNameAsString() | String | 获取列族名称 |
| getTimeToLive() | int | 获取列族数据有效时长 |
| isInMemory() | boolean | 获取列族数据在内存中是否优先 |

构建列族时，先创建 ColumnFamilyDescriptorBuilder 实例，进行列族属性设置，然后构建 ColumnFamilyDescriptor 对象。

（8）Table

要对表中数据进行操作，首先需要创建一个与表对应的 Table 实例，用于与 HBase 进行通信。Table 接口中提供了众多的表数据操作方法，可以用于数据的插入、查询、删除等操作。表 5-22 列出了 Table 接口常用的方法。

表 5-22 Table 接口常用的方法

| 方法 | 返回值 | 描述 |
|---|---|---|
| delete(Delete delete) | void | 删除指定的行或单元值 |
| delete(List\<Delete\> deletes) | void | 批量删除指定的行或单元格 |
| exists(Get get) | boolean | 判断 Get 指定的列是否存在 |
| get(Get get) | Result | 根据 Get 条件获取一行中特定单元值 |
| getDescriptor() | TableDescriptor | 获取表的描述信息 |
| getScanner(byte[] family) | ResultScanner | 获取所有行的指定列族包含的数据 |
| getScanner(byte[] family, byte[] qualifier) | ResultScanner | 获取所有行的指定列族限定符包含的数据 |
| getScanner(Scan scan) | ResultScanner | 返回满足 Scan 条件的数据 |
| put(Put put) | void | 向表中插入一行数据 |
| put(List\<Put\> puts) | void | 向表中批量插入数据 |

Table 实例的创建是通过调用 Connection 实例的 getTable()方法实现的。操作完成后，需要

调用 close()方法关闭 Table 实例以释放资源。例如：

```
Table table=conn.getTable();
table.close();
```

（9）TableName

TableName 接口是对表名进行封装的 POJO（Plain Ordinary Java Object）对象，表名完整形式为"命名空间:表名"。如果没有指定命名空间，系统会使用 HBase 默认的命名空间 default。调用 TableName 接口的 valueOf()方法可以将表名封装成 TableName 对象。例如：

```
TableName tableName=TableName.valueOf("table");
```

（10）Put

如果要进行数据插入，需要构造 Put 实例，然后设置要插入到表中的值。最后以 Put 对象作为参数，调用 Table 对象的 put()方法实现数据的插入。表 5-23 列出了 Put 类常用的方法。

表 5-23 Put 类常用的方法

| 方法 | 返回值 | 描述 |
| --- | --- | --- |
| Put(byte[] row) | Put | 根据行键创建 Put 实例 |
| addColumn(byte[] family, byte[] qualifier, byte[] value) | Put | 指定要插入值的列族、列族限定符和值 |
| addColumn(byte[] family, byte[] qualifier, long ts, byte[] value) | Put | 指定要插入值的列族、列族限定符和值，以及作为单元值版本的时间戳 |
| setPriority(int priority) | Put | 设置插入值的优先级 |
| setTimestamp(long timestamp) | Put | 设置插入值的时间戳 |
| setTTL(long ttl) | Put | 设置插入值的有效时长，单位为 ms |

例如，向表中插入一个行键为 2018710104、score 列族中 math 单元值为 90 的数据。

```
Put put = new Put(Bytes.getBytes("2018710104"));
put.addColumn(Bytes.toBytes("score"), Bytes.toBytes("math"),Bytes.toBytes(90));
table.put(put);
```

需要注意的是，HBase 中数据都是以未经解析的二进制字节数组方式存储的，因此需要将行键、列族、列族限定符、值都转换为字节数组类型。

（11）Delete

如果要进行删除表中的数据，需要构造 Delete 实例，然后指定要删除的数据。可以删除最新版本的单元值、所有版本的单元值、列族、一行数据等。最后以 Delete 对象作为参数，调用 Table 对象的 delete()方法实现数据的删除。表 5-24 列出了 Delete 类常用的方法。

表 5-24 Delete 类常用的方法

| 方法 | 返回值 | 描述 |
| --- | --- | --- |
| Delete(byte[] row) | Delete | 根据行键创建 Delete 实例 |
| addColumn(byte[] family, byte[] qualifier) | Delete | 删除指定单元格中最新版本的值 |
| addColumn(byte[] family, byte[] qualifier, long timestamp) | Delete | 删除指定单元格中特定版本的值 |
| addColumns(byte[] family, byte[] qualifier) | Delete | 删除指定单元格中所有版本的值 |
| addFamily(byte[] family) | Delete | 删除指定列族中所有列的所有版本值 |
| addFamily(byte[] family, long timestamp) | Delete | 删除指定列族中所有数据 |

例如，删除表中行键为 2018710104 的行中的 score 列族、info 列族下的 age 限定符对应的

单元格最新值和 info 列族下 email 限定符对应的单元格的所有值。

```
Delete delete= new Delete(Bytes.toBytes("2018710104"));
delete.addFamily(Bytes.toBytes("score")）
delete.addColumn(Bytes.toBytes("info"), Bytes.toBytes("age"));
delete.addColumns(Bytes.toBytes("info"), Bytes.toBytes("email"));
table.delete(delete);
```

（12）Get

如果要进行表中单行数据的查询，需要构造 Get 实例，然后指定要查询的数据。可以查询一行中特定单元格的新版本单元值、指定版本单元值、一个或多个列族、一行完整数据等。最后以 Get 对象作为参数，调用 Table 对象的 get()方法实现单行数据的查询。表 5-25 列出了 Get 类常用的方法。

表 5-25　Get 类常用的方法

| 方法 | 返回值 | 描述 |
| --- | --- | --- |
| Get (byte[] row) | Get | 根据行键创建 Get 实例 |
| addColumn(byte[] family, byte[] qualifier) | Get | 查询指定列族中特定列族限定符的值 |
| addFamily(byte[] family) | Get | 查询指定列族中的值 |
| readAllVersions() | Get | 查询单元格所有版本值 |
| readVersions(int versions) | Get | 查询单元格指定个数的版本值 |
| setTimeRange(long minStamp, long maxStamp) | Get | 查询指定时间戳范围内的单元格版本值 |
| setTimestamp(long timestamp) | Get | 查询指定时间戳对应的单元格版本值 |

例如，查询表中行键为 2018710104 的行中的 score 列族下所有值和 info 列族下的 name 列族限定符对应的值。

```
Get get= new Get(Bytes.toBytes("2018710104"));
get.addFamily(Bytes.toBytes("score"))
get.addColumn(Bytes.toBytes("info"), Bytes.toBytes("name"));
Result result=table.get(get);
```

（13）Result

利用 Table 实例的 get()方法进行数据查询时，返回的一行数据保存在 Result 实例中。Result 实例中，查询结果以<key,value>的格式被封装成一个 NavigableMap 对象。Result 对象的内部结构如图 5-21 所示。

图 5-21　Result 对象的内部结构

由图 5-21 可以发现，Result 对象内部本质是 Cell 对象数组，每个 Cell 对象代表 HBase 表中的一个单元值，由行键、列族、列族限定符、版本号（时间戳）限定。表 5-26 列出了 Result 类常用的方法。

表 5-26 Result 类常用的方法

| 方法 | 返回值 | 描述 |
|---|---|---|
| containsColumn(byte[]family, byte[] qualifier) | boolean | 检查返回结果中指定的列族限定符是否存在 |
| getFamilyMap(byte[] family) | NavigableMap<byte[],byte[]> | 返回指定列族下所有列族限定符与最新版本值的键值对 |
| getNoVersionMap() | NavigableMap<byte[],NavigableMap<byte[], byte[]>> | 返回所有列族下所有列族限定符与最新版本值的键值对 |
| getMap() | NavigableMap<byte[], NavigableMap<byte[], NavigableMap<Long,byte[]>>> | 返回所有列族下所有列族限定符对应单元格的所有版本值 |
| getRow() | byte[] | 返回 Result 对象对应的行键 |
| getValue(byte[] family, byte[] qualifier) | byte[] | 返回指定列族中的特定列族限定符最新版本值 |
| value() | byte[] | 返回第一个列标识符的值 |
| listCells() | List<Cell> | 返回 Result 对象中所有单元的有序列表 |
| rawCells() | Cell[] | 返回 Result 对象所有单元的数组 |

例如，获取 result 查询结果中的行键、info 列族中 name 列族限定符的单元值，并将 result 中的单元值转换为 Cells 列表。

```
Result result=table.get(get)
String sno=Bytes.toString(result.getRow())
String name=Bytes.toString(result.getValue(Bytes.toBytes("info"),Bytes.toBytes("name")) 
List<Cell> cells=result.ListCells()
```

（14）Scan

如果要进行表中多行数据的查询，需要构造 Scan 实例，然后设定查询条件。可以指定返回一个或多个列族、列族限定符数据，可以指定返回单元值版本数量、返回行数、起始行键等。最后以 Scan 对象作为参数，调用 Table 对象的 getScanner()方法实现多行数据的查询。表 5-27 列出了 Scan 类常用的方法。

表 5-27 Scan 类常用的方法

| 方法 | 返回值 | 描述 |
|---|---|---|
| Scan() | Scan | 构造 Scan 实例 |
| addColumn(byte[] family, byte[] qualifier) | Scan | 查询指定列族中的特定列族限定符的单元值 |
| addFamily(byte[] family) | Scan | 查询指定的列族中所有单元格的值 |
| setColumnFamilyTimeRange(byte[] cf, long minStamp, long maxStamp) | Scan | 查询指定列族中特定时间范围内的单元值 |
| setFilter(Filter filter) | Scan | 设置查询过滤器 |
| setLimit(int limit) | Scan | 设置返回的行数 |
| setTimeRange(long minStamp, long maxStamp) | Scan | 设置返回指定时间范围内的单元数据 |
| withStopRow(byte[] stopRow) | Scan | 查询数据的起始行键 |
| withStartRow(byte[] stopRow) | Scan | 查询数据的终止行键（结果不包括该行） |
| setCaching(int caching) | Scan | 设置缓存行的数量 |
| readAllVersions() | Scan | 设置读取单元格中所有版本数据 |
| readVersions(int versions) | Scan | 设置读取单元格中数据的版本数 |

例如，查询表中行键从 2018710101 到 2018710103 的行的 info 列族的所有值。

```
Scan scan=new Scan();
scan.addFamily(Bytes.toByte("info");
scan.setStartRow(Bytes.toBytes("2018710101"));
scan.setStopRow(Bytes.toBytes("2018710104"));
ResultScanner rs=table.getScanner(scan);
```

（15）ResultScanner

以 Scan 对象为参数，调用 Table 对象的 getScanner()方法时，返回的是 ResultScanner 实例，是 Result 对象的集合。

## 5.6.2 Hbase 表管理程序设计

下面通过一个 HBase 工程，介绍如何利用 HBase Java API 进行 HBase 表管理开发。

### 1. 在 Eclipse 中创建 Maven 工程

工程名为 HBase，过程与 MapReduce 工程创建类似。

### 2. 编辑 pom.xml 文件

在工程的 pom.xml 文件中添加下列 HBase 依赖信息和调试依赖信息。

```xml
<dependencies>
    <dependency>
        <groupId>org.apache.hbase</groupId>
        <artifactId>hbase-client</artifactId>
        <version>2.2.6</version>
    </dependency>
    <dependency>
        <groupId>junit</groupId>
        <artifactId>junit</artifactId>
        <version>3.8.2</version>
        <scope>test</scope>
    </dependency>
</dependencies>
```

### 3. 在 HBase 工程中创建一个名为 HBaseDDL 的类

（1）引入程序需要的内置类

```java
import java.io.IOException;
import java.util.ArrayList;
import java.util.List;
import org.apache.hadoop.conf.Configuration;
import org.apache.hadoop.hbase.HBaseConfiguration;
import org.apache.hadoop.hbase.TableName;
import org.apache.hadoop.hbase.client.Admin;
import org.apache.hadoop.hbase.client.ColumnFamilyDescriptor;
import org.apache.hadoop.hbase.client.ColumnFamilyDescriptorBuilder;
import org.apache.hadoop.hbase.client.Connection;
import org.apache.hadoop.hbase.client.ConnectionFactory;
import org.apache.hadoop.hbase.client.TableDescriptor;
import org.apache.hadoop.hbase.client.TableDescriptorBuilder;
```

（2）在 HBaseDDL 类中定义类属性、初始化方法 init()和释放资源方法 close()

由于任何对表的管理操作都需要首先连接 HBase 数据库，创建一个 Admin 对象，因此可以在类中声明 Configuration、Connection、Admin 三个公有属性，并定义 init()方法创建这些对象实例，定义一个 close()方法在程序结束时释放资源。

```java
public class HBaseDDL {
    public static Configuration conf;
    public static Connection conn;
    public static Admin admin;
    //创建Configuration对象、Connection对象和Admin对象
    public static void init() {
        conf = HBaseConfiguration.create();         //创建Configuration对象
        conf.set("hbase.rootdir", "hdfs://master:9000/hbase");   //配置信息设置
        try {
            conn = ConnectionFactory.createConnection(conf);   //创建Connection对象
            admin = conn.getAdmin();        //获取Admin对象
        } catch (IOException e) {
            e.printStackTrace();
        }
    }
    //释放Admin对象和Connection对象
    public static void close() {
        try {
            if (admin != null) {
                admin.close();
            }
            if (conn != null) {
                conn.close();
            }
        } catch (Exception e) {
            e.printStackTrace();
        }
    }
    public static void main(String[] args) {
    }
}
```

（3）创建表

创建一个方法，以表名和列族名称数组为参数创建表。代码如下：

```java
public static void createTable(String myTableName, String[] colFamily) {
    //调用init()方法创建admin对象
    init();
    //基于表名参数创建TableName对象
    TableName tableName = TableName.valueOf(myTableName);
    //检查表是否存在，如果已经存在，返回
    try {
        if (admin.tableExists(tableName)) {
            System.out.println("The table is exists !");
            return;
        } else {
            //基于TableName对象创建一个表描述生成器
            TableDescriptorBuilder tableDescBuilder = TableDescriptorBuilder.newBuilder(tableName);
```

```java
            //创建一个包含所有列族描述的列表
            List<ColumnFamilyDescriptor> colFamilyList = new ArrayList<>();
            for (String col : colFamily) {
                //基于列族名称创建列族生成器
                ColumnFamilyDescriptorBuilder columnFamilyDescBuiler =
                    ColumnFamilyDescriptorBuilder.newBuilder(col.getBytes());
                //进行列族设置
                columnFamilyDescBuiler.setMaxVersions(5);
                //构建列族描述对象
                ColumnFamilyDescriptor columnFamilyDesc = columnFamilyDescBuiler.build();
                //将列族描述对象添加到列族描述列表中
                colFamilyList.add(columnFamilyDesc);
            }
            //将所有列族描述对象添加到表描述生成器中
            tableDescBuilder.setColumnFamilies(colFamilyList);
            //构建表描述对象
            TableDescriptor tableDesc = tableDescBuilder.build();
            //基于表描述对象创建表
            admin.createTable(tableDesc);
            System.out.println("Create "+myTableName+" successful!");
            //释放资源
            close();
        }
    } catch (IOException e) {
        e.printStackTrace();
    }
}
```

然后在 main()函数中调用该方法:

```java
public static void main(String[] args) {
    createTable("newStudent", new String[] { "info", "score" });
}
```

程序执行完成后，可以通过 HBase Shell 命令查看表创建情况，如图 5-22 所示。

```
hbase(main):007:0> list
TABLE
newStudent
student
2 row(s)
Took 0.0159 seconds
=> ["newStudent", "student"]
hbase(main):008:0> desc 'newStudent'
Table newStudent is ENABLED
newStudent
COLUMN FAMILIES DESCRIPTION
{NAME => 'info', VERSIONS => '5', EVICT_BLOCKS_ON_CLOSE => 'false', NEW_VERSION_BEHAVIOR => 'false', KEEP_DELETED_CELLS => 'FALSE', CACHE_DAT
A_ON_WRITE => 'false', DATA_BLOCK_ENCODING => 'NONE', TTL => 'FOREVER', MIN_VERSIONS => '0', REPLICATION_SCOPE => '0', BLOOMFILTER => 'ROW',
CACHE_INDEX_ON_WRITE => 'false', IN_MEMORY => 'false', CACHE_BLOOMS_ON_WRITE => 'false', PREFETCH_BLOCKS_ON_OPEN => 'false', COMPRESSION =>
'NONE', BLOCKCACHE => 'true', BLOCKSIZE => '65536'}

{NAME => 'score', VERSIONS => '5', EVICT_BLOCKS_ON_CLOSE => 'false', NEW_VERSION_BEHAVIOR => 'false', KEEP_DELETED_CELLS => 'FALSE', CACHE_DA
TA_ON_WRITE => 'false', DATA_BLOCK_ENCODING => 'NONE', TTL => 'FOREVER', MIN_VERSIONS => '0', REPLICATION_SCOPE => '0', BLOOMFILTER => 'ROW',
CACHE_INDEX_ON_WRITE => 'false', IN_MEMORY => 'false', CACHE_BLOOMS_ON_WRITE => 'false', PREFETCH_BLOCKS_ON_OPEN => 'false', COMPRESSION =>
'NONE', BLOCKCACHE => 'true', BLOCKSIZE => '65536'}
```

图 5-22　利用 Java API 创建表

（4）添加列族

创建一个方法，以表名和列族名称数组为参数，向表中添加列族。代码如下：

```java
public static void addColumnFamily(String myTableName, String[] familyNames)    {
    init();
    TableName tableName = TableName.valueOf(myTableName);
```

```java
        try {
            if(admin.tableExists(tableName)) {
                for(String col :familyNames) {
                    ColumnFamilyDescriptorBuilder columnFamilyDescBuiler =
                        ColumnFamilyDescriptorBuilder.newBuilder(col.getBytes());
                    ColumnFamilyDescriptor colFamilyDes=columnFamilyDescBuiler.build();
                    //将要添加的列族对应的列族描述对象为参数，向表中添加列族
                    admin.addColumnFamily(tableName, colFamilyDes);
                    System.out.println("add "+col+" successful!");
                }
                close();
            }
        }catch (IOException e) {
            e.printStackTrace();
        }
    }
```

然后在 main()函数中调用该方法：

```java
public static void main(String[] args) {
    addColumnFamily("newStudent", new String[] { "address"});
}
```

执行完成后，可以通过 HBase Shell 命令查看添加的列族信息。

（5）删除列族

创建一个方法，以表名和列族名称数组为参数，删除表中指定的列族。代码如下：

```java
public static void removeColumnFamily(String myTableName, String[] familyNames) {
    init();
    TableName tableName = TableName.valueOf(myTableName);
    try {
        for (String col : familyNames) {
            //以列族名称为参数，删除一个列族
            admin.deleteColumnFamily(tableName, col.getBytes());
            System.out.println("delete " + col + " successful!");
        }
        close();
    } catch (IOException e) {
        e.printStackTrace();
    }
}
```

然后在 main()函数中调用该方法：

```java
public static void main(String[] args) {
    removeColumnFamily("newStudent", new String[] {"address"});
}
```

（6）修改列族

创建一个方法，以表名和列族名称数组为参数，修改表中参数指定的列族。代码如下：

```java
public static void modifyColumnFamily(String myTableName, String[] familyNames) {
    init();
    TableName tableName = TableName.valueOf(myTableName);
    try {
        for (String col : familyNames) {
```

```
                ColumnFamilyDescriptorBuilder columnFamilyDescBuiler =
                    ColumnFamilyDescriptorBuilder.newBuilder(col.getBytes());
                //修改列族设置
                columnFamilyDescBuiler.setMaxVersions(3);
                columnFamilyDescBuiler.setTimeToLive(360000);
                //构建新的列族描述对象
                ColumnFamilyDescriptor colFamilyDes = columnFamilyDescBuiler.build();
                //用修改后的列族描述对象修改列族
                admin.modifyColumnFamily(tableName, colFamilyDes);
                System.out.println("modify " + col + " successful!");
            }
            close();
        } catch (IOException e) {
            e.printStackTrace();
        }
    }
```

然后在 main()函数中调用该方法：

```
public static void main(String[] args) {
    modifyColumnFamily("newStudent", new String[] { "score", "info" });
}
```

（7）查询 HBase 中所有表名称

创建一个方法，显示 HBase 中所有表名称。代码如下：

```
public static void listTables() {
    init();
    try {
        //获取包含HBase中当前所有表名对象数组
        TableName[] tableNames = admin.listTableNames();
        for (TableName name : tableNames) {
            System.out.println(name.toString());
        }
        close();
    } catch (IOException e) {
        e.printStackTrace();
    }
}
```

然后在 main()函数中调用该方法：

```
public static void main(String[] args) {
    listTables();
}
```

（8）显示 HBase 中所有表的描述信息

创建一个方法，获取 HBase 中所有表的描述信息。代码如下：

```
public static void describeTables() {
    init();
    try {
        //获取包含HBase中当前所有表的描述对象列表
        List<TableDescriptor> tableDescs = admin.listTableDescriptors();
        //遍历每个表的描述对象，获取表的所有列族信息
        for (TableDescriptor tableDesc : tableDescs) {
```

```java
            //获取一个表的所有列族描述数组
            ColumnFamilyDescriptor[] columnFamilyDescs = tableDesc.getColumnFamilies();
            String columnNames = "";
            //遍历每个列族描述对象，获取列族限定符信息
            for (ColumnFamilyDescriptor colFamilyDesc : columnFamilyDescs) {
                columnNames = colFamilyDesc.getNameAsString() + " " + columnNames;
            }
            System.out.println(tableDesc.getTableName().getNameAsString() + " " + columnNames);
        }
        close();
    } catch (IOException e) {
        e.printStackTrace();
    }
}
```

然后在 main()函数中调用该方法：

```java
public static void main(String[] args) {
    describeTables();
}
```

(9) 修改表

创建一个方法，以表名为参数，修改表的设置。代码如下：

```java
public static void modifyTable(String myTableName) {
    init();
    TableName tableName = TableName.valueOf(myTableName);
    try {
        TableDescriptor tableDesc = admin.getDescriptor(tableName);
        TableDescriptorBuilder tableDescBuilder = TableDescriptorBuilder.newBuilder(tableDesc);
        //设置表描述信息
        tableDescBuilder.setMaxFileSize(134217728);
        tableDescBuilder.setReadOnly(true);
        //使用修改后的表描述对象修改表
        admin.modifyTable(tableDescBuilder.build());   //修改表
        System.out.println("modify "+myTableName+" successful !");
        close();
    } catch (Exception e) {
        e.printStackTrace();
    }
}
```

然后在 main()函数中调用该方法：

```java
public static void main(String[] args) {
    modifyTable("newStudent");
}
```

(10) 删除表

创建一个方法，以表名为参数，删除表。代码如下：

```java
public static void dropTable(String myTableName) {
    init();
    TableName tableName = TableName.valueOf(myTableName);
    try {
        if (admin.tableExists(tableName)) {
```

```
                admin.disableTable(tableName);    //禁用表
                admin.deleteTable(tableName);     //删除表
                System.out.println("drop " + myTableName + " successful !");
                close();
            } else {
                System.out.println("the table does not exist !");
            }
        } catch (Exception e) {
            e.printStackTrace();
        }
    }
```

然后在 main()函数中调用该方法：

```
public static void main(String[] args) {
    dropTable("newStudent");
}
```

### 5.6.3 HBase 数据操作程序设计

利用 HBaseJava API 进行数据操作，主要包括数据插入、单行数据查询、多行数据查询、数据删除等。

在 5.6.2 节创建的 HBase 工程中创建一个名为 HBaseDML 的类。

（1）引入程序需要的内置类

```
package com.bigdata.example;
import java.io.IOException;
import java.util.ArrayList;
import java.util.List;
import org.apache.hadoop.conf.Configuration;
import org.apache.hadoop.hbase.Cell;
import org.apache.hadoop.hbase.CellUtil;
import org.apache.hadoop.hbase.HBaseConfiguration;
import org.apache.hadoop.hbase.TableName;
import org.apache.hadoop.hbase.client.Connection;
import org.apache.hadoop.hbase.client.ConnectionFactory;
import org.apache.hadoop.hbase.client.Delete;
import org.apache.hadoop.hbase.client.Get;
import org.apache.hadoop.hbase.client.Put;
import org.apache.hadoop.hbase.client.Result;
import org.apache.hadoop.hbase.client.ResultScanner;
import org.apache.hadoop.hbase.client.Scan;
import org.apache.hadoop.hbase.client.Table;
import org.apache.hadoop.hbase.util.Bytes;
```

（2）在 HBaseDML 类中定义类属性、初始化方法 init()和释放资源方法 close()

由于数据操作需要连接 HBase 数据库，然后基于不同表创建 Table 对象，因此可以在类中定义 Configuration、Connection 两个公有属性，并定义 init()方法创建这些对象实例，定义一个 close()方法在程序结束时释放资源。

```
public class HBaseDML {
    public static Configuration conf;
    public static Connection conn;
```

```java
public static void init() {
    conf = HBaseConfiguration.create();
    conf.set("hbase.rootdir", "hdfs://master:9000/hbase");
    try {
        conn = ConnectionFactory.createConnection(conf);
    } catch (IOException e) {
        e.printStackTrace();
    }
}
public static void close() {
    if (conn != null) {
        try {
            conn.close();
        } catch (IOException e) {
            e.printStackTrace();
        }
    }
}
public static void main(String[] args) {

}
}
```

（3）数据插入

数据插入是根据行键创建一个 Put 对象，将要插入的数据封装到 Put 对象中，然后调用 Table 对象的 put()方法实现数据插入。可以进行单行数据插入，也可以创建 Put 对象列表实现批量插入。

例如，创建一个 put()方法，以表名为单位进行批量数据插入。

```java
public static void put(String tableName) {
    init();
    try {
        Table table = conn.getTable(TableName.valueOf(tableName));
        Put put1 = new Put(Bytes.toBytes("2018710101"));
        put1.addColumn(Bytes.toBytes("info"), Bytes.toBytes("name"), Bytes.toBytes("liuYan"));
        put1.addColumn(Bytes.toBytes("info"), Bytes.toBytes("age"), Bytes.toBytes("20"));
        put1.addColumn(Bytes.toBytes("info"), Bytes.toBytes("email"), Bytes.toBytes("liu@qq.com"));
        put1.addColumn(Bytes.toBytes("score"), Bytes.toBytes("math"), Bytes.toBytes("90"));
        put1.addColumn(Bytes.toBytes("score"), Bytes.toBytes("english"), Bytes.toBytes("98"));
        put1.addColumn(Bytes.toBytes("score"), Bytes.toBytes("database"), Bytes.toBytes("80"));

        Put put2 = new Put(Bytes.toBytes("2018710104"));
        put2.addColumn(Bytes.toBytes("info"), Bytes.toBytes("name"), Bytes.toBytes("CaoBO"));
        put2.addColumn(Bytes.toBytes("info"), Bytes.toBytes("age"), Bytes.toBytes("19"));
        put2.addColumn(Bytes.toBytes("info"), Bytes.toBytes("email"), Bytes.toBytes("cao@qq.com"));
        put2.addColumn(Bytes.toBytes("info"), Bytes.toBytes("email"), Bytes.toBytes("bo@qq.com"));
        put2.addColumn(Bytes.toBytes("score"), Bytes.toBytes("math"), Bytes.toBytes("80"));
        put2.addColumn(Bytes.toBytes("score"), Bytes.toBytes("english"), Bytes.toBytes("90"));
        put2.addColumn(Bytes.toBytes("score"), Bytes.toBytes("database"), Bytes.toBytes("95"));
        List<Put> puts = new ArrayList<Put>();
```

```
            puts.add(put1);
            puts.add(put2);
            table.put(puts);
            table.close();
            close();
        } catch (IOException e) {
            e.printStackTrace();
        }
    }
```

然后在 main()函数中调用该方法：

```
public static void main(String[] args) {
    put("newStudent");
}
```

程序执行完成后，可以通过 HBase Shell 命令查看数据插入情况，如图 5-23 所示。

```
hbase(main):002:0> scan 'newStudent'
ROW                    COLUMN+CELL
 2018710101            column=info:age, timestamp=1612183692670, value=20
 2018710101            column=info:email, timestamp=1612183692670, value=liu@qq.com
 2018710101            column=info:name, timestamp=1612183692670, value=liuYan
 2018710101            column=score:database, timestamp=1612183692670, value=80
 2018710101            column=score:english, timestamp=1612183692670, value=98
 2018710101            column=score:math, timestamp=1612183692670, value=90
 2018710104            column=info:age, timestamp=1612183692670, value=19
 2018710104            column=info:email, timestamp=1612183692670, value=bo@qq.com
 2018710104            column=info:name, timestamp=1612183692670, value=CaoBO
 2018710104            column=score:database, timestamp=1612183692670, value=95
 2018710104            column=score:english, timestamp=1612183692670, value=90
 2018710104            column=score:math, timestamp=1612183692670, value=80
2 row(s)
Took 0.5354 seconds
```

图 5-23　利用 Java API 数据插入

重载（Overload）是面向对象程序设计中的重要特性，因此在 Java 程序设计中，可以根据参数不同创建多个方法。例如，在数据插入中，还可以创建 put()方法，以表名、列族名、列族限定符、值为参数。

```
public static void put(String tableName, String rowKey, String colFamily, String qualifier, String value) {
    init();
    try {
        Table table = conn.getTable(TableName.valueOf(tableName));
        Put put = new Put(rowKey.getBytes());
        put.addColumn(colFamily.getBytes(), qualifier.getBytes(), value.getBytes());
        table.put(put);
        table.close();
        close();
    } catch (IOException e) {
        e.printStackTrace();
    }
}
```

（4）单行数据查询

单行数据查询是根据行键创建一个 Get 对象,将查询条件封装到 Get 对象中,然后调用 Table 对象的 get()方法返回查询结果。

可以根据表名、行键、列族和列族限定符为参数，返回一个单元格的值。例如：

```
public static void get(String tableName,String rowKey,String colFamily,String qualifier){
```

```
        init();
        Table table;
        try {
            table = conn.getTable(TableName.valueOf(tableName));
            Get get=new Get(rowKey.getBytes());
            get.addColumn(colFamily.getBytes(), qualifier.getBytes());
            get.readVersions(2);
            Result result =table.get(get);
            for(Cell cell :result.listCells()){
                System.out.println(
                        "RowKey: "+Bytes.toString(CellUtil.cloneRow(cell))+
                        " Family: "+Bytes.toString(CellUtil.cloneFamily(cell))+
                        " qualifier: " +Bytes.toString(CellUtil.cloneQualifier(cell)) +
                        " value:"+Bytes.toString(CellUtil.cloneValue(cell)) +
                        " timestamp: "+ cell.getTimestamp());
            }
            table.close();
            close();
        } catch (IOException e) {
            e.printStackTrace();
        }
    }
```

然后在 main()函数中调用该方法：
```
public static void main(String[] args) {
    get("newStudent","2018710101","score","math");
}
```

单行数据查询结果如图 5-24 所示。

```
RowKey: 2018710101 Family: score qualifier: math value:90 timestamp: 1612183692670
```

图 5-24　单行数据查询结果

可以创建一个用于显示 Result 结果的方法 show()，例如：
```
public static void show(Result result){
    List<Cell> cells=result.listCells();
    for(Cell cell: cells){
        System.out.println(
                "RowKey: "+Bytes.toString(CellUtil.cloneRow(cell))+
                " Family: "+Bytes.toString(CellUtil.cloneFamily(cell))+
                " qualifier: " +Bytes.toString(CellUtil.cloneQualifier(cell)) +
                " value:"+Bytes.toString(CellUtil.cloneValue(cell)) +
                " timestamp: "+ cell.getTimestamp());
    }
}
```

可以根据表名、行键、列族为参数，返回指定列族的所有的值。例如：
```
public static void get(String tableName,String rowKey,String[] colFamily){
    init();
```

```
        Table table;
        try {
            table = conn.getTable(TableName.valueOf(tableName));
            Get get=new Get(rowKey.getBytes());
            for(String col:colFamily) {
                get.addFamily(col.getBytes());
            }
            Result result =table.get(get);
            show(result);
            table.close();
            close();
        } catch (IOException e) {
            e.printStackTrace();
        }
    }
```

可以根据表名、行键为参数，返回完整一行数据。例如：

```
public static void get(String tableName,String rowKey){
    init();
    Table table;
    try {
        table = conn.getTable(TableName.valueOf(tableName));
        Get get=new Get(rowKey.getBytes());
        Result result =table.get(get);
        show(result);
        table.close();
        close();
    } catch (IOException e) {

        e.printStackTrace();
    }
}
```

（5）多行数据查询

多行数据查询是创建一个 Scan 对象，将查询条件封装到 Scan 对象中，然后调用 Table 对象的 getScanner()方法返回查询结果。

```
public static void scan(String tableName){
    init();
    try {
        Table table=conn.getTable(TableName.valueOf(tableName));
        Scan scan=new Scan();
        scan.readVersions(2);
        scan.setCacheBlocks(true);
        scan.setBatch(1000);
        ResultScanner rs=table.getScanner(scan);
        for(Result result: rs){
            show(result);
        }
        rs.close();
        table.close();
```

```
        close();
    } catch (IOException e) {
        e.printStackTrace();
    }
}
```

可以通过参数指定查询的特定列族限定符的值，例如：

```
public static void scan(String tableName,String colFamily,String qualifier){
    init();
    try {
        Table table=conn.getTable(TableName.valueOf(tableName));
        Scan scan=new Scan();
        scan.addColumn(Bytes.toBytes(colFamily),Bytes.toBytes(qualifier));
        ResultScanner rs=table.getScanner(scan);
        for(Result result: rs){
            show(result);
        }
        rs.close();
        table.close();
        close();
    } catch (IOException e) {
        e.printStackTrace();
    }
}
```

可以通过参数指定查询的特定列族的值，例如：

```
public static void scan(String tableName,String colFamily){
    init();
    try {
        Table table=conn.getTable(TableName.valueOf(tableName));
        Scan scan=new Scan();
        scan.addFamily(Bytes.toBytes(colFamily));
        ResultScanner rs=table.getScanner(scan);
        for(Result result: rs){
            show(result);
        }
        rs.close();
        table.close();
        close();
    } catch (IOException e) {
        e.printStackTrace();
    }
}
```

（6）数据删除

数据删除是根据行键创建一个 Delete 对象，将删除条件封装到 Delete 对象中，然后调用 Table 对象的 delete()方法实现数据删除。可以进行单行数据删除，也可以创建 Delete 对象列表实现批量删除。

可以删除参数指定单元格中的数据。例如：

```
public static void delete(String tableName, String rowKey,
        String colFamily, String qualifier) {
```

```java
        init();
        try {
            Table table = conn.getTable(TableName.valueOf(tableName));
            Delete delete = new Delete(rowKey.getBytes());
            delete.addColumn(colFamily.getBytes(), qualifier.getBytes());
            table.delete(delete);
            table.close();
            close();
        } catch (IOException e) {
            e.printStackTrace();
        }
    }
```

可以删除参数指定的列族对应的数据。例如：

```java
public static void delete(String tableName, String rowKey,String colFamily) {
    init();
    try {
        Table table = conn.getTable(TableName.valueOf(tableName));
        Delete delete = new Delete(rowKey.getBytes());
        delete.addFamily(colFamily.getBytes());
        table.delete(delete);
        table.close();
        close();
    } catch (IOException e) {
        e.printStackTrace();
    }
}
```

可以删除参数指定的一行数据。例如：

```java
public static void delete(String tableName, String rowKey) {
    init();
    try {
        Table table = conn.getTable(TableName.valueOf(tableName));
        Delete delete = new Delete(rowKey.getBytes());
        table.delete(delete);
        table.close();
        close();
    } catch (IOException e) {
        e.printStackTrace();
    }
}
```

### 5.6.4 HBase Filter API

与 HBase Shell scan 和 get 命令中使用的 Filter 相对应，HBase Java API 也提供了大量的过滤器。HBase 过滤器在服务器端与 Scan、Get 对象一起过滤掉不需要的数据，以减少在服务器端与客户端的数据传输量，所有的过滤器均继承自抽象类 org.apache.hadoop.hbase.filter.Filter，如图 5-25 所示。

在 Java HBase 2.x API 中，CompareOperator 类中定义了比较运算符，包括 EQUAL(等)、GREATER（大于）、GREATER_OR_EQUAL（大于或等于）、LESS（小于）、LESS_OR_EQUAL

（小于或等于）、NOT_EQUAL（不等于）和 NO_OP（排除所有）。

```
○ org.apache.hadoop.hbase.filter.Filter
    ○ org.apache.hadoop.hbase.filter.FilterBase
        ○ org.apache.hadoop.hbase.filter.ColumnCountGetFilter
        ○ org.apache.hadoop.hbase.filter.ColumnPaginationFilter
        ○ org.apache.hadoop.hbase.filter.ColumnPrefixFilter
        ○ org.apache.hadoop.hbase.filter.ColumnRangeFilter
        ○ org.apache.hadoop.hbase.filter.ColumnValueFilter
        ○ org.apache.hadoop.hbase.filter.CompareFilter
            ○ org.apache.hadoop.hbase.filter.DependentColumnFilter
            ○ org.apache.hadoop.hbase.filter.FamilyFilter
            ○ org.apache.hadoop.hbase.filter.QualifierFilter
            ○ org.apache.hadoop.hbase.filter.RowFilter
            ○ org.apache.hadoop.hbase.filter.ValueFilter
        ○ org.apache.hadoop.hbase.filter.FilterList
        ○ org.apache.hadoop.hbase.filter.FilterListBase
            ○ org.apache.hadoop.hbase.filter.FilterListWithAND
            ○ org.apache.hadoop.hbase.filter.FilterListWithOR
        ○ org.apache.hadoop.hbase.filter.FirstKeyOnlyFilter
            ○ org.apache.hadoop.hbase.filter.FirstKeyValueMatchingQualifiersFilter
        ○ org.apache.hadoop.hbase.filter.FuzzyRowFilter
        ○ org.apache.hadoop.hbase.filter.InclusiveStopFilter
        ○ org.apache.hadoop.hbase.filter.KeyOnlyFilter
        ○ org.apache.hadoop.hbase.filter.MultipleColumnPrefixFilter
        ○ org.apache.hadoop.hbase.filter.MultiRowRangeFilter
        ○ org.apache.hadoop.hbase.filter.PageFilter
        ○ org.apache.hadoop.hbase.filter.PrefixFilter
        ○ org.apache.hadoop.hbase.filter.RandomRowFilter
        ○ org.apache.hadoop.hbase.filter.SingleColumnValueFilter
            ○ org.apache.hadoop.hbase.filter.SingleColumnValueExcludeFilter
        ○ org.apache.hadoop.hbase.filter.SkipFilter
        ○ org.apache.hadoop.hbase.filter.TimestampsFilter
        ○ org.apache.hadoop.hbase.filter.WhileMatchFilter
    ○ org.apache.hadoop.hbase.filter.FilterWrapper
```

图 5-25　Java Filter API

Java HBase API 也提供了实现多样化目标匹配效果的比较器，包括：
- BinaryComparator，匹配完整字节数组；
- BinaryPrefixComparator，匹配字节数组前缀；
- BitComparator，匹配位；
- NullComparator，匹配空值；
- RegexStringComparator，匹配正则表达式；
- SubstringComparator，匹配子字符串。

它们都是 ByteArrayComparable 的子类。其中，BitComparator、RegexStringComparator、SubstringComparator 这 3 种比较器只能与 EQUAL 和 NOT EQUAL 运算符搭配使用，通过 compareTo()方法进行比较，匹配时返回 0，不匹配时返回 1。RegexStringComparator、SubstringComparator 基于字符串比较，需要将给定的值转化为 String 类型，因此运行较慢，较消耗资源。

应用过滤器时，按行键过滤时可以使用 RowFilter，按列族过滤时可以使用 FamliyFilter，按列族限定符过滤时可以使用 QualifierFilter，按值过滤时可以使用 ValueFilter。可以使用 SkipFilter 和 WhileMatchFilter 封装来添加更多控制，也可以使用 FilterList 组合多个过滤器。

（1）RowFilter

RowFilter 过滤器支持基于行键过滤数据，可以执行精确匹配、子字符串匹配或正则表达式匹配，筛选出匹配的所有的行。例如：

```
Filter filter = new RowFilter(CompareOperator.EQUAL, new BinaryComparator(Bytes.toBytes("10005")));
```

Filter filter=new RowFilter(CompareOperator.LESS_OR_EQUAL,new BinaryComparator(Bytes.toBytes ("1000")));

（2）FamilyFilter

FamilyFilter 过滤器支持基于列族名称过滤数据，可以匹配等值条件，也可以匹配小于、大于、不等于等条件。例如：

FamilyFilter filter= new FamilyFilter(CompareOperator.EQUAL, new BinaryPrefixComparator("info".getBytes()));

（3）QualifierFilter

QualifierFilter 过滤器支持基于列族限定符名称过滤数据，可以进行等值、大于、小于、不等于等条件的匹配。例如：

Filter filter = new QualifierFilter(CompareOperator.EQUAL, new RegexStringComparator("name"));

（4）ValueFilter

ValueFilter 过滤器支持基于单元值过滤数据，可以进行等值、大于、小于、不等于等条件的匹配。例如：

Filter filter = new ValueFilter(CompareOperator.EQUAL, new BinaryComparator(Bytes.toBytes("80")));

（5）PrefixFilter

PrefixFilter 过滤器是 RowFilter 过滤器的一种特例，基于行键的前缀值进行过滤，它相当于给扫描构造函数 Scan(byte[] startRow, byte[] stopRow)提供了一个停止键，不需要自己计算停止键。例如：

Filter filter = new PrefixFilter(Bytes.toBytes("2018"));

（6）KeyOnlyFilter

KeyOnlyFilter 过滤器只会返回每行的行键、列族、列族限定符，而不返回单元值。这对于不需要值的应用场景来说非常实用，减少了值的传递。例如：

Filter filter = new KeyOnlyFilter();

（7）FirstKeyOnlyFilter

FirstKeyOnlyFilter 过滤器返回每一行的第一列族限定符的单元值。例如：

Filter filter = new FirstKeyOnlyFilter();

（8）ColumnPrefixFilter

ColumnPrefixFilter 过滤器是按照列名的前缀来扫描单元格的，只会返回符合条件的列数据。例如：

Filter filter = new ColumnPrefixFilter(Bytes.toBytes("LiuYun"));

（9）SingleColumnValueFilter

SingleColumnValueFilter 过滤器根据指定列族、列族限定符对应单元值是否匹配指定的条件，决定是否将该数据行返回。由于 HBase 支持动态模式，因此有些数据行可能不包含指定的列族限定符，可以设置 filter.setFilterIfMissing(true)，查询结果不会返回不包含指定列族限定符的数据行，如果为 false（默认），则会返回所有的行信息。例如：

SingleColumnValueFilter filter = new SingleColumnValueFilter(Bytes.toBytes("info"), Bytes.toBytes("name"), CompareOperator.EQUAL.EQUAL, new SubstringComparator("Liu"));
filter.setFilterIfMissing(true);

（10）ColumnValueFilter

ColumnValueFilter 过滤器与 SingleColumnValueFilter 过滤器类似，不同点在于 ColumnValueFilter 过滤器只返回匹配的单元，而不返回整个行。例如：

SingleColumnValueFilter filter = new ColumnValueFilter(Bytes.toBytes("info"), Bytes.toBytes("name"), CompareOperator.EQUAL, new SubstringComparator("Liu"));

（11）TimestampsFilter

TimestampsFilter 过滤器允许针对返回给客户端的单元版本进行更细粒度的控制，可以提供一个返回的时间戳列表，只有与时间戳匹配的单元才可以返回。例如：

List<Long> timestamps = new ArrayList<Long>();
timestamps.add(1611923503132L);
timestamps.add(1611923504132L);
TimestampsFilter filter = new TimestampsFilter(timestamps);

（12）WhileMatchFilter

WhileMatchFilter 是包装过滤器，包装过滤器通过包装其他过滤器以实现某些拓展的功能。当 WhileMatchFilter 过滤器包装的过滤器条件不满足时，WhileMatchFilter 结束本次扫描，返回已经扫描到的结果。

例如，当遇到单元值等于 100 时返回结果，结束扫描：

Filter filter = new ValueFilter(CompareOperator.NOT_EQUAL, new BinaryComparator(Bytes.toBytes("100")));
Filter filter2 = new WhileMatchFilter(filter1);

WhileMatchFilter 过滤器的执行类似 while 循环语句，当条件不满足即结束。

（13）SkipFilter

SkipFilter 过滤器与 WhileMatchFilter 类似，也是包装过滤器。不同之处在于，SkipFilter 遇到满足条件的数据时，跳过该满足过滤器条件的行。

例如，跳过单元值等于 100 的行：

Filter filter1= new ValueFilter(CompareOperator.NOT_EQUAL,new BinaryComparator(Bytes.toBytes("100")));
Filter filter2 = new SkipFilter(filter1);

（14）FilterList

FilterList 是一个过滤器列表，可以包含一组即将应用于目标数据集的过滤器，过滤器可以是"与"关系（FilterList.Operator.MUST_PASS_ALL）或"或"关系（FilterList.Operator.MUST_PASS_ONE）。例如：

Filter rowFilter = new RowFilter(CompareOperator.EQUAL, new BinaryComparator(Bytes.toBytes("10001")));
Filter valueFilter = new ValueFilter(CompareOperator.EQUAL, new BinaryPrefixComparator(Bytes.toBytes("80")));
List<Filter> filters = new ArrayList<Filter>();
filters.add(rowFilter);
filters.add(valueFilter);
FilterList fl = new FilterList(FilterList.Operator.MUST_PASS_ALL, filters);

下面是一个过滤器应用的综合示例，使用一个或多个过滤器进行数据的过滤。

```
public class HBaseFilter {
    public static Configuration conf;
    public static Connection conn;
    public static void main(String[] args) throws Exception {
        conf=HBaseConfiguration.create();
        conf.set("hbase.rootdir","hdfs://master:9000/hbase");
        conn=ConnectionFactory.createConnection(conf);
        Table table=conn.getTable(TableName.valueOf("newStudent"));
        //查询行健中包含2018的行
        Filter filter1=new RowFilter(CompareOperator.EQUAL,new SubstringComparator("2018"));
        //查询info列族信息
        Filter filter2=new FamilyFilter(CompareOperator.EQUAL,new BinaryComparator(Bytes.toBytes("info")));
```

```
        //查询name列族限定符对应的单元值
        Filter filter3=new QualifierFilter(CompareOperator.LESS_OR_EQUAL,new BinaryComparator(Bytes.toBytes("name")));
        //查询单元值大于80的信息
        Filter filter=new ValueFilter(CompareOperator.GREATER_OR_EQUAL,new BinaryComparator(Bytes.toBytes(80)));
        //多个过滤器组合
        List<Filter> filters = new ArrayList<Filter>();
        filters.add(filter1);
        filters.add(filter2);
        filters.add(filter3);
        FilterList filterList = new FilterList(FilterList.Operator.MUST_PASS_ALL, filters);
        Scan scan=new Scan();
        scan.setFilter(filterList);
        ResultScanner rs=table.getScanner(scan);
        for(Result result :rs) {
            System.out.print(result);
        }
        rs.close();
        table.close();
        conn.close();
    }
}
```

## 5.7 HBase 与 MapReduce 融合

### 5.7.1 HBase 与 MapReduce 融合概述

**1. HBase 与 MapReduce 融合的必要性**

HBase 与 MapReduce 融合包括 3 种形式：HBase 作为 MapReduce 作业数据输入的来源、HBase 作为 MapReduce 作业处理结果输出的目的地、HBase 同时作为 MapReduce 作业数据输入和输出的目的地。这样，既利用了 MapReduce 分布式计算的优势，也利用了 HDFS 海量存储的特点，特别是利用了 HBase 对海量数据实时访问的特点。

除将 HBase 作为 MapReduce 作业数据的输入和输出外，HBase 与 MapReduce 融合还可以实现下列任务：

① 对 HBase 中的数据进行实时性的统计分析。HBase 适合进行键值对查询，但默认 HBase 不带聚合函数，如果没有实时性要求，可以使用 MapReduce 来完成统计分析。

② 对 HBase 表中数据进行分布式计算。HBase 的目标是在海量数据中快速定位数据并访问，但 HBase 只支持基于行键的查询，如果要进行业务逻辑扩展，可以将业务逻辑放到 MapReduce 计算框架中。

③ 在多个 MapReduce 作业间使用 HBase 作为中间存储的介质，HBase 可以同时作为多个 MapReduce 作业的数据源和数据存储的目的地。

**2. 配置 Hadoop 环境**

为了保证 MapReduce 程序可以访问操作 HBase，需要预先进行 Hadoop 环境的设置。

① 将 hbase-site.xml 复制到 Hadoop 配置文件目录下，以便使 MapReduce 任务在运行时可

以连接到 Zookeeper。

cp $HBASE_HOME/conf/hbase-site.xml $HADOOP_HOME/etc/hadoop/

② 在 Hadoop 环境变量中引入 HBase，以便使 MapReduce 作业可以访问 HBase 的相关类。

gedit $HADOOP_HOME/etc/hadoop/hadoop-env.sh

在 hadoop-env.sh 文件中添加下列内容：

export HADOOP_CLASSPATH=$HADOOP_CLASSPATH:/usr/local/hbase/lib/*

③ 重启 Hadoop 和 HBase，然后验证安装。

例如，验证 HBase 中 student 表中数据的行数。

hadoop jar /usr/local/hbase/lib/hbase-mapreduce-2.2.6.jar rowcounter student

执行结果如图 5 26 所示。

```
HBaseCounters
        BYTES_IN_REMOTE_RESULTS=0
        BYTES_IN_RESULTS=46
        MILLIS_BETWEEN_NEXTS=51
        NOT_SERVING_REGION_EXCEPTION=0
        NUM_SCANNER_RESTARTS=0
        NUM_SCAN_RESULTS_STALE=0
        REGIONS_SCANNED=1
        REMOTE_RPC_CALLS=0
        REMOTE_RPC_RETRIES=0
        ROWS_FILTERED=0
        ROWS_SCANNED=1
        RPC_CALLS=1
        RPC_RETRIES=0
org.apache.hadoop.hbase.mapreduce.RowCounter$RowCounterMapper$Counters
        ROWS=1
```

图 5-26 验证表中数据行数的执行结果

### 5.7.2 HBase MapReduce Java API

HBase Java API 对 MapReduce API 进行了扩展，提供了基于 HBase 表数据输入和输出操作的 API。对 HBase 表的输入、输出格式提供了 TableInputFormat 和 TableOutputFormat 类，同时提供了 TableMapper 和 TableReducer 类用于编写 MapReduce 程序。

（1）TableMapper 类

TableMapper 类继承自 Mapper 类，其定义为：

public abstract class TableMapper<KEYOUT,VALUEOUT>extends
org.apache.hadoop.mapreduce.Mapper<ImmutableBytesWritable,Result,KEYOUT,VALUEOUT>{
}

可以看出 TableMapper 类没有实现任何功能，是一个不提供 KEYIN 和 VALUEIN 参数的 Mapper 类，即 TableMapper 类限制了每一行输入数据的键值类型，分别为 ImmutableBytesWritable 和 Result，而输出的键值类型由参数 KEYOUT 和 VALUEOUT 指定。其中，ImmutableBytesWritable 类型表示行键，Result 类型表示 HBase 中的一行数据。

（2）TableReducer 类

TableReducer 类继承自 Reducer 类，其定义为：

public abstract class TableReducer<KEYIN,VALUEIN,KEYOUT>
extends org.apache.hadoop.mapreduce.Reducer<KEYIN,VALUEIN,KEYOUT,Mutation>{
}

可以看出，TableReducer 类需要指定 KEYIN、VALUEIN 和 KEYOUT 三个参数。

KEYIN：输入键类型，对应 Map 任务的输出键类型。

VALUEIN：输入值类型，对应 Map 任务输出值类型。

KEYOUT：输出键类型。

而 TableReducer 类默认的输出值类型为 Mutation，与 TableOutputFormat 类的输出值类型保持一致。

（3）TableMapReduceUtil 类

在利用 MapReduce 程序操作 HBase 时，为了解决第三方 JAR 文件依赖问题，需要调用 TableMapReduceUtil 类的静态方法 initTableMapperJob 来指定作为数据输入来源的 HBase 表名和自定义的 Mapper 类，调用 TableMapReduceUtil 类的静态方法 initTableReduceJob 来指定作为数据输出目的的 HBase 表名和自定义的 Reducer 类。

initTableMapperJob 方法与 TableMapReduceUtil 方法的原型为：

● static void initTableMapperJob(String table, Scan scan,Class<?extends TableMapper> mapper, Class<?> outputKeyClass, Class<?> outputValueClass, org.apache.hadoop.mapreduce.Job job)

● static void initTableReducerJob(String table, Class<? extends TableReducer> reducer, org.apache.hadoop.mapreduce.Job job)

### 5.7.3 HBase MapReduce 程序设计

#### 1. 读取 HDFS 数据，将处理结果写入 HBase

下面以统计词频数为例，读取 4.4.1 节的 HDFS 文件，计算每个单词出现的次数，结果写入 HBase 表中。

（1）在 HBase 中创建一个名为 wordcount 的表用于接收 MapReduce 程序的输出结果。

```
hbase>create 'wordcount','result'
```

（2）在 Eclipse 中创建一个名为"HBase_MapReduce"的 Maven 工程。

（3）编辑工程 pom.xml 文件，添加下列依赖。

```xml
<dependencies>
    <dependency>
        <groupId>org.apache.hadoop</groupId>
        <artifactId>hadoop-client</artifactId>
        <version>3.1.2</version>
    </dependency>
    <dependency>
        <groupId>org.apache.hbase</groupId>
        <artifactId>hbase-mapreduce</artifactId>
        <version>2.2.6</version>
    </dependency>
    <dependency>
        <groupId>org.apache.hbase</groupId>
        <artifactId>hbase-client</artifactId>
        <version>2.2.6</version>
    </dependency>
    <dependency>
        <groupId>org.apache.hadoop</groupId>
        <artifactId>hadoop-auth</artifactId>
        <version>3.1.2</version>
    </dependency>
</dependencies>
```

（4）创建一个名为 HDFSToHBase 的类。

① 在 HDFSToHBase 类中创建一个继承 Mapper 的内嵌类，重写 map()函数，读取 HDFS

文件，输出键值对。

```java
public static class WriteHBaseMapper extends Mapper<LongWritable, Text, Text, IntWritable> {
    private final static IntWritable one = new IntWritable(1);
    private Text word = new Text();
    public void map(LongWritable key,Text value,Context context)throws IOException,InterruptedException {
        String line=value.toString();
        StringTokenizer token = new StringTokenizer(line);
        while (token.hasMoreTokens()) {
            word.set(token.nextToken());
            context.write(word, one);
        }
    }
}
```

② 在 HDFSToHBase 类中创建一个继承 TableReducer 类的内嵌类，重写 reduce()函数，接收 map()函数输出，将处理结果写入 HBase。

```java
public static class WriteHBaseReducer extends TableReducer<Text, IntWritable,Text> {
    public void reduce(Text key, Iterable<IntWritable> values,
            Context context) throws IOException, InterruptedException {
        int sum = 0;
        for (IntWritable val : values) {
            sum += val.get();
        }
        Put put=new Put(Bytes.toBytes(key.toString()));
        put.addColumn(Bytes.toBytes("result"),Bytes.toBytes("count"),Bytes.toBytes(String.valueOf(sum)));
        context.write(key, put);
    }
}
```

③ 编写 HDFSToHBase 类的 main()函数，进行 MapReduce 作业设置。

```java
public static void main(String[] args) throws Exception{
    Configuration conf = HBaseConfiguration.create();
    conf.set("fs.defaultFS", "hdfs://master:9000/");
    conf.set("hbase.rootdir", "hdfs://master:9000/hbase");
    conf.set("hbase.zookeeper.quorum", "master");
    String[] pathArgs = new String[] {"/mapreduce/wordcount/input"};
    Job job = Job.getInstance(conf," Data From HDFS to HBaseData From HDFS to HBase");
    job.setJarByClass(HDFSToHBase.class);
    job.setMapperClass(WriteHBaseMapper.class);
    TableMapReduceUtil.initTableReducerJob("wordcount", WriteHBaseReducer.class, job);
    job.setMapOutputKeyClass(Text.class);
    job.setMapOutputValueClass(IntWritable.class);
    job.setOutputKeyClass(Text.class);
    job.setOutputValueClass(Put.class);
    FileInputFormat.addInputPath(job, new Path(pathArgs[0]));
    System.exit(job.waitForCompletion(true) ? 0 : 1);
}
```

（5）调试 MapReduce 程序，可以查看 wordcount 表中的数据。

```
hbase>scan 'wordcount'
```

如图 5-27 所示。

```
hbase(main):006:0> scan 'wordcount'
ROW                    COLUMN+CELL
 a                     column=result:count, timestamp=1614069631357, value=4
 count                 column=result:count, timestamp=1614069631357, value=1
 data                  column=result:count, timestamp=1614069631357, value=1
 example               column=result:count, timestamp=1614069631357, value=4
 hadoop                column=result:count, timestamp=1614069631357, value=1
 is                    column=result:count, timestamp=1614069631357, value=4
 mapreduce             column=result:count, timestamp=1614069631357, value=1
 sort                  column=result:count, timestamp=1614069631357, value=1
 that                  column=result:count, timestamp=1614069631357, value=2
 this                  column=result:count, timestamp=1614069631357, value=2
 word                  column=result:count, timestamp=1614069631357, value=1
11 row(s)
```

图 5-27　HBase MapReduce 写入 HBase 数据

（6）将 MapReduce 程序导出，部署到集群并执行。

```
$ hadoop jar   ~/HDFSToHBase.jar com.bigdata.example.HDFSToHBase
```

### 2．读取 HBase 数据，将处理结果写入 HDFS

下面以统计 HBase 中 student 表中 info:name 出现的次数为例，介绍如何读取 HBase 数据，并将统计结果写入 HDFS。

读取 HBase 时，需要在 main() 函数中定义一个 Scan 对象，作为 TableMapReduceUtil.initTableMapperJob() 方法的参数。在 TableMapper 类的 map 方法中，每次读取根据 Scan 查询条件返回一条记录。

（1）在工程 HBase_MapReduce 中创建一个名为 HBaseToHDFS 的类。

（2）在 HBaseToHDFS 类中创建一个继承 TableMapper 类的内嵌类，重写 map() 函数，读取 HBase 数据，输出键值对。

```
public static class ReadHBaseMapper extends TableMapper<Text, IntWritable> {
    public void map(ImmutableBytesWritable key, Result value, Context context)
            throws IOException, InterruptedException {
        // 取出每行中的所有单元，实际上只有一列 info:name
        List<Cell> cells = value.listCells();
        for (Cell cell : cells) {
            context.write(new Text(Bytes.toString(CellUtil.cloneValue(cell))), new IntWritable(1));
        }
    }
}
```

（3）在 HBaseToHDFS 类中创建一个继承 Reducer 类的内嵌类，重写 reduce() 函数，接收 map() 函数的输出，将处理结果写入 HDFS。

```
public static class ReadHBaseReducer extends Reducer<Text, IntWritable, Text, IntWritable> {
    public void reduce(Text key, Iterable<IntWritable> values, Context context)
            throws IOException, InterruptedException {
        int sum = 0;
        for (IntWritable val : values) {
            sum += val.get();
        }
        context.write(key, new IntWritable(sum));
    }
}
```

（4）编写 HBaseToHDFS 类的 main() 函数，进行 MapReduce 作业设置。

```
public static void main(String[] args) throws Exception {
```

```java
        Configuration conf = HBaseConfiguration.create();
        conf.set("fs.defaultFS", "hdfs://master:9000/");
        conf.set("hbase.rootdir", "hdfs://master:9000/hbase");
        conf.set("hbase.zookeeper.quorum", "master");
        String[] pathArgs = new String[] {"/hbase/wordcount/output"};
        Job job = Job.getInstance(conf, "Data From HDFS to HBase");
        job.setJarByClass(HBaseToHDFS.class);
        job.setReducerClass(ReadHBaseReducer.class);
        Scan scan=new Scan();
        scan.addColumn(Bytes.toBytes("info"),Bytes.toBytes("name"));
        TableMapReduceUtil.initTableMapperJob("student",scan,ReadHBaseMapper.class,Text.class,IntWritable.class,job,false);
        job.setMapOutputKeyClass(Text.class);
        job.setMapOutputValueClass(IntWritable.class);
        job.setOutputKeyClass(Text.class);
        job.setOutputValueClass(IntWritable.class);
        FileOutputFormat.setOutputPath(job, new Path(pathArgs[0]));
        System.exit(job.waitForCompletion(true) ? 0 : 1);
    }
```

（5）调试 MapReduce 程序，可以查看 HDFS 中的输出结果。

```
$ hdfs dfs -ls /hbase/wordcount/output
$ hdfs dfs -cat /hbase/wordcount/output/part-r-00000
```

如图 5-28 所示。

```
hadoop@master:~$ hdfs dfs -ls /hbase/wordcount/output
Found 2 items
-rw-r--r--   3 hadoop supergroup          0 2021-02-23 21:29 /hbase/wordcount/output/_SUCCESS
-rw-r--r--   3 hadoop supergroup         20 2021-02-23 21:29 /hbase/wordcount/output/part-r-00000
hadoop@master:~$ hdfs dfs -cat /hbase/wordcount/output/part-r-00000
LiuYuan 1
YeShuai 2
```

图 5-28　HBase MapReduce 写入 HDFS 数据

（6）将 MapReduce 程序导出，部署到集群并执行。

```
$ hadoop jar ~/HBaseToHDFS.jar com.bigdata.example.HBaseToHDFS
```

## 5.8　HBase 学生成绩分析

### 5.8.1　任务描述

一个存放学生多门课程成绩的文件 score.txt，内部每行对应一个学生的学号和多门课程成绩，各列之间使用制表符分隔，如图 5-29 所示。要求将 score.txt 中的数据导入 HBase 的 scorelist 表中，然后利用 HBase MapReduce 程序统计每个学生的平均成绩，并将统计结果写入 HBase 的 scoreaverage 表中。

| | | | | | |
|---|---|---|---|---|---|
| 18001050101 | 76 | 81 | 83 | 81 | 91 |
| 18001050102 | 75 | 79 | 69 | 90 | 86 |
| 18001050104 | 72 | 85 | 60 | 77 | 87 |
| 18001050105 | 76 | 76 | 78 | 83 | 88 |
| 18001050106 | 70 | 76 | 40 | 73 | 84 |

图 5-29　score.txt 数据格式

## 5.8.2 导入原始数据到 HBase

（1）首先在 HBase 中用 scorelist 表来存放原始数据，然后创建 scoreaverage 表来存放统计结果。

执行以下 HBase Shell 命令：
```
hbase> create 'scorelist','score'
hbase> create 'scoreaverage','average'
```

（2）将原始数据上传到 HDFS。
```
$ hdfs dfs -mkdir -p /mapreduce/hbase/score/input
$ hdfs dfs -put ~/data/score.txt /mapreduce/hbase/score/input
```

（3）利用 HBase 提供的批量导入工具 ImportTsv 和 completebulkload 将 HDFS 文件中的数据导入 HBase 表中。

① 使用 ImportTsv 工具按 HBase 的内部数据格式生成一个临时的 HFile 文件。
```
$ /usr/local/hbase/bin/hbase org.apache.hadoop.hbase.mapreduce.ImportTsv  -Dimporttsv.bulk.output=tmp
-Dimporttsv.columns=HBASE_ROW_KEY,score:math,score:chinese,score:english,score:database,score:computer
scorelist /mapreduce/hbase/score/input
```

参数说明：

ImportTsv：一个 MapReduce 的应用名称，也是工具名称。

-Dimporttsv.bulk.output=tmp：表示生成的 HFile 文件的临时存储目录。

-Dimporttsv.columns=HBASE_ROW_KEY：表示行键。

score:math…：表示 score 列族下的列族限定符。

scorelist：表示 HBase 中的表名。

/mapreduce/hbase/score/input：表示原始数据在 HDFS 中的存储位置。

执行结果如图 5-30 所示，其中 Bad Lines=0 表示原始数据全部导入 HFile 文件。

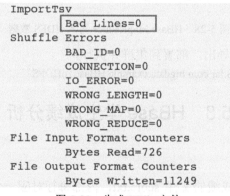

图 5-30　生成 HFile 文件

② 使用 completebulkload 工具将临时 HFile 文件加载到 HBase 的 Region 中。

使用 completebulkload 工具之前，需要先把 HBase 安装目录下的 lib 子目录的 JAR 文件复制到 Hadoop 的 common 目录下：
```
$ cp /usr/local/hbase/lib/* /usr/local/hadoop/share/hadoop/common/lib/
```
然后执行下面命令将临时文件中的数据加载到 HBase 表中：
```
hadoop jar /usr/local/hbase/lib/hbase-mapreduce-2.2.6.jar completebulkload /user/hadoop/tmp scorelist
```
参数说明：

completebulkload：工具名称。

tmp：HFile 文件所在目录名，对应 ImportTsv 中的-D importtsv.bulk.output 参数值。
scorelist：接收数据的 HBase 表名称。
数据加载成功后，可通过 HBase Shell 命令查看表中数据，验证数据导入结果。

hbase>scan 'scorelist'

执行结果如图 5-31 所示。

```
hbase(main):016:0> scan 'scorelist'
ROW                      COLUMN+CELL
 18001050101             column=score:chinese, timestamp=1614138832932, value=81
 18001050101             column=score:computer, timestamp=1614138832932, value=91
 18001050101             column=score:database, timestamp=1614138832932, value=81
 18001050101             column=score:english, timestamp=1614138832932, value=83
 18001050101             column=score:math, timestamp=1614138832932, value=76
 18001050102             column=score:chinese, timestamp=1614138832932, value=79
 18001050102             column=score:computer, timestamp=1614138832932, value=86
 18001050102             column=score:database, timestamp=1614138832932, value=90
 18001050102             column=score:english, timestamp=1614138832932, value=69
 18001050102             column=score:math, timestamp=1614138832932, value=75
 18001050104             column=score:chinese, timestamp=1614138832932, value=85
 18001050104             column=score:computer, timestamp=1614138832932, value=87
 18001050104             column=score:database, timestamp=1614138832932, value=77
 18001050104             column=score:english, timestamp=1614138832932, value=60
 18001050104             column=score:math, timestamp=1614138832932, value=72
 18001050105             column=score:chinese, timestamp=1614138832932, value=76
 18001050105             column=score:computer, timestamp=1614138832932, value=88
 18001050105             column=score:database, timestamp=1614138832932, value=83
 18001050105             column=score:english, timestamp=1614138832932, value=78
 18001050105             column=score:math, timestamp=1614138832932, value=76
```

图 5-31　数据导入执行结果

### 5.8.3　统计学生平均成绩

可以利用 HBase MapReduce 程序读取 HBase 中 scorelist 表中数据，统计每个学生的平均成绩，并将统计结果写入 scoreaverage 表中。

（1）在 Eclipse 中创建一个名为"HBase_MapReduce"的 Maven 工程。
（2）编辑工程 pom.xml 文件，添加依赖，参见 5.7.3 节介绍。
（3）创建一个名为 ScoreStatistics 的类。
（4）在 ScoreStatistics 类中创建一个继承 TableMapper<ImmutableBytesWritable, Text>的内嵌类，重写 map(ImmutableBytesWritable key, Result value,Context context)函数。

```
static class ScoreMapper extends TableMapper<ImmutableBytesWritable, Text> {
    protected void map(ImmutableBytesWritable key, Result value,
            Context context) throws IOException, InterruptedException {
        // 取出每行中的所有单元,行键与每个单元值组成键值对输出
        List<Cell> cells = value.listCells();
        for (Cell cell : cells) {
            context.write(key,new Text(Bytes.toString(CellUtil.cloneValue(cell))));
        }
    }
}
```

（5）在 ScoreStatistics 类中创建一个继承 TableReducer<ImmutableBytesWritable, Text, ImmutableBytesWritable>的内嵌类，重写 reduce(ImmutableBytesWritable key, Iterable<Text> values,Context context)函数。

```
static class ScoreReducer extends TableReducer<ImmutableBytesWritable, Text, ImmutableBytesWritable> {
    protected void reduce(ImmutableBytesWritable key, Iterable<Text> values, Context context)
            throws IOException, InterruptedException {
```

```
            int count = 0;
            int sum=0;
            int average=0;
            for (Text grade : values) {
                sum += Integer.parseInt(grade.toString());
                count++;
            }
            average=(int)sum/count;
            Put put = new Put(key.get());
            put.addColumn(Bytes.toBytes("average"), Bytes.toBytes("grade"), Bytes.toBytes(String.valueOf(average)) );
            context.write(key, put);
        }
    }
```

（6）编写 ScoreStatistics 类的 main()函数，进行 MapReduce 作业设置。创建 Scan 对象，设置要读取的数据。通过 TableMapReduceUtil.initTableMapperJob()方法设置 Map 任务读取的数据源所在表、过滤条件、Mapper 类等信息。通过 TableMapReduceUtil.initTableReducerJob()方法设置 Reduce 任务将数据写入的表、Reducer 类等信息。

```
    public static void main(String[] args) throws IOException,ClassNotFoundException, InterruptedException {
        Configuration conf = HBaseConfiguration.create();
        conf.set("fs.defaultFS", "hdfs://master:9000/");
        conf.set("hbase.rootdir", "hdfs://master:9000/hbase");
        conf.set("hbase.zookeeper.quorum", "master");
        Job job = Job.getInstance(conf, "average score");
        job.setJarByClass(ScoreStatistics.class);
        Scan scan = new Scan();
        scan.addFamily(Bytes.toBytes("score"));
        TableMapReduceUtil.initTableMapperJob("scorelist", scan, ScoreMapper.class,ImmutableBytesWritable.class, Text.class, job);
        TableMapReduceUtil.initTableReducerJob("scoreaverage", ScoreReducer.class, job);
        job.waitForCompletion(true);
    }
```

（7）调试 MapReduce 程序，可以查看 scoreaverage 表中的数据。

```
hbase>scan 'scoreaverage'
```

结果如图 5-32 所示。

```
hbase(main):026:0> scan 'scoreaverage'
ROW                           COLUMN+CELL
 18001050101                  column=average:grade, timestamp=1614150906082, value=82
 18001050102                  column=average:grade, timestamp=1614150906082, value=79
 18001050104                  column=average:grade, timestamp=1614150906082, value=76
 18001050105                  column=average:grade, timestamp=1614150906082, value=80
 18001050106                  column=average:grade, timestamp=1614150906082, value=68
 18001050107                  column=average:grade, timestamp=1614150906082, value=80
 18001050108                  column=average:grade, timestamp=1614150906082, value=81
 18001050109                  column=average:grade, timestamp=1614150906082, value=73
 18001050110                  column=average:grade, timestamp=1614150906082, value=77
 18001050111                  column=average:grade, timestamp=1614150906082, value=82
```

图 5-32 平均成绩统计结果

（8）将 MapReduce 程序导出，部署到集群并执行。

```
$ hadoop jar ~/scorestatistics.jar com.bigdata.example.ScoreStatistics
```

# 本 章 小 结

本章首先系统介绍了分布式数据库 HBase 的数据模型、体系结构、工作机制等理论知识，然后详细介绍了 HBase Shell 命令和基于 Java 的 HBase 程序设计，最后通过一个学生成绩分析案例介绍了 HBase 与 MapReduce 相融合的应用开发。通过本章的学习，读者可以了解 Hbase 的基本结构及工作机制，掌握利用 Java 语言实现对 HBase 的应用开发，利用 MapReduce 程序操作 HBase，从而实现大数据的存储、读取与数据分析。

# 思考题与习题

1. 简答题

（1）HBase 架构由哪些组件构成？每种组件有什么功能？
（2）HBase 数据库有哪些特性？
（3）简述 HBase 的概念视图和物理视图的结构。
（4）HBase 运行模式有哪些？有何不同？
（5）简述 HBase 与 MapReduce 相融合如何进行程序设计。

2. 选择题

（1）Base 数据库与传统关系数据库比较，下列说法不正确的是（  ）。
A．存储数据量大，可达数十亿行、数百万列
B．数据类型简单，数据被存储为字节数组
C．列式存储，同一列族数据集中存储在几个文件中
D．数据操作简单，支持数据的增、删、改、查
（2）下列与 HBase 特性不符的是（    ）。
A．稠密表　　　　　B．列式存储　　　　C．数据多版本　　　D．横向扩展性
（3）关于 HBase 数据模型描述不正确的是（    ）。
A．采用表来组织数据，表由行和列族组成
B．每个列族都可以有多个列族限定符
C．表中数据由行组成，每个行都有一个行键，行键可以重复
D．每个单元值都对应一个时间戳
（4）在 HBase 系统架构中，系统的协调管理器是（    ）。
A．HMaster　　　B．HRegionServer　　　C．Zookeeper　　　D．Chubby
（5）关于 HBase 客户端描述不正确的是（    ）。
A．客户端包含 HBase 访问接口
B．执行管理操作时，客户端与 HMaster 进行交互
C．执行数据读写操作时，客户端与 HregionServer 进行交互
D．执行数据修改操作时，客户端与 Hmaster 进行交互
（6）在 HBase 系统架构中，Zookeeper 的功能描述不正确的是（    ）。
A．监控 HMaster 状态　　　　　　B．保存 HBase:meta
C．监控 HRegionServer 状态　　　D．保证系统任何时刻都有一个 HMaster 运行
（7）在 HBase 系统架构中，HMaster 的描述不正确的是（    ）。

A. 执行用户对表的管理操作

B. 实现不同 HRegionServer 之间的负载均衡

C. 对发生故障失效的 HRegionServer 进行数据迁移

D. 执行用户的数据读写操作

（8）在 HBase 系统架构中，HRegionServer 的描述不正确的是（　　）。

A. 响应用户的 I/O 请求，执行 HDFS 数据读写操作

B. 管理一系列 HLog 对象，每一个 HLog 对应一个 Region 对象

C. 管理一系列 Region 对象，每一个 Region 对象对应一个 Table 表的一个 Region

D. 一台服务器上只能运行一个 HRegionServer

（9）关于用户读写数据，不正确的是（　　）。

A. 用户写入数据时，先写 MemStore，然后写入 HLog

B. 用户写入数据时，被分配到相应的 HRegionServer 上执行

C. 只有当操作写入 HLog 后，用户操作才会被提交

D. 用户读取数据时，首先访问 MemStore 缓存

（10）关于用户数据写入 HBase，下列描述不正确的是（　　）。

A. 首先定位要写入的 HRegionServer　　B. 数据先写入相应 Store 的 Memstore

C. 每个 Store 存储一个或多个列族的数据　　D. MemStore 写满后会刷入一个 StoreFile 中

### 3．实训题

（1）利用 Shell 命令完成下列操作。

① 在 HBase 中创建一个客户表，名为 customer，列族名为 info，最大版本数为 3。

② 将 customer 表 info 列族的 TTL 改为 30 天。

③ 向 customer 表中插入一行数据（cust101，2020-05-04，124.64.242.30，1），各字段分别表示 ID（rowkey）、注册日期（reg_date）、IP 地址（reg_ip）和状态（reg_status）。

④ 查询 ID 为 cust101 的客户的注册日期。

⑤ 查询 ID 为 cust101 的客户的所有信息。

（2）编写 Java 程序完成下列任务。

① 编写程序向 customer 表中写入两行数据：（cust102，2010-05-05，117.136.0.172，1）、（cust103，2010-05-06，114.94.44.230，0）。

② 编程查询 ID 为 cust103 的客户的注册信息，并输出。

③ 编程查询状态为 1 的客户的注册信息，并输出。

# 第 6 章　数据仓库 Hive

Hive 是运行在 Hadoop 集群上的数据仓库，利用 HDFS 进行数据存储、利用 MapReduce 引擎进行数据计算。Hive 将 HDFS 文件映射为表，利用 HiveQL 语句进行表的操作。本章将介绍 Hive 的理论体系与管理开发。

- Hive 基础：Hive 系统架构、工作原理、数据存储模型、数据类型及数据存储格式。
- Hive 安装与配置：MySQL 安装与配置、伪分布式 Hive 安装与配置。
- Beeline：Beeline 简介、Beeline 基本操作。
- Hive 操作：DDL 操作、DML 操作、数据查询操作。
- Hive 函数：内置 UDF、内置 UDAF、内置 UDTF。
- Hive 高级应用：Hive 用户自定义函数、Hive 与 HBase 整合。
- Hive 程序设计：利用 Java API 操作 Hive。

## 6.1　Hive 基础

### 6.1.1　Hive 简介

Hive 是基于 Hadoop 的一个数据仓库工具，是 Facebook 公司开发的一个开源项目，可以将结构化的数据文件映射为一张数据库表，并提供简单的类 SQL 查询功能，可以将类 SQL 语句转换为 MapReduce 任务进行运行。Hive 提供了一系列的工具，可以用来进行数据提取、转化、加载（ETL），是一种可以存储、查询和分析存储在 Hadoop 中的大规模数据的机制。

**1．Hive 广泛应用的原因**

Hive 在数据分析中得到广泛应用的原因在于：

① Hive 定义了简单的类 SQL 查询语言，称为 HiveQL，熟悉 SQL 的用户都可以进行数据查询。

② 通过 HiveQL 语句可以快速实现简单的 MapReduce 统计，不必开发专门的 MapReduce 应用，十分适合数据仓库的统计分析。

③ HiveQL 语言允许熟悉 MapReduce 的开发者开发自定义的 Mapper 类和 Reducer 类来处理内置的 Mapper 和 Reducer 无法完成的复杂的分析工作。

**2．Hive 特点**

运行在 Hadoop 架构之上的 Hive 具有下列特点：

① 可以自由扩展集群的规模，一般情况下不需要重启服务，具有很好的可扩展性。

② 支持用户自定义函数（UDF），用户可以根据自己的需求来实现自定义的函数，具有很好的延展性。

③ 具有良好的容错性，即使节点出现问题，HiveQL 语句仍可完成执行。

**3．Hive 的应用场景**

由于 Hive 运行在 Hadoop 架构之上，因此 Hive 适用于特定的应用场景：

① Hive 构建在基于静态批处理的 Hadoop 之上，Hadoop 通常都有较高的延迟并且在作业提交和调度时需要大量的开销。因此，Hive 不适合需要低延迟的应用。例如，Hive 不能够在大

规模数据集上实现低延迟快速的查询。

② Hive 不是为联机事务处理而设计的，Hive 不提供实时的查询和基于行级的数据更新操作。

③ Hive 的最佳使用场合是大数据集的批处理作业，如网络日志分析。

**4．Hive 与传统数据库的比较**

Hive 作为大数据环境下的数据仓库工具，和传统数据库虽然有很多相似之处，但是还是有很多不同之处。

① 数据存储格式不同：Hive 没有专门的数据存储格式，只需要在创建表时指定数据的列分隔符和行分隔符，Hive 就可以解析数据；传统数据库的数据存储格式由系统预先定义。

② 数据验证不同：Hive 在数据加载过程中不进行数据验证，而是在数据查询时才进行验证；传统数据库在数据加载时进行验证，因此 Hive 加载数据比传统数据库快。

③ DML 操作不同：Hive 不支持数据更新操作，支持批量数据导入；传统数据库支持各种 DML 操作，支持数据更新、单条或批量数据导入。

④ 延迟性不同：Hive 操作延迟性高，不适合低延迟操作；传统数据库延迟性低，适合低延迟操作。

⑤ 数据规模不同：Hive 数据存储在 HDFS 中，利用 MapReduce 进行并行计算，适合大规模数据操作；传统数据库主要采用本地文件系统存储数据，存在容量上限，在本地运行，数据处理能力有限。

## 6.1.2 Hive 系统架构

Hive 的系统架构如图 6-1 所示，由 4 部分组成：User Interface（用户接口）、MetaStore（元数据存储数据库）、Driver（驱动模块）和 Hadoop 集群。

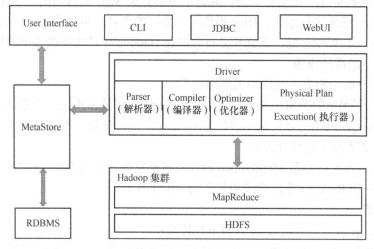

图 6-1　Hive 架构

（1）User Interface

用户接口用于与 Hive 交互，主要包括 CLI、JDBC 和 Web UI。CLI 通过控制台以命令行的方式访问 Hive，JDBC 通过 Java API 访问 Hive，Web UI 通过 Web 访问 Hive。

（2）MetaStore

Hive 中表的名字、列名、分区、表属性等信息称为 Hive 元数据，这些元数据存储在数据库中。由于 Hive 元数据可能面临不断更新、修改和读取，因此不适合使用 Hadoop 文件系统进行存储。目前，Hive 将元数据存储在关系数据库中，如 MySQL、Derby，其中 Derby 是存储于本地磁盘的 Hive 自带的数据库，Hive 的内嵌模式安装就是使用 Derby。但由于使用 Derby 数据库的内嵌模式只能允许一个会话连接，因此在实际应用中通常采用 MySQL 数据库存储 Hive 元数据，支持多用户会话。

（3）Driver

驱动模块包括 HiveQL 的解析器、编译器、优化器、执行器。解析器、编译器、优化器完成 HiveQL 语句从词法分析、语法分析、编译、优化和查询计划的生成。生成的查询计划存储在 HDFS 中，并在随后由 MapReduce 调用执行。

（4）Hadoop 集群

Hive 的数据存储在 Hadoop 的 HDFS 中，大部分的查询由 MapReduce 完成。

Hive 中元数据和业务数据的存储位置在参数文件 hive-site.xml 中设置。属性"hive.metastore.warehouse.dir"设置了业务数据在 HDFS 上的存储主目录，属性"javax.jdo.option.ConnectionURL"设置了 Hive 元数据存储的关系数据库的 URL。

在 Hive 中创建一个数据库时，系统自动在 Hive 的存储主目录下创建一个目录，目录命名为数据库名.db；在 Hive 数据库中创建表（外部表除外）时，系统自动在 Hive 的存储目录下创建与表同名的目录；向表中插入数据时，系统自动在表对应的目录下创建存储文件。

例如，在 Hive 中创建一个名为 test 的数据库，test 数据库中创建一个名为 student 的表，并向表中插入数据，此时可以看到在 Hadoop 集群中，在 Hive 的存储主目录下创建了 test.db 文件夹，在该文件夹中创建了 student 文件夹，在 student 文件夹中创建了存储数据的文件，如图 6-2 所示。

```
hadoop@master:~$ hdfs dfs -ls /user/hive/warehouse/test.db/student
Found 1 items
-rw-r--r--   1 hadoop supergroup          2 2021-03-07 21:18 /user/hive/warehouse/test.db/student/000000_0
```

图 6-2　Hive 中对象存储位置

与此同时，在 Hive 中创建的库、表的元数据信息都被存储到了关系数据库中，如 MySQL 数据库。在 MySQL 的对应库中，生成多个存储 Hive 元数据的表。其中，DBS 表保存了 Hive 中的库信息，TBLS 表保存 Hive 表的信息。通过查询这两个表，可以得到 Hive 中的库信息和表信息，如图 6-3 所示。

```
mysql> use hive;
mysql> show tables;
+---------------------------+
| Tables_in_hive            |
+---------------------------+
| AUX_TABLE                 |
| BUCKETING_COLS            |
| CDS                       |
| COLUMNS_V2                |
| COMPACTION_QUEUE          |
| COMPLETED_COMPACTIONS     |
| COMPLETED_TXN_COMPONENTS  |
| CTLGS                     |
| DATABASE_PARAMS           |
| DBS                       |
| DB_PRIVS                  |

mysql> SELECT DB_ID,NAME FROM DBS;
+-------+---------+
| DB_ID | NAME    |
+-------+---------+
|     1 | default |
|     6 | test    |
+-------+---------+
2 rows in set (0.00 sec)

mysql> SELECT tbl_id,db_id,tbl_name FROM TBLS;
+--------+-------+----------+
| tbl_id | db_id | tbl_name |
+--------+-------+----------+
|      1 |     1 | s        |
|      6 |     6 | student  |
+--------+-------+----------+
```

图 6-3　Hive 元数据信息

## 6.1.3　Hive 工作原理

用户输入 HiveQL 语句后，其执行过程如图 6-4 所示。

① 由 Hive 驱动模块中的编译器对用户输入的 HiveQL 语言进行词法和语法解析，将 HiveQL 语句转化为抽象语法树。

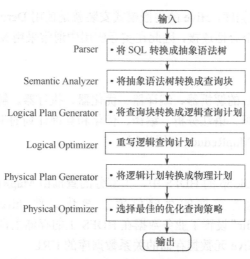

图 6-4  HiveQL 执行过程

② 由于抽象语法树的结构仍很复杂，因此，把抽象语法树转化为查询块。
③ 把查询块转换成逻辑查询计划，里面包含了许多逻辑操作符。
④ 重写逻辑查询计划，进行优化，合并多余操作，减少 MapReduce 任务数量。
⑤ 将逻辑操作符转换成需要执行的具体 MapReduce 任务。
⑥ 对生成的 MapReduce 任务进行优化，生成最终的 MapReduce 任务执行计划。
⑦ 由 Hive 驱动模块中的执行器对最终的 MapReduce 任务进行执行输出。

### 6.1.4 Hive 数据存储模型

Hive 中包含数据库（database）、内部表（Table）、外部表（External Table）、分区表（Partition）和桶（Bucket）5 种数据存储模型。

① Hive 中每个数据库对应 HDFS 中一个文件夹。

② Hive 中的内部表和传统数据库中的表在概念上是类似的，每一个内部表在 Hive 中都有一个相应的同名目录存储数据。向内部表中加载数据时，实际数据文件移动到内部表对应的文件夹中，之后对数据的访问将会直接在表对应的文件夹中完成。删除内部表时，表中的数据和元数据将会被同时删除。

③ Hive 中外部表和内部表在元数据的组织上是相同的，但实际数据的存储则不同。外部表的创建与数据加载同时完成，并不会移动数据到 Hive 的存储目录中，只是与外部数据建立了一个链接。当删除外部表时，只是删除了该链接，并不会删除实际的外部数据。

④ Hive 中的分区是对表中数据进行粗略划分的机制。表中的一个分区对应于表存储目录下的一个子目录，所有分区的数据都存储在对应的目录中。Hive 把表进行分区存储，可以加快分区的查询速度。例如，互联网应用每天都要存储大量的日志文件，日志文件必然有表示日志产生日期的属性，可以将日志文件根据日期进行分区，将每天的日志保存到一个目录中，从而可以提高特定条件的日志查询。

⑤ 桶是对表中指定列进行 hash 运算，根据 hash 值切分数据，目的是为了并行，每一个桶对应一个文件。数据加载到桶时，会对指定列取 hash 值，然后与桶的数量取模，把数据放到相应目录中。桶是对表或分区的数据进行的细粒度划分，可以获得更高的查询处理效率和更高效的取样。

## 6.1.5 Hive 数据类型

在 Hive 中创建表时，表中每个字段对应一个数据类型。Hive 数据类型的构成与传统关系数据库有所不同：传统关系数据库的字段类型设计得比较仔细，大多都有最大长度设置，方便字段数据索引映射关系的建立，加快查找速度；Hive 中所有的数据类型都是针对 Java 接口实现的，因此这些类型的具体行为细节和 Java 中对应的类型完全一致。

Hive 中的数据类型分为基本数据类型和复合数据类型。基本数据类型如表 6-1 所示。

表 6-1 Hive 基本数据类型

| 分类 | 类型 | 描述 | 示例 |
| --- | --- | --- | --- |
| 数值类型 | TINYINT | 1 字节（8 位）有符号整数 | 1 |
| | SMALLINT | 2 字节（16 位）有符号整数 | 1 |
| | INT | 4 字节（32 位）有符号整数 | 1 |
| | BIGINT | 8 字节（64 位）有符号整数 | 1 |
| | FLOAT | 4 字节（32 位）单精度浮点数 | 1.0 |
| | DOUBLE | 8 字节（64 位）双精度浮点数 | 1.0 |
| | DECIMAL | 任意精度的带符号小数 | 1.0 |
| 布尔类型 | BOOLEAN | 布尔类型，true/false | True |
| 字符类型 | CHAR | 固定长度字符串 | "hello", 'world' |
| | VARCHAR | 变长字符串 | "hello", 'world' |
| | STRING | 变长字符串 | "hello" |
| 日期/时间类型 | TIMESTAMP | 时间戳，纳秒（ns）精度 | 122327497689 |
| | DATE | 日期 | '2019-4-23' |
| 二进制类型 | BINARY | 字节数组 | [0,1,0,1,0,1,0,1] |

需要注意的是，TIMESTAMP 数据类型有 3 种取值。
① 整数：距离 UNIX 新纪元时间（1970 年 1 月 1 日零点）的秒数。
② 浮点数：距离 UNIX 新纪元时间的秒数，精确到纳秒（小数点后保留 9 位）。
③ 字符串：JDBC 所约定的时间字符串格式，格式为 YYYY-MM-DD hh:mm:ss:ffffffff。
BINARY 数据类型用于存储变长的二进制数据。
复合数据类型主要包括 ARRAY、MAP、STRUCT 和 UNIONTYPE，如表 6-2 所示。

表 6-2 Hive 复合数据类型

| 类型 | 描述 | 示例 | 元素访问方式 |
| --- | --- | --- | --- |
| ARRAY<data_type> | 一系列相同类型的数据组成 | Array(10,20) | array[index] |
| MAP<primitive_type,data_type> | 一组无序的键值对组成 | Map('a',1,'b',2) | map['key'] |
| STRUCT<col_name:data_type [COMMENT col_comment],···> | 一组不同类型的数据组成 | Struct('a',1,1,0) | struct.fieldname |
| UNIONTYPE<data_type, data_type, ···> | 在给定的任何一个时间点，保存指定数据类型中的任意一种 | UNIONTYPE< int,string> | |

### 6.1.6 Hive 数据存储格式

Hive 中所有数据都存储在 HDFS 中，没有专门的数据存储格式，支持 TEXTFILE、SEQUENCEFILE、RCFILE、ORC、PARQUET、AVRO、JSONFILE 等存储格式。其中，TEXTFILE 是默认文件格式，采用行存储方式；SEQUENCEFILE 是二进制键值对序列化文件格式，采用行存储方式；RCFILE、ORC、PARQUET 采用列存储方式。在 Hive 中创建表时，只需告诉 Hive 数据文件中列分隔符和行分隔符，Hive 就可以解析数据。

## 6.2 Hive 安装与配置

由于 Hive 是运行在 Hadoop 集群上的，因此安装 Hive 之前应保证 Hadoop 集群已经安装并启动。此外，安装的 Hive 版本需要与 Hadoop 版本相匹配。本书采用 Hive 3.1.2，与 Hadoop 3.1.2 完全兼容。

### 6.2.1 安装 MySQL

由于 Hive 默认使用内嵌的 Derby 数据库存储元数据信息，而 Derby 数据库一次只能打开一个会话，无法进行多用户并发访问，因此需要安装、使用 MySQL 数据库存储 Hive 元数据。在安装、配置 Hive 之前，需要首先安装、配置 MySQL 数据库。

（1）首先执行下面 3 条命令，在线安装 MySQL

```
$ sudo apt-get install mysql-server
$ sudo apt-get install mysql-client
$ sudo apt-get install libmysqlclient-dev
```

（2）安装成功后，可以通过下面的命令测试是否安装成功

```
$ sudo netstat -tap | grep mysql
```

出现如图 6-5 所示信息，证明安装成功。

```
hadoop@master:~$ sudo netstat -tap | grep mysql
tcp        0      0 localhost.localdo:mysql 0.0.0.0:*               LISTEN      12091/mysqld
```

图 6-5　查询 MySQL 服务

（3）连接 MySQL 数据库

在 MySQL 5.7 中，如果直接使用 root 用户登录，将被拒绝，如图 6-6 所示。

```
hadoop@master:~$ mysql -uroot -p
Enter password:
ERROR 1698 (28000): Access denied for user 'root'@'localhost'
```

图 6-6　root 用户登录失败

① 使用 sudo 命令、root 用户登录。

```
$ sudo mysql -uroot -p
```

此时不需要密码，按回车键就可以了。如图 6-7 所示。

```
hadoop@master:~$ sudo mysql -u root -p
Enter password:
Welcome to the MySQL monitor.  Commands end with ; or \g.
Your MySQL connection id is 4
Server version: 5.7.33-0ubuntu0.18.04.1 (Ubuntu)

Copyright (c) 2000, 2021, Oracle and/or its affiliates.

Oracle is a registered trademark of Oracle Corporation and/or its
affiliates. Other names may be trademarks of their respective
owners.

Type 'help;' or '\h' for help. Type '\c' to clear the current input statement.
```

图 6-7　以 sudo 命令形式连接 MySQL 数据库

② 修改 root 用户密码：

mysql> alter user 'root'@'localhost' identified by 'root';

③ 修改 plugin 字段内容为 mysql_native_password：

mysql>UPDATE mysql.user SET plugin='mysql_native_password' WHERE user='root' and host = 'localhost';

④ 刷新权限：

mysql>FLUSH PRIVILEGES；

⑤ 退出 MySQL，以 root 用户重新登录。

mysql> exit;
$ mysql -uroot -proot

（4）创建用户

mysql>CREATE USER 'hive'@'%' IDENTIFIED BY  'hive';   #创建hive用户
mysql>CREATE DATABASE hive;        #创建hive数据库
mysql>GRANT ALL ON hive.* TO'hive'@'%' IDENTIFIED BY 'hive';
mysql> FLUSH PRIVILEGES；
mysql> exit        #退出mysql
$ mysql -uhive -phive #验证hive用户

（5）修改 MySQL 配置文件，允许远程访问

编辑 MySQL 配置文件/etc/mysql/mysql.conf.d/mysqld.cnf，将 bind-address = 127.0.0.1 行注释掉，如图 6-8 所示。

$ sudo gedit /etc/mysql/mysql.conf.d/mysqld.cnf

```
[mysqld]
#
# * Basic Settings
#
user            = mysql
pid-file        = /var/run/mysqld/mysqld.pid
socket          = /var/run/mysqld/mysqld.sock
port            = 3306
basedir         = /usr
datadir         = /var/lib/mysql
tmpdir          = /tmp
lc-messages-dir = /usr/share/mysql
skip-external-locking
#
# Instead of skip-networking the default is now to listen only on
# localhost which is more compatible and is not less secure.
#bind-address           = 127.0.0.1
#
# * Fine Tuning
#
```

图 6-8  允许远程访问 MySQL 数据库

（6）重启 MySQL 服务

$ sudo service mysql restart

## 6.2.2  Hive 安装与配置过程

### 1．Hive 安装

（1）软件下载

从 Hive 官网（http://hive.apache.org/downloads.html）下载 apache-hive-3.1.2-bin.tar.gz，并上传到本地~/Downloads 目录中。

（2）解压 Hive 安装文件

$ sudo tar -zxvf    ~/Downloads/apache-hive-3.1.2-bin.tar.gz -C /usr/local

（3）修改文件夹名称与所有者

将解压后的文件夹重命名为 Hive，并修改所有者为用户 hadoop。

```
$ cd /usr/local/
$ sudo mv ./apache-hive-3.1.2-bin/ ./hive
$ sudo chown -R hadoop:hadoop ./hive
```

（4）配置环境变量，并使其生效

① 编辑~/.bashrc 文件：

```
$ gedit ~/.bashrc
```

写入下列信息：

```
export HIVE_HOME=/usr/local/hive
export PATH=$PATH:$HIVE_HOME/bin
```

② 使修改的配置文件立即生效。

```
$ source ~/.bashrc
```

2. Hive 配置

（1）添加 MySQL 驱动

下载 Hive 的 JDBC 驱动程序 mysql-connector-java-8.0.23.jar 到本地的~/Downloads 目录中，然后将该驱动程序移动到 Hive 安装目录的 lib 子目录中，命令如下：

```
$ mv ~/Downloads/mysql-connector-java-8.0.23.jar /usr/local/hive/lib/
```

（2）创建、配置 hive-site.xml 文件

```
$ cd    /usr/local/hive/conf
$ mv    hive-default.xml.template    ./hive-site.xml
$ gedit    ./hive-site.xml
```

在文件末尾的</configuration>标签之前添加下列内容：

```
<property>
        <name>hive.metastore.warehouse.dir</name>
        <value>/user/hive/warehouse</value>
</property>
<property>
        <name>javax.jdo.option.ConnectionURL</name>
        <value>jdbc:mysql://master:3306/hive?createDatabaseIfNotExist=true&amp;useUnicode=true&amp;characterEncoding=UTF-8&amp;</value>
<property>
        <name>javax.jdo.option.ConnectionDriverName</name>
        <value>com.mysql.jdbc.Driver</value>
</property>
<property>
        <name>javax.jdo.option.ConnectionPassword </name>
        <value>hive</value>
</property>
<property>
        <name>javax.jdo.option.ConnectionUserName</name>
        <value>hive</value>
</property>
```

（3）修改 hive-site.xml 文件中的配置项

将 hive-site.xml 文件中含有"system:java.io.tmpdir"的配置项的值修改为/usr/local/hive/tmp，共有 4 处，如图 6-9 所示。

```
<property>
  <name>hive.exec.local.scratchdir</name>
  <value>${system:java.io.tmpdir}/${system:user.name}</value>
  <description>Local scratch space for Hive jobs</description>
</property>
```

```
<property>
  <name>hive.exec.local.scratchdir</name>
  <value>/usr/local/hive/tmp</value>
  <description>Local scratch space for Hive jobs</description>
</property>
```

图 6-9　修改 hive-site.xml 中的配置项

（4）创建、配置 hive-env.sh 文件

```
$ cd    /usr/local/hive/conf
$ mv hive-env.sh.template    hive-env.sh
$ gedit    hive-env.sh
```

在 hive-env.sh 文件中加入下列配置信息：

```
export JAVA_HOME=/usr/lib/jvm/jdk1.8.0_231
export HIVE_HOME=/usr/local/hive
export HADOOP_HOME=/usr/local/hadoop
```

（5）修改 Hadoop 集群配置，并重启集群

由于 Hive 是基于 Hadoop 集群运行的，因此 Hive 运行用户要具有 Hadoop 集群运行用户的部分访问权限，可以在 Hadoop 集群配置文件 core-site.xml 中设置代理，代码如下：

```
<property>
    <name>hadoop.proxyuser.hadoop.groups</name>
    <value>*</value>
</property>
<property>
    <name>hadoop.proxyuser.hadoop.hosts</name>
    <value>*</value>
</property>
```

修改完 Hadoop 集群配置文件 core-site.xml 后，需要重新启动集群。

```
$ stop-all.sh
$ start-all.sh
```

（6）Hive 初始化

```
$ schematool -dbType mysql  –initSchema
```

（7）如果要远程访问 Hive，需要启动远程服务 HiveServer 2

```
$ hiveserver2 &
```

（8） 打开控制台命令行工具

① 在 Hive 2.1 之前使用 HiveCLI（将被淘汰）：

```
$ hive
```

② 在 Hive 2.1 之后使用 Beeline：

```
$ beeline -u jdbc:hive2://master:10000 -n hadoop -p hadoop
```

连接成功，如图 6-10 所示。

```
hadoop@master:~$ beeline -u jdbc:hive2://master:10000/ -n hadoop -p hadoop
SLF4J: Class path contains multiple SLF4J bindings.
SLF4J: Found binding in [jar:file:/usr/local/hive/lib/log4j-slf4j-impl-2.10.0.jar!/org/slf4j/impl/
StaticLoggerBinder.class]
SLF4J: Found binding in [jar:file:/usr/local/hadoop/share/hadoop/common/lib/slf4j-log4j12-1.7.25.j
ar!/org/slf4j/impl/StaticLoggerBinder.class]
SLF4J: See http://www.slf4j.org/codes.html#multiple_bindings for an explanation.
SLF4J: Actual binding is of type [org.apache.logging.slf4j.Log4jLoggerFactory]
Connecting to jdbc:hive2://master:10000/
Connected to: Apache Hive (version 3.1.2)
Driver: Hive JDBC (version 3.1.2)
Transaction isolation: TRANSACTION_REPEATABLE_READ
Beeline version 3.1.2 by Apache Hive
0: jdbc:hive2://master:10000/>
```

图 6-10　Beeline 连接 Hive

# 6.3 Beeline

## 6.3.1 Beeline 简介

在 Hive 2.1 之前版本中，使用的控制台命令行工具是 HiveCLI，命令为：

```
$ HIVE_HOME/bin/hive
```

由于 HiveCLI 只能操作本地 Hive 服务，且不支持多用户并发操作，因此 HiveServer 2 引入了新的控制台命令行工具 Beeline。Beeline 是基于 SQLLine CLI 的 JDBC 客户端，通过 JDBC 连接远程服务，并支持多用户、安全控制。Hive 2.1 以后版本中使用 Beeline 替代 HiveCLI，并会废弃 HiveCLI 客户端工具。因此，本书使用 Beeline 作为 Hive 的控制台命令行工具，连接并操作 Hive。

由于 Beeline 基于 JDBC 远程访问 Hive，因此，使用 Beeline 需要两个前提条件：

① 在 Hadoop 的配置文件 core-site.xml 中设置了代理，并重启了 Hadoop 集群；

② 启动了 HiveServer2 远程服务。

关于 core-site.xml 文件配置详见 6.2.2 节中的介绍。HiveServer 2 启动有下列两种方式：

```
$ hive --service  hiveserver2 &
$ hiveserver2 &
```

&表示 hiveserver2 作为守护进程在后台运行。HiveServer 2 的默认端口为 10000，可以启动时设置端口：

```
$ nohup hiveserver2 --hiveconf hive.server2.thrift.port=10010 &
```

## 6.3.2 Beeline 基本操作

### 1. Beeline 命令参数

可以通过 Beeline --help 查看 Beeline 命令的参数信息，如图 6-11 所示。

```
hadoop@master:~$ beeline --help
  -u <database url>               the JDBC URL to connect to
  -c <named url>                  the named JDBC URL to connect to,
                                  which should be present in beeline-site.xml
                                  as the value of beeline.hs2.jdbc.url.<namedUrl>
  -r                              reconnect to last saved connect url (in conjunction with !save)
  -n <username>                   the username to connect as
  -p <password>                   the password to connect as
  -d <driver class>               the driver class to use
  -i <init file>                  script file for initialization
  -e <query>                      query that should be executed
  -f <exec file>                  script file that should be executed
  -w (or) --password-file <password file>  the password file to read password from
  --hiveconf property=value       Use value for given property
  --hivevar name=value            hive variable name and value
                                  This is Hive specific settings in which variables
                                  can be set at session level and referenced in Hive
                                  commands or queries.
  --property-file=<property-file> the file to read connection properties (url, driver, user, password)
  --color=[true/false]            control whether color is used for display
  --showHeader=[true/false]       show column names in query results
  --escapeCRLF=[true/false]       show carriage return and line feeds in query results as escaped \r an
  --headerInterval=ROWS;          the interval between which heades are displayed
  --fastConnect=[true/false]      skip building table/column list for tab-completion
  --autoCommit=[true/false]       enable/disable automatic transaction commit
  --verbose=[true/false]          show verbose error messages and debug info
  --showWarnings=[true/false]     display connection warnings
  --showDbInPrompt=[true/false]   display the current database name in the prompt
  --showNestedErrs=[true/false]   display nested errors
  --numberFormat=[pattern]        format numbers using DecimalFormat pattern
```

图 6-11 Beeline 命令的参数信息

Beeline 命令的常用参数及其示例如表 6-3 所示。

例如，连接 Hive 服务器，并查询数据。

```
$ beeline -u jdbc:hive2://master:10000/default  -n hadoop -p hadoop -e "select * from test.student"
```

表 6-3　Beeline 命令的常用参数及其示例

| 参数名称 | 描述 | 示例 |
| --- | --- | --- |
| -u &lt;database url&gt; | 用于 JDBC URL 连接 | beeline -u db_URL |
| -r | 重新连接到最近使用过的 URL | beeline -r |
| -n &lt;username&gt; | 连接时使用的用户名 | beeline -n valid_user |
| -p &lt;password&gt; | 连接时使用的密码 | beeline -p valid_password |
| -d &lt;driver class&gt; | 配置使用的驱动类 | beeline -d driver_class |
| -e &lt;query&gt; | 应执行的查询。查询语句两端用单引号或双引号 | beeline -e "query_string" |
| -f &lt;exec file&gt; | 需要被执行的脚本文件 | beeline -f filepath |
| -w 或--password-file &lt;password file&gt; | 从文件中读取密码 | |
| --hiveconf property=value | 为给定的配置属性赋值 | beeline --hiveconf prop1=value1 |
| hivevar name=value | Hive 的变量名和变量值 | beeline --hivevar var1=value1 |

## 2．Beeline 管理命令

启动 Beeline 后，可以执行一些管理命令，如连接、中断、退出等。执行 Beeline 管理命令需要以"！"开始，不需要终止符。常用的 Beeline 管理命令及其描述如表 6-4 所示。

表 6-4　常用的 Beeline 管理命令及其描述

| 命令 | 描述 | 命令 | 描述 |
| --- | --- | --- | --- |
| !all | 在所有的连接上执行指定的 SQL 语句 | !autocommit | 打开自动提交模式 |
| !batch | 开始或执行一个批处理语句 | !brief | 关闭 verbose 模式 |
| !close | 关闭当前到数据库的连接 | !closeall | 关闭所有到数据库的连接 |
| !columns | 显示指定表的所有列 | !commit | 提交当前事务 |
| !connect | 打开一个到数据库的新连接 | !dbinfo | 返回数据库的元数据信息 |
| !delimiter | 设置查询语句分隔符，默认为"；" | !describe | 描述一个表 |
| !dropall | 删除当前数据库中的所有表 | !help | 打印所有命令的汇总信息 |
| !indexes | 显示指定表的所有索引 | !list | 显示所有连接 |
| !metadata | 获取元数据信息 | !quit | 退出 Shell |
| !reconnect | 重建到数据库的连接 | !record | 将输出记录写入指定文件中 |
| !set | 设置 beeline 变量 | !sql | 执行一个 SQL 命令 |
| !tables | 显示当前数据库的所有表 | !verbose | 显示查询追加的明细 |

例如，可以执行!help 查看 Beeline 的管理命令，如图 6-12 所示。

```
0: jdbc:hive2://master:10000/test> !help
!addlocaldriverjar  Add driver jar file in the beeline client side.
!addlocaldrivername Add driver name that needs to be supported in the beeline
                    client side.
!all                Execute the specified SQL against all the current connections
!autocommit         Set autocommit mode on or off
!batch              Start or execute a batch of statements
!brief              Set verbose mode off
!call               Execute a callable statement
!close              Close the current connection to the database
!closeall           Close all current open connections
!columns            List all the columns for the specified table
!commit             Commit the current transaction (if autocommit is off)
!connect            Open a new connection to the database.
```

图 6-12　查看 Beeline 的管理命令

## 3. Beeline Hive 命令

Beeline 连接到 Hive 服务器后，可以执行 Hive 命令，命令以分号";"结束。
常用的 Beeline Hive 命令及其描述如表 6-5 所示。

表 6-5 常用的 BeelineHive 命令及其描述

| 命令 | 描述 | 命令 | 描述 |
| --- | --- | --- | --- |
| add | 向分布式缓存的资源列表中添加文件等 | reset | 重设配置项的默认值 |
| delete | 从分布式缓存中删除资源 | set <key>=<value> | 设置指定的配置变量值 |
| list | 显示所有添加到分布式缓存中的资源 | dfs <dfs command> | 执行 DFS 命令 |
| set -v | 打印所有 Hadoop 和 Hive 配置变量 | <query string> | 执行 Hive 查询语句 |
| set | 打印所有被重写的配置变量 | | |

例如，可以执行 dfs 命令操作 HDFS，执行 HiveQL 语句操作 Hive，如图 6-13 所示。

```
0: jdbc:hive2://master:10000> dfs -ls / ;
+----------------------------------------------------+
|                     DFS Output                     |
+----------------------------------------------------+
| Found 2 items                                      |
| drwx-wx-wx   - hadoop supergroup   0 2021-03-07 14:55 /tmp  |
| drwxr-xr-x   - hadoop supergroup   0 2021-03-07 17:18 /user |
+----------------------------------------------------+
3 rows selected (0.043 seconds)
0: jdbc:hive2://master:10000> show databases;
+----------------+
| database_name  |
+----------------+
| default        |
| test           |
+----------------+
2 rows selected (1.667 seconds)
```

图 6-13 BeelineHive 命令应用示例

## 6.4 Hive DDL 操作

HiveQL 是 Hive 提供的类似 SQL 的语言，与大部分 SQL 语法兼容，但是并不完全支持 SQL 标准，例如，子查询、连接查询、事务处理、数据更新等都有较多限制，这是由于 Hive 底层依赖于 Hadoop 平台的特性决定的。HiveQL 使得不熟悉 Hive，但熟练掌握 SQL 的开发人员可以快速使用 Hive 进行应用开发。

与 SQL 语句类似，HiveQL 也分为 DDL、DML、DQL、UDF 等几类。其中，DDL（Data Definition Language）用于数据库、表、视图等对象的创建与管理；DML（Data Manipulation Language）用于数据加载、插入、更新、删除等操作；DQL（Data Select Language）用于数据查询；UDF（User-Defined Functions）包括内置函数（Built-in Functions）、UDAF（Built-in Aggregate Functions）、UDTF（Built-in Table-Generating Functions）和用户自定义函数，用于辅助数据查询。

### 6.4.1 Hive 数据库管理

#### 1. 创建数据库

在 Hive 中使用 CREATE DATABASE 语句创建数据库，语法为：

CREATE DATABASE [IF NOT EXISTS] database_name

[COMMENT database_comment]
[LOCATION hdfs_path]
[WITH DBPROPERTIES (property_name=property_value, …)]

参数说明：

**IF NOT EXISTS**：如果同名数据库已经存在，则放弃创建数据库操作，从而避免因为要创建的数据库已经存在而抛出异常。

**COMMENT**：设置新建数据库的描述信息。

**LOCATION**：设置数据库数据的存储位置。默认在 Hive 的存储主目录下新建一个目录，目录命名为"数据库名.db"。

**WITH DBPROPERTIES**：设置数据库的属性。

例如，创建一个名为 dbhive 的数据库。

```
0: jdbc:hive2://master:10000> create database if not exists dbhive
                            > comment 'This a hive test database!'
                            > with dbproperties('creator'='hive','date'='2021-3-12');
```

### 2．查看数据库信息

数据库创建后，可以使用 DESCRIBE 语句查看数据库信息，包括数据库的名称、描述和存储位置信息等。如果使用 EXTENDED 关键字，还可以查看数据库的属性信息。DESCRIBE 语句的语法为：

DESCRIBE DATABASE [EXTENDED]database_name

例如，查看 dbhive 数据库的信息：

DESCRIBE DATABASE EXTENDED dbhive;

查询结果如图 6-14 所示。

图 6-14　查看数据库信息

### 3．查看 Hive 中所有数据库

可以使用 SHOW DATABASES 语句查看 Hive 中有哪些数据库，可以通过正则表达式查询满足条件的数据库。SHOW DATABASES 语法为：

SHOW DATABASES [LIKE 'identifier_with_wildcards'];

例如，查看名字以字母 d 开头的数据库：

0: jdbc:hive2://master:10000> SHOW DATABASES LIKE 'd.*';

### 4．修改数据库

数据库创建后，可以使用 ALTER DATABASE…SET 语句修改数据库属性、存储位置等信息，语法为：

ALTER DATABASE database_name SET DBPROPERTIES (property_name=property_value, …)
ALTER DATABASE database_name SET LOCATION hdfs_path;

### 5．使用数据库

如果要对某个数据库进行操作，需要使用 USE 语句切换到相应的数据库，默认使用的是 DEFAULT 数据库。

例如，切换到 dbhive 数据库：

0: jdbc:hive2://master:10000> use dbhive;

### 6. 查看当前使用的数据库

可以通过 current_database() 函数返回当前使用的数据库名称。例如：

0: jdbc:hive2://master:10000> SELECT current_database();

### 7. 删除数据库

如果要删除某个数据库，可以使用 DROP DATABASE 语句，语法为：

DROP DATABASE [IF EXISTS] database_name [RESTRICT|CASCADE]

参数说明：

**IF EXISTS**：如果要删除的数据库不存在，则不执行，避免删除不存在的数据库时系统抛出异常。

**RESTRICT**：默认选项。如果数据库非空，则删除操作失败。

**CASCADE**：删除数据库及其所有对象。

## 6.4.2 Hive 表管理

### 1. 创建表

在 Hive 中使用 CREATE TABLE 语句创建表，语法为：

CREATE [TEMPORARY] [EXTERNAL] TABLE [IF NOT EXISTS] [db_name.]table_name
[(col_name data_type [column_constraint_specification] [COMMENT col_comment], …
constraint_specification] )]
[COMMENT table_comment]
[PARTITIONED BY (col_name data_type [COMMENT col_comment], …)]
[CLUSTERED BY (col_name, col_name, …) [SORTED BY (col_name [ASC|DESC], …)] INTO num_buckets BUCKETS]
[ROW FORMAT row_format]
[STORED AS file_format]
[LOCATION hdfs_path]
[TBLPROPERTIES (property_name=property_value, …)]
[AS select_statement]
[LIKE existing_table_or_view_name]

参数说明：

**TEMPORARY**：创建临时表。临时表仅对当前会话可见，临时表的数据将暂时存储在用户的暂存目录中，并在会话结束后删除。

**EXTERNAL**：默认创建的是内部表，指定 EXTERNAL 关键字创建一个外部表。外部表在创建时必须同时指定一个指向实际数据的路径，由 LOCATION 关键字指定。

**IF NOT EXISTS**：如果指定名称的表不存在，则创建该表；如果该表已经存在，则语句不执行，避免系统抛出异常。

**data_type**：列的数据类型，详见 6.1.5 节中的介绍。

**COMMENT**：为表字段或者表添加注释说明。

**constraint_specification**：设置列或表约束，语法为：

[, PRIMARY KEY (col_name, …) DISABLE NOVALIDATE RELY/NORELY ]
[, PRIMARY KEY (col_name, …) DISABLE NOVALIDATE RELY/NORELY ]
[, CONSTRAINT constraint_name FOREIGN KEY (col_name, …) REFERENCES table_name(col_name, …) DISABLE NOVALIDATE
[, CONSTRAINT constraint_name UNIQUE (col_name, …) DISABLE NOVALIDATE RELY/NORELY ]

[, CONSTRAINT constraint_name CHECK [check_expression] ENABLE|DISABLE NOVALIDATE RELY/NORELY ]

PARTITIONED BY：创建分区表，指定分区字段，多个字段列之间用逗号分隔，相互之间是从属关系。

CLUSTERED BY：创建桶，指定将表或分区进一步组织成桶时所用的列。

SORTED BY：是一个高效的归并排序，按指定列进行升序或降序排序。

INTO…BUCKETS：指定将表或分区划分的桶数。

ROW FORMAT row_format：指定表对应的数据行的格式。row_format 语法结构为：

DELIMITED [FIELDS TERMINATED BY char [ESCAPED BY char]]
[COLLECTION ITEMS TERMINATED BY char]
[MAP KEYS TERMINATED BY char]
[LINES TERMINATED BY char]
[NULL DEFINED AS char]
|SERDE serde_name [WITH SERDEPROPERTIES (property_name=property_value, property_name=property_value, …)]

- FIELDS TERMINATED BY：指定字段之间的分隔符。
- ESCAPED BY：定义转移字符。
- COLLECTION ITEMS TERMINATED BY：指定集合类型中元素分隔符。
- MAP KEYS TERMINATED BY：指定 MAP 类型中键与值的分隔符。
- LINES TERMINATED BY：指定行分隔。
- NULL DEFINED AS：指定空值的表示。
- SERDE：设置表序列化方式。

STORED AS file_format：指定表数据的存储格式。file_format 语法为：

SEQUENCEFILE | TEXTFILE | RCFILE | ORC | PARQUET | AVRO | JSONFILE
| INPUTFORMAT input_format_classname OUTPUTFORMAT output_format_classname

LOCATION：指定表中数据对应的存储位置，对外部表是必需的。

TBLPROPERTIES：用于设置表的属性。

AS：根据子查询的结果来创建表。

LIKE：基于已有表或视图的结构创建表，即新建的表与已有表或视图结构完全相同。

例如，有一个保存员工信息的文件 empinfo.txt，数据格式如图 6-15 所示，字段间以 Tab 键分隔。

```
姓名    工资    爱好                     电话                                    住址
Tom     5000    football,running,movie   tel:84832283,mobile:16740947707 softpark,dalian,china,116023
Marry   6000    watching TV,table tennis  tel:84832267,mobile:16723489760 redroad,shenyang,china,110011
Jack    7000    football                 tel:84832283,mobile:16740947707 softpark,dalian,china,116034
```

图 6-15 文件中数据格式

可以创建一个与文件对应的表，以加载文件中的数据：

CREATE TABLE employees (
  name    string,
  salary  double,
  hobby   array<string>,
  phone   map<string,string>,
  address struct<street:string,city:string,state:string,zip:int> )
ROW FORMAT DELIMITEDFIELDS TERMINATED BY  '\t'
COLLECTION ITEMS TERMINATED BY ','MAP KEYS TERMINATED BY ':'
LINES TERMINATED BY  '\n';

文本格式文件是以逗号或制表符分隔的文本文件。Hive 支持这两种文件格式，但是需要注意文本文件中的那些不需要作为分隔符处理的逗号或制表符。因此，Hive 默认使用了如表 6-6 所示文本文件分隔符，这些字符很少出现在字段值中。

表 6-6 Hive 文本文件分隔符

| 分隔符 | 描述 |
| --- | --- |
| \n | 针对文本文件，每行都是一条记录，换行符可以分隔记录 |
| ^A(Ctrl+A) | 用于分隔字段（列）。在 CREATE TABLE 语句中使用八进制编码\001 表示 |
| ^B | 用于分隔 ARRAY 或 STRUCT 中的元素，或用于 MAP 中键和值之间的分隔。在 CREATE TABLE 语句中使用八进制编码\002 表示 |
| ^C | 用于 MAP 中键和值之间的分隔。在 CREATE TABLE 语句中使用八进制编码\003 表示 |

例如，文本文件格式如图 6-16 所示。

```
姓名      工资        爱好                                电话                         住址
Tom^A5000^Afootball^Brunning^Bmovie^Atel^C84832283^Bmobile^C16740947707^Asoftpark^Bdalian^Bchina^B116023
Marry^A6000^Awatching TV^Btable tennis^Atel^C84832267^Bmobile^C16723489760^Aredroad^Bshenyang^Bchina^B110011
Jack^A7000^Afootball^Atel^C84832283^Bmobile^C16740947707^Asoftpark^Bdalian^Bchina^B116034
```

图 6-16 文本文件格式

此时，可以按下列语句创建 Hive 表。

```
CREATE TABLE employees2 (
  name      string,
  salary    double,
  hobby     array<string>,
  phone     map<string,string>,
  address struct<street:string,city:string,state:string,zip:int> )
ROW FORMAT DELIMITED FIELDS TERMINATED BY '\001'
COLLECTION ITEMS TERMINATED BY '\002'
MAP KEYS TERMINATED BY '\003'
LINES TERMINATED BY   '\n' STORED AS TEXTFILE;
```

（1）创建内部表

创建内部表时会在默认存储目录中创建一个与表同名的文件夹，加载数据时数据文件移动到与表同名的目录中。内部表的创建与数据加载是两个独立的过程。内部表与文件是一体的，删除表时，数据文件也一起删除。

例如，创建一个保存员工信息的内部表 inner_emp 和保存部门信息的 inner_dept 表。

```
USER dbhive;
CREATE TABLE IF NOT EXISTS inner_emp(
  empno int COMMENT '员工号',
  ename char(20) COMMENT '员工名',
  job string COMMENT '职位',
  mgr int COMMENT '领导员工号',
  hiredate string COMMENT '入职日期',
  sal float COMMENT '工资',
  comm float COMMENT '奖金',
  deptno int COMMENT '部门号')
ROW FORMAT DELIMITED FIELDS TERMINATED BY ',' STORED AS TEXTFILE;

CREATE TABLE IF NOT EXISTS inner_dept(
```

```
    deptno int COMMENT '部门号',
    dname string COMMENT '部门名',
    manager_id   int COMMENT '部门领导号码')
    ROW FORMAT DELIMITED FIELDS TERMINATED BY ',' STORED AS TEXTFILE;
```

（2）创建外部表

创建外部表只是创建一个外部文件的链接，并不移动数据文件。外部表的创建与数据加载是同时完成的。当删除一个外部表时，仅删除表自身的元数据及与外部文件的链接，数据文件不受影响。创建外部表时，需要通过 location 参数指定外部文件所在的目录。

例如，创建一个外部表 outet_emp。

```
CREATE EXTERNAL TABLE IF NOT EXISTS oxt_emp(
    empno int ,ename char(20),job string,mgr int,hiredate string,sal float,comm float,deptno int)
    ROW FORMAT DELIMITED FIELDS TERMINATED BY ','
    LOCATION '/user/hive/warehouse/ext_data';
```

（3）创建分区表

分区表是将数据根据特定条件分散存储在不同的目录中，每个分区对应一个存储目录。分区列不是表中的一个实际字段，而是一个或多个伪列，在表的数据文件中实际上并不保存分区列的信息与数据。

例如，创建一个分区表 part_emp，根据员工工资等级进行分区。

```
CREATE TABLE IF NOT EXISTS part_emp(
    empno int ,ename char(20),job string,mgr int,hiredate string,sal float,comm float,deptno int)
    PARTITIONED BY (salary_level string)
    ROW FORMAT DELIMITED FIELDS TERMINATED BY ',' STORED AS TEXTFILE;
```

（4）创建桶

创建桶是通过对指定列进行 hash 运算实现的，通过 hash 值将一个列名下的数据切分为一组桶，并使每个桶对应于该列名下的一个存储文件。

例如，创建一个桶 bucket_emp，将员工信息根据员工号分散存储到 5 个桶中。

```
CREATE TABLE IF NOT EXISTS bucket_emp(
    empno int ,ename char(20),job string,mgr int,hiredate string,sal float,comm float,deptno int)
    CLUSTERED BY(empno) SORTED BY(empno ASC ) INTO 5 BUCKETS
    ROW FORMAT DELIMITED FIELDS TERMINATED BY ',' STORED AS TEXTFILE;
```

（5）利用子查询创建表

利用子查询创建表是基于查询结果创建表，注意不能指定列名，也不能创建外部表。

例如，查询 inner_emp，将结果保存到一个新表中。

```
CREATE TABLE as_emp
AS
SELECT empno,ename,sal,deptno FROM inner_emp;
```

（6）利用已存在表创建表

可以基于已有表的结构创建一个新表，新表中没有数据。例如，创建一个与 inner_emp 结构完全相同的表：

```
CREATE TABLE like_emp LIKE inner_emp;
```

2．查看表的列信息

表创建后，可以使用 DESCRIBE 语句查看表的列信息，语法为：

```
DESCRIBE [EXTENDED | FORMATTED][db_name.]table_name
[PARTITION partition_spec] [col_name ( [.field_name] | [.'$elem$'] | [.'$key$'] | [.'$value$'] )* ];
```

参数说明：

EXTENDED：以 Thrift 序列形式显示表的所有元数据。

FORMATTED：以 tabular 形式显示表的所有元数据。

例如：

0: jdbc:hive2://master:10000> DESCRIBE EXTENDED inner_emp;

0: jdbc:hive2://master:10000> DESCRIBE EXTENDED part_emp PARTITION (salary_level='low');

3．查看表信息

可以使用 SHOW 语句查看表、列、分区的元数据信息。

① 查看当前数据库中的所有表：

SHOW TABLES ['identifier_with_wildcards'];

② 查看表分区：

SHOW PARTITIONS [db_name.]table_name [PARTITION(partition_spec)];

③ 查看表分区详细信息：

SHOW TABLE EXTENDED LIKE 'identifier_with_wildcards' [PARTITION(partition_spec)];

④ 查看表定义信息：

SHOW CREATE TABLE ([db_name.]table_name|view_name);

例如：

0: jdbc:hive2://master:10000> SHOW TABLES;

0: jdbc:hive2://master:10000> SHOW PARTITIONS part_emp PARTITION(salary_level='low');

0: jdbc:hive2://master:10000> SHOW TABLE EXTENDED like 'part_emp' PARTITION(salary_level='low');

0: jdbc:hive2://master:10000> SHOW CREATE TABLE part_emp;

4．修改表

表创建后，可以使用 ALTER TABLE 语句修改表，包括修改表名、修改约束、修改分区等。为了方便介绍修改表的操作，先创建一个名为 alter_example 的表。

CREATE TABLE alter_example(sno string,sname char(10),sage int,brithdate date)
PARTITIONED BY(class_id int)
ROW FORMAT DELIMITED FIELDS TERMINATED BY ','STORED AS TEXTFILE;

（1）重命名表

语法：

ALTER TABLE table_name RENAME TO new_table_name;

示例：

ALTER TABLE alter_example RENAME TO alterExample;

（2）修改表属性

语法：

ALTER TABLE table_name SET TBLPROPERTIES (property_name = property_value, … )

示例：

ALTER TABLE alterExample SET TBLPROPERTIES('editor'='sfd','date'='2021-3-14');

（3）修改表约束

语法：

● ALTER TABLE table_name ADD CONSTRAINT constraint_name PRIMARY KEY (column, …) DISABLE NOVALIDATE;

● ALTER TABLE table_name ADD CONSTRAINT constraint_name FOREIGN KEY (column, …) REFERENCES table_name(column, …) DISABLE NOVALIDATE RELY;

● ALTER TABLE table_name DROP CONSTRAINT constraint_name;

示例：
- ALTER TABLE alterExample ADD CONSTRAINT pk PRIMARY KEY(sno) DISABLE NOVALIDATE;
- ALTER TABLE alterExample DROP CONSTRAINT pk;

（4）为表添加分区

语法：

ALTER TABLE table_name ADD [IF NOT EXISTS] PARTITION partition_spec
[LOCATION 'location'][, PARTITION partition_spec [LOCATION 'location'], …];

示例：

ALTER TABLE alterExample ADD PARTITION(class_id=5)
LOCATION '/user/hive/warehouse/alterExample/class_id=5';

（5）重命名分区

语法：

ALTER TABLE table_name PARTITION partition_spec RENAME TO PARTITION partition_spec;

示例：

ALTER TABLE alterExample PARTITION(class_id=5) RENAME TO PARTITION(class_id=6);

（6）移动分区

移动分区是指将一个表中一个或多个分区中的数据移动到另一个具有相同模式的表中。

语法：

ALTER TABLE table_name_2 EXCHANGE PARTITION (partition_spec, partition_spec2, …)
WITH TABLE table_name_1;

示例：

ALTER TABLE alterExample2 EXCHANGE PARTITION(class_id=6) WITH TABLE alterExample;

（7）删除分区

语法：

ALTER TABLE table_name DROP [IF EXISTS]
PARTITION partition_spec[, PARTITION partition_spec, …] [PURGE];

示例：

ALTER TABLE alterExample DROP IF EXISTS PARTITION(class_id=6);

（8）添加列

语法：

ALTER TABLE table_name ADD COLUMNS (col_name data_type [COMMENT col_comment], …)

示例：

ALTER TABLE alterExample
ADD COLUMNS (address string COMMENT 'home address infomation');

（9）修改列

语法：

ALTER TABLE table_name CHANGE [COLUMN] col_old_name  col_new_name  column_type
[COMMENT col_comment] [FIRST|AFTER column_name] [CASCADE|RESTRICT];

示例：

ALTER TABLE alterExample CHANGE sno student_id string;
ALTER TABLE alterExample CHANGE sage student_age string AFTER student_id;
ALTER TABLE alterExample CHANGE sname student_name string FIRST;

（10）删除或替换列

语法：

ALTER TABLE table_name
REPLACE COLUMNS (col_name    data_type [COMMENT col_comment], …)

示例：

ALTER TABLE alterExample REPLACE COLUMNS(student_age string);
ALTER TABLE alterExample REPLACE COLUMNS(student_age string,sage int);

5．删除表

可以使用 DROP TABLE 语句删除表及其元素。语法为：

DROP TABLE [IF EXISTS] table_name [PURGE];

如果没有指定 PURGE 关键字，表中数据被转移到 Trash/Current 目录中，数据可以恢复；如果指定了 PURGE 关键字，数据被删除，无法恢复。如果删除的是外部表，那么文件系统中的数据不会被删除，只有表的元数据被删除。为了避免删除不存在的表导致系统抛出异常，可以使用 IF EXISTS 关键字。

### 6.4.3 视图管理

#### 1．创建视图

与关系数据库类似，Hive 中也支持视图的应用。可以使用 CREATE VIEW 语句创建视图，语法为：

CREATE VIEW [IF NOT EXISTS] [db_name.]view_name
[(column_name [COMMENT column_comment], …) ]
[COMMENT view_comment]
[TBLPROPERTIES (property_name = property_value, …)]
AS SELECT …;

创建视图时，可以指定字段名称及其描述，但不能指定数据类型。如果不指定名称，则采用子查询的字段名称。

例如，创建一个统计各个部门员工人数与平均工资的视图。

CREATE VIEW IF NOT EXISTS emp_view
AS
SELECT deptno,avg(sal) avgsal,count(*) num FROM emp GROUP BY deptno;

#### 2．修改视图

视图创建后，可以使用 ALTER VIEW 语句修改视图，包括视图属性修改、视图定义修改等。ALTER VIEW 语法为：

ALTER VIEW [db_name.]view_name
SET TBLPROPERTIES (property_name = property_value, property_name = property_value, …);
ALTER VIEW [db_name.]view_name AS select_statement;

#### 3．删除视图

可以使用 DROP VIEW 语句删除视图，语法为：

DROP VIEW [IF EXISTS] [db_name.]view_name;

## 6.5　Hive DML 操作

Hive DML 用于 Hive 表中数据加载、插入、更新、删除、合并等操作。

## 1. 加载数据

在 Hive 中，可以在创建表的同时加载数据，也可以在表创建后加载数据。可以使用 Load 命令将数据加载到 Hive 表中，语法为：

LOAD DATA [LOCAL] INPATH 'filepath'
[OVERWRITE] INTO TABLE tablename [PARTITION (partcol1=val1, partcol2=val2 …)]

参数说明：

LOCAL：如果使用该参数，说明要加载的数据文件是本地文件；如果没有指定该参数，说明要加载的是 HDFS 文件。

INPATH：指明要加载的文件名或文件所在的目录。如果是目录，则加载该目录中的所有数据文件。

OVERWRITE：如果使用该参数，在数据加载时先清空表中原有数据，然后进行加载，即数据覆盖；如果没有指定该参数，采用数据追加方式加载数据。

PARTITION：如果要将数据加载到分区表中，需要使用 PARTITION 参数指明数据要加载的分区名称。

Load 命令只是进行简单的复制或移动操作，将数据文件移动到 Hive 表对应的 HDFS 存储目录中。

**例 1**：将本地系统/home/hadoop/hivedata 目录中的 emp.txt 文件加载到 inner_emp 表中，将 dept.txt 文件加载到 inner_dept 表中，将 empinfo.txt 文件加载到 employees 表中。

LOAD DATA LOCAL INPATH '/home/hadoop/hivedata/emp.txt' INTO TABLE inner_emp;
LOAD DATA LOCAL INPATH '/home/hadoop/hivedata/dept.txt' INTO TABLE inner_dept;
LOAD DATA LOCAL INPATH '/home/hadoop/hivedata/empinfo.txt' INTO TABLE employees;

数据加载成功后，可以查看表对应存储目录中的文件信息，如图 6-17 所示。

```
hadoop@master:~$ hdfs dfs -ls /user/hive/warehouse/dbhive.db/inner_emp
Found 1 items
-rw-r--r--   1 hadoop supergroup       5326 2021-03-15 14:18 /user/hive/warehouse/dbhive.db/inner_emp/emp.txt
hadoop@master:~$ hdfs dfs -ls /user/hive/warehouse/dbhive.db/inner_dept;
Found 1 items
-rw-r--r--   1 hadoop supergroup        568 2021-03-15 14:18 /user/hive/warehouse/dbhive.db/inner_dept/dept.txt
hadoop@master:~$ hdfs dfs -ls /user/hive/warehouse/dbhive.db/employees;
Found 1 items
-rw-r--r--   1 hadoop supergroup    9500127 2021-03-19 17:01 /user/hive/warehouse/dbhive.db/employees/empinfo.txt
```

图 6-17 内部表加载数据后的存储结构

**例 2**：将本地系统/home/hadoop/hivedata 目录中的 emp_low.txt 文件加载到 part_emp 表的 salary_level=low 分区，将 emp_high.txt 文件加载到 part_emp 表的 salary_level=high 分区。

LOAD DATA LOCAL INPATH '/home/hadoop/hivedata/emp_low.txt' OVERWRITE INTO TABLE part_emp PARTITION(salary_level='low');
LOAD DATA LOCAL INPATH '/home/hadoop/hivedata/emp_high.txt' OVERWRITE INTO TABLE part_emp PARTITION(salary_level='high');

数据加载成功后，可以查看分区表对应存储目录中的文件信息，如图 6-18 所示。

```
:~$ hdfs dfs -ls -R /user/hive/warehouse/dbhive.db/part_emp
drwxr-xr-x   - hadoop supergroup          0 2021-03-15 14:45 /user/hive/warehouse/dbhive.db/part_emp/salary_level=high
-rw-r--r--   1 hadoop supergroup       1657 2021-03-15 14:45 /user/hive/warehouse/dbhive.db/part_emp/salary_level=high/emp_high.txt
drwxr-xr-x   - hadoop supergroup          0 2021-03-15 14:45 /user/hive/warehouse/dbhive.db/part_emp/salary_level=low
-rw-r--r--   1 hadoop supergroup       3811 2021-03-14 11:44 /user/hive/warehouse/dbhive.db/part_emp/salary_level=low/emp_low
-rw-r--r--   1 hadoop supergroup       3811 2021-03-15 14:45 /user/hive/warehouse/dbhive.db/part_emp/salary_level=low/emp_low.txt
```

图 6-18 分区表加载数据后的存储结构

**例 3**：将 HDFS 中/user/hadoop/empinfo.txt 文件加载到 employees 表中。

HDFS DFS -put /home/hadoop/hivedata/empinfo.txt /home/hadoop
LOAD DATA INPATH '/user/hadoop/empinfo.txt' INTO TABLE employees;

## 2. 利用子查询插入数据

除使用 Load 命令直接加载数据文件外，还可以采用 INSERT 命令将查询的结果插入一个或多个 Hive 表中。

(1) 单表插入

利用 INSERT 语句将查询结果插入一个 Hive 表中，语法为：

INSERT OVERWRITE | INTO TABLE tablename1 [PARTITION (partcol1=val1, partcol2=val2 …) ] select_statement1 FROM from_statement;

参数说明：

OVERWRITE：覆盖插入，首先删除表或分区中的原有数据，然后进行数据插入。

INTO：追加插入，保留原有数据。

例 4：将内部表 inner_emp 中的员工号、员工名及部门号信息插入外部表 ext_emp 中。

INSERT INTO TABLE ext_emp( empno,ename,deptno) SELECT empno,ename,deptno FROM inner_emp;

例 5：将 inner_emp 表中工资范围为 5000～10000 元的员工信息插入分区表 part_emp 的一个名为 salary_level=middle 的分区中，如果该分区有数据，则覆盖原有数据。

INSERT INTO TABLE part_emp PARTITION(salary_level='middle')
SELECT * FROM inner_emp WHERE sal>5000 and sal<10000;

(2) 多表插入

Hive 支持多表插入操作，即将一个表的查询结果插入多个表中，只需要将 FROM 短语放在 INSERT 关键字之前指明数据来源表即可。多表插入的语法为：

FROM from_statement
INSERT OVERWRITE |INTO TABLE tablename1 [PARTITION (partcol=val …)] select_statement1
[INSERT OVERWRITE | INTO TABLE tablename2 [PARTITION …] select_statement2]
[INSERT OVERWRITE | INTO TABLE tablename2 [PARTITION …] select_statement2] …;

例 6：查询内部表 inner_emp，将 20 号部门员工信息插入外部表 ext_emp 中，将工资低于 3000 元的员工信息插入分区表 part_emp 的 salary_level=low 的分区中，分别采用覆盖插入方式和追加插入方式进行。

采用覆盖插入方式，语句为：

FROM inner_emp
INSERT OVERWRITE TABLE ext_emp SELECT * WHERE deptno=20
INSERT OVERWRITE TABLE part_emp PARTITION(salary_level='low') SELECT * WHERE sal<3000;

采用追加插入方式，语句为：

FROM inner_emp
INSERT INTO TABLE ext_emp SELECT * WHERE deptno=20
INSERT INTO TABLE part_emp PARTITION(salary_level='low') SELECT * WHERE sal<3000;

(3) 动态分区插入

往 Hive 分区表中插入数据时，如果需要创建的分区很多，则需要执行多个插入操作，每次指定一个要插入的分区名，这样效率非常低。由于 Hive 是批处理系统，因此 Hive 提供了一个动态分区功能，基于查询语句参数的位置去推断分区的名称，从而建立分区。通常，将查询语句最后一个字段作为分区列，值作为分区名称。

在 Hive 中有几个与动态分区相关的参数，需要进行相应的设置。

hive.exec.dynamic.partition：开启动态分区功能，默认值为 true。

hive.exec.dynamic.partition.mode：默认值为 strict，表示只有当分区表中创建一个静态分区后才可以动态创建其他分区。当参数值为 nonstrict 时，允许所有分区都是动态创建的。

hive.exec.max.dynamic.partitions.pernode：每个 Mapper 或 Reducer 可以动态创建的最大分区数量，默认值为 100。

hive.exec.max.dynamic.partitions：可以创建的最大动态分区个数，默认值为 1000。

hive.exec.max.created.files：在一个 MapReduce 作业中，所有 Mapper 和 Reducer 可以创建的 HDFS 文件的最大数量。

动态分区插入与静态数据插入的语法类似，不同点在于 PARTITON 短语中可以只指定分区列名（子查询最后的列），而无须指定完整的分区名称。

INSERT OVERWRITE | INTO TABLE tablename PARTITION (partcol1, partcol2 …)
select_statement FROM from_statement;

**例 7**：创建一个分区表，并实现数据动态分区插入。

① 创建一个分区表 multi_part_insert，以 deptno 列为分区列。

CREATE TABLE   IF NOT EXISTS dynamic_part_insert (empno int,ename string,sal double)
PARTITIONED BY(deptno int) ROW FORMAT DELIMITED FIELDS TERMINATED BY ','
STORED AS TEXTFILE;

② 设置参数，启动 Hive 的动态分区功能，允许分区表的所有分区可以动态创建。

SET hive.exec.dynamic.partition=true;
SET hive.exec.dynamic.partition.mode=nonstrict;

③ 利用 INSERT INTO TABLE 语句进行动态分区数据插入。

INSERT INTO TABLE dynamic_part_insert PARTITION(deptno)
SELECT empno,ename,sal,deptno FROM inner_emp;

④ 数据插入完成后，使用 SHOW PARTITIONS 命令，可以看到该表自动分区情况。

SHOW PARTITIONS dynamic_part_insert;

结果如图 6-19 所示。

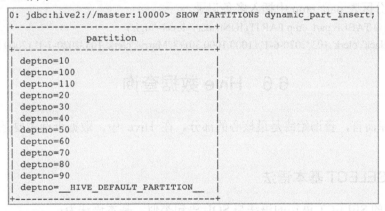

图 6-19  动态分区创建结果

### 3．将查询结果写入文件

在 Hive 中，可以使用 INSERT…DIRECTORY 语句将查询结果写入一个或多个目录中，可以是本地目录，也可以是 HDFS 目录。可以通过 STORED AS 短语设置文件的存储格式，通过 ROW FORMAT 短语设置文件中行的格式。

（1）将查询结果写入一个文件

将查询结果写入一个目录的 INSERT…DIRECTORY 语句的语法为：

INSERT OVERWRITE [LOCAL] DIRECTORY directory1
[ROW FORMAT row_format] [STORED AS file_format]
SELECT … FROM …

**例 8**：将 inner_emp 表中工资大于 5000 元的员工信息导出到本地/home/hadoop/emp_output 目录中。

INSERT OVERWRITE LOCAL DIRECTORY '/home/hadoop/hivedata/emp_output'
SELECT * FROM inner_emp WHERE sal>5000;

（2）将查询结果写入多个文件

将查询结果写入多个目录的 INSERT…DIRECTORY 语句的语法为：

FROM from_statement
INSERT OVERWRITE [LOCAL] DIRECTORY directory1 select_statement1
[INSERT OVERWRITE [LOCAL] DIRECTORY directory2 select_statement2] …

**例 9**：将分区表 part_emp 中分区名为 salary_level='high'和 salary_level='low'的数据分别导入 HDFS 的当前目录中的 highsalary 和 lowsalay 中。

FROM part_emp
INSERT OVERWRITE DIRECTORY '/user/hadoop/highsalary'
ROW FORMAT DELIMITED FIELDS TERMINATED BY '\t'SELECT * WHERE salary_level='high'
INSERT OVERWRITE DIRECTORY '/user/hadoop/lowsalary'
ROW FORMAT DELIMITED FIELDS TERMINATED BY '\t'SELECT * WHERE salary_level='low';

**4．利用 INSERT…VALUES 语句插入数据**

可以使用 INSERT INTO TABLE…VALUES 语句向表或分区中插入一条或多条数据。语法为：

INSERT INTO TABLE tablename [PARTITION (partcol1[=val1], partcol2[=val2] …)]
VALUES ( value [, value …] ) [( value [, value …] ) …]

**例 10**：向 inner_emp 表中插入一条记录。

INSERT INTO TABLE inner_emp VALUES(1,'TOM','clerk',100,'2020-5-1',6900,0,20);

**例 11**：向分区表 part_emp 中插入多条记录。

INSERT INTO TABLE part_emp PARTITION(salary_level='high')
VALUES(2,'Jack','clerk',103,'2020-6-1',11000,1000,30),(3,'Marry','clerk',103,'2020-7-1',12000,500,40);

## 6.6 Hive 数据查询

对于数据库而言，查询始终是最核心的部分。在 Hive 中，数据查询主要是通过 SELECT 语句进行的。

### 6.6.1 Hive SELECT 基本语法

HiveQL 中的 SELECT 语句的语法与 SQL 语句类似，基本语法为：

SELECT [ALL | DISTINCT] select_expr, select_expr, …
FROM table_reference
[WHERE where_condition]
[GROUP BY col_list [HAVING condition]]
[SORT | ORDER BY col_list]
[CLUSTER BY col_list | [DISTRIBUTE BY col_list] [SORT BY col_list]]
[LIMIT [offset,] rows]

参数说明：

DISTINCT：去除查询结果中的重复值，默认为 ALL，保留所有结果。

FROM：指定从哪个表或哪些表中查询。

WHERE：指定查询表中数据的过滤条件。

GROUP BY：对查询的行数据进行分组。

HAVING：对分组后形成的组进行筛选。

ORDER BY：对查询结果进行全局排序，可以升序（ASC）或降序（DESC）排序。

DISTRIBUTE BY：将具有相同键（key）值的记录分发到同一个 Reducer 上进行处理。

SORT BY：对进入 Reducer 的数据进行局部排序，保证每个 Reducer 输出是有序的。

CLUSTER BY：除具有 DISTRIBUTE BY 的功能外，还对局部数据按指定字段进行升序排序。因此，如果 DISTRIBUTE BY 与 SORT BY 使用的字段相同，可以使用 CLUSTER BY=DISTRIBUTE BY+SORT BY。

LIMIT：指定查询返回的行数，可以通过 offset 设定偏移量，即从哪一行开始返回指定数量的数据。

与 SQL 语句类型，HiveQL 的查询语句分为无条件查询、有条件查询、分组查询、嵌套查询、连接查询等多种形式。

## 6.6.2 无条件查询

在 SELECT 语句中，SELECT 子句后的目标列可以是表中的所有列、部分列，也可以是表达式，包括算术表达式、字符串常量、函数等。

（1）查询所有列

如果查询表或视图中所有列的数据，可以用 "*" 表示目标列。

例 12：查询所有员工的信息。

SELECT * FROM inner_emp;

（2）查询指定列

如果查询特定列的数据，可以在目标列中列出相应列名，用逗号分隔。

例 13：查询员工的员工号、员工名及其部门号。

SELECT empno,ename,deptno FROM inner_emp;

（3）使用列别名

可以为查询的目标列或目标表达式起别名。别名紧跟在列名后，或者在列名和别名之间加入关键字 AS。

例 14：查询员工号、员工名、工资，并为输出列起别名。

SELECT empno employee_id,ename AS employee_name,deptno AS department_name FROM inner_emp;

（4）使用表达式

如果需要对查询目标列进行计算，那么可以在目标列表达式中使用表达式。表达式中可以使用的算术运算符如表 6-7 所示。

表 6-7 算术运算符

| 运算符 | 说明 | 运算符 | 说明 | 运算符 | 说明 | 运算符 | 说明 | 运算符 | 说明 |
| --- | --- | --- | --- | --- | --- | --- | --- | --- | --- |
| + | 加 | - | 减 | * | 乘 | / | 除 | % | 求余 |
| DIV | 取整 | & | 位与 | \| | 位或 | ^ | 异或 | ~ | 位非 |

例 15：查询员工工资与奖金的和。

SELECT empno,sal+commtotal FROM inner_emp;

（5）使用字符常量

可以在查询结果中加入字符串，用单引号（''）引起来。

例 16：查询员工号与员工工资，在工资列之前加入 "salary is:"。

SELECT empno, 'salary is:', sal FROM inner_emp;

（6）使用函数

可以在目标列表达式中采用函数对查询结果进行运算。

**例 17**：查询员工号、员工名及员工入职年份。

SELECT empno,ename,year(hiredate) FROM inner_emp;

（7）使用连接字符串

可以使用"||"运算符将查询的目标列或目标表达式连接起来。

**例 18**：查询员工号、员工名、员工值，结果以一个字符串输出。

SELECT  '员工号：'||empno ||'员工名'：||ename||'工资：sal   FROM inner_emp;

（7）消除重复行

如果不希望在查询结果中出现重复记录，可以使用 DISTINCT 语句。

**例 19**：查询有员工的部门号。

SELECT DISTINCT deptno FROM inner_emp;

（8）限制返回的行数

可以使用 LIMIT 关键字限制返回的行数。Limit 后包含两个参数，第一个参数指定偏移量，第二个参数指定返回的函数，如果没有指定偏移量，则从第一条记录开始返回指定的函数。

**例 20**：返回第 3 个到第 7 个员工的信息。

SELECT * FROM inner_emp LIMIT 2,5;

### 6.6.3 有条件查询

在 HiveQL 的 SELECT 语句中，可以通过 WHERE 子句进行数据行的筛选。在 WHERE 子句中使用的运算符如表 6-8 所示。

表 6-8　WHERE 子句中使用的运算符

| 查询条件 | 运算符 |
| --- | --- |
| 关系运算 | =，==，>，<，>=，<=，<>，!= |
| 确定范围 | BETWEEN… AND，NOT BETWEEN… AND |
| 确定集合 | IN，NOT IN |
| 字符匹配 | LIKE，NOT LIKE |
| 空值判断 | IS NULL，IS NOT NULL |
| 正则表达式 | RLIKE |
| 逻辑操作 | NOT，AND，OR |

（1）关系运算

在查询条件中可以使用简单的关系运算符。

**例 21**：查询 10 号部门之外的其他部门的员工号、员工名及员工工资的信息。

SELECT empno,ename,sal FROM inner_empWHERE deptno!= 10;

**例 22**：查询工资大于 5000 元的员工的员工号、员工名及员工工资的信息。

SELECT empno,ename,sal FROM inner_empWHERE sal> 5000;

（2）确定范围

在 WHERE 条件表达式中可以使用 BETWEEN…AND 指定特定的范围（包括边界），也可以用 NOT BETWEEN…AND 指定在特定范围之外（不包括边界）。

**例 23**：查询工资大于或等于 5000 元，并且小于或等于 12000 元的员工信息。

SELECT * FROM inner_emp WHERE sal BETWEEN 5000 AND 12000;

例 24：查询工资小于 5000 元，或者工资大于 12000 元的员工信息。

SELECT * FROM inner_emp WHERE sal NOT BETWEEN 5000 AND 12000;

（3）确定集合

如果查询条件中涉及多个等于或不等于运算，可以使用 IN 或 NOT IN 运算符。

例 25：查询 10、20、30、50 号部门的员工的员工号、员工姓名及员工工资信息。

SELECT empno,ename,sal FROM inner_empWHERE deptno IN(10,20,30,50);

例 26：查询 50 和 90 号部门之外的其他部门的员工号、员工姓名及员工工资信息。

SELECT empno,ename,sal FROM inner_empWHERE deptno NOT IN(50,90);

（4）字符匹配

如果要进行模糊查询，可以在查询条件中使用 LIKE、NOT LIKE、RLIKE 运算符。RLIKE 是 LIKE 的功能扩展，可以通过正则表达式来指定匹配条件进行查询。

为了实现模糊查询，HiveQL 使用 "%"（百分号）和 "_"（下画线）两个通配符。

%：代表 0 个或多个字符。

_：代表任意一个字符。

例 27：查询名字中含有 "S" 的员工信息。

SELECT * FROM inner_emp WHERE ename LIKE '%S%';

例 28：查询名字的第二个字母为 "a" 的员工信息。

SELECT * FROM inner_emp WHERE ename LIKE '_a%';

例 29：查询名字以字母 M 开头，以字母 s 结尾的员工信息。

SELECT * FROM inner_emp WHERE ename RLIKE '^M.*s$';

（5）空值判断

如果要判断列或表达式的结果是否为空，则需要使用 IS NULL 或 IS NOT NULL 运算符，切记不能使用 "=" 进行判断。

例 30：查询没有奖金的员工的信息。

SELECT * FROM inner_emp WHERE comm IS NULL;

例 31：查询有奖金的员工的信息。

SELECT * FROM inner_emp WHERE comm IS NOT NULL;

（6）逻辑操作

在进行数据过滤时，有时会有多个条件，如果是或的关系，需要使用 OR 连接起来；如果是与的关系，需要使用 AND 连接起来。需要注意的是，AND 的优先级高于 OR，如果需要，可以用括号改变运算的优先级。

例 32：查询 10 号部门中工资高于 1400 元的员工信息。

SELECT * FROM inner_emp WHERE deptno=10 AND sal >1400;

例 33：查询工资高于 1400 元的 10 号部门和 20 号部门的员工信息。

SELECT * FROM inner_empWHERE (deptno=10 OR deptno=20) AND sal>1400;

### 6.6.4 查询统计

在数据查询过程中，经常涉及对查询信息的统计。对查询信息的统计通常使用内置的聚集函数实现。表 6-9 列出了 Hive 中常用的聚集函数。

表 6-9 Hive 中常用的聚集函数

| 函数 | 格式 | 功能 |
| --- | --- | --- |
| COUNT | COUNT([DISTINCT] *) | 返回结果集中记录总个数（包含空值） |
| COUNT | COUNT([DISTINCT] expr) | 返回结果集中非空记录个数 |
| AVG | AVG([DISTINCT] column) | 返回组中列平均值 |
| MAX | MAX(column) | 返回组中列最大值 |
| MIN | MIN(column) | 返回组中列最小值 |
| SUM | SUM([DISTINCT] column) | 返回组中列总和 |
| STDDEV | STDDEV(column) | 返回列的标准差 |
| VARIANCE | VARIANCE(column) | 返回列的方差 |

使用聚集函数时，需要注意以下几点。

① 除 COUNT(*)函数外，其他的聚集函数都不考虑返回值为 NULL 的情况。

② 聚集函数只能出现在查询结果表达式、ORDER BY 子句、HAVING 子句中，不能出现在 WHERE 子句和 GROUP BY 子句中。

③ 默认对组中所有行的指定列进行统计，包括重复值；如果要统计不重复的信息，则可以使用 DISTINCT 选项。

④ 如果对查询结果进行了分组，则聚集函数的作用范围为各个组，否则聚集函数作用于整个返回结果。

**例 34**：统计 50 号部门员工的人数、平均工资、最高工资、最低工资。

SELECT count(*),avg(sal),max(sal),min(sal) FROM inner_empWHERE deptno=50;

**例 35**：统计所有员工的平均工资和工资总额。

SELECT avg(sal),sum(sal) FROM inner_emp;

**例 36**：统计有员工的部门的个数。

SELECT count(DISTINCT deptno) FROM inner_emp;

**例 37**：统计员工工资的方差和标准差。

SELECT variance(sal),stddev(sal) FROM inner_emp;

在分组统计执行过程中，可以通过参数 hive.map.aggr 控制聚合操作如何进行。hive.map.aggr 默认值为 FALSE。如果为 TRUE，则在执行过程中首先在 Map 端进行聚合操作，这样会提高分组聚合的效率，但会消耗更多的内存资源。

**例 38**：统计不同部门员工人数，采用 Map 端聚合。

SET hive.map.aggr=TRUE;
SELECT deptno,count(*) num FROM inner_emp GROUP BY deptno;

### 6.6.5 分组查询

如果需要对查询的数据进行分组统计，可以使用 GROUP BY 子句，可以基于一列或多列进行分组。分组查询的基本语法为：

SELECT expression [, expression…)] FROM src
GROUP BY groupByExpression [, groupByExpression…]
HAVING groupCondition

**例 39**：查询不同部门不同职位的员工人数与平均工资。

SELECT deptno,job,count(*),avg(sal) FROM inner_empGROUP BY deptno,job;

进行分组查询时，需要注意下列事项：

① GROUP BY 子句用于指定分组列或分组表达式。

② 聚集函数用于对分组进行统计。如果未对查询结果分组，则聚集函数将作用于整个查询结果；如果对查询结果分组，则聚集函数对每个组进行一次统计。

③ 在分组查询中，SELECT 子句后面的所有列名或表达式要么是分组列，要么是分组表达式，要么是聚集函数。

如果需要对分组后的数据进一步筛选，即对组进行筛选，可以使用 HAVING 子句，只有满足条件的组才会返回。

**例 40**：查询平均工资高于 7000 元的部门号及部门平均工资。

SELECT deptno,avg(sal) FROM inner_emp GROUP BY deptno HAVING avg(sal)>7000;

使用 HAVING 子句，需要注意下列事项：

① HAVING 子句用于限制分组的返回结果。

② WHERE 子句对表中的记录进行过滤，而 HAVING 子句对分组后的组进行过滤。

可以将统计分析的结果保存到多个表或 HDFS 文件中。

**例 41**：将 inner_emp 表中各个部门的部门号和部门人数信息写入 dept_sum 表中，将不同职位的名称及员工数写入 HDFS 文件/user/hadoop/emp/job_sum。

FROM inner_emp
INSERT OVERWRITE TABLE dept_sum
　　SELECT inner_emp.deptno, count(DISTINCT inner_emp.empno)
　　GROUP BY inner_emp.deptno
INSERT OVERWRITE DIRECTORY '/user/hadoop/emp/job_sum'
　　SELECT inner_emp.job, count(DISTINCT inner_emp.empno)
　　GROUP BY inner_emp.job;

### 6.6.6　子查询

Hive 只支持在 FROM 短语和 WHERE 子句中使用子查询。

（1）FROM 短语中的子查询

在 FROM 短语中使用子查询时，子查询必须起名字，同时子查询中列也要有唯一的列名。

**例 42**：查询工资与奖金总和大于 10000 元的员工信息。

SELECT * FROM (SELECT empno,ename,job,sal+comm total,deptno FROM emp)new_emp
WHERE total>10000;

（2）WHERE 子句中的子查询

WHERE 子句中的子查询可以作为 IN、NOT IN 语句中的常量处理。

**例 43**：查询工资与 10 号部门某个员工工资相同的员工的信息。

SELECT * FROM inner_emp WHERE job IN (SELECT job FROM inner_emp WHERE deptno=10);

如果 WHERE 子句需要判断子查询是否有返回结果，可以使用 EXISTS 或 NOT EXISTS 关键词。如果子查询返回结果，EXISTS 返回 TRUE，NOT EXISTS 返回 FALSE；如果子查询没有返回结果，NOT EXISTS 返回 TRUE，EXISTS 返回 FALSE。

**例 44**：查询有员工的部门的信息。

SELECT * FROM inner_dept WHERE EXISTS (
SELECT 1 FROM inner_emp WHERE deptno=inner_dept.deptno);

### 6.6.7 连接查询

HiveQL 支持连接操作,包括内连接、外连接、笛卡儿积连接等,基本语法为:
table_reference [INNER] JOIN table_factor [join_condition]
| table_reference {LEFT|RIGHT|FULL} [OUTER] JOIN table_reference join_condition
| table_reference CROSS JOIN table_reference [join_condition]

(1)内连接

内连接是将连接的表中符合连接条件的数据查询出来,通过 INNER JOIN 关键字连接表,INNER 可以省略,默认为内连接。

**例 45**:查询工资大于 5000 元的员工及其部门信息。
SELECT e.empno,e.ename,e.deptno,d.dname
FROM inner_emp e JOIN inner_dept d ON e.deptno=d.deptno WHERE sal>5000;

只有满足连接条件的行才发生内连接。有时希望得到符合连接条件的数据外,还希望得到部分不符合连接条件的数据,为此引入了外连接。

(2)外连接

外连接的结果集是内连接结果集的超集。根据额外增加数据的来源不同,外连接分为左外连接(LEFT [OUTER]JOIN)、右外连接(RIGHT [OUTER]JOIN)和全外连接(FULL [OUTER]JOIN)。

左外连接:在内连接的基础上将 JOIN 关键字左侧表中的所有数据加入结果集中,右侧对应字段值为空。

右外连接:在内连接的基础上将 JOIN 关键字右侧表中的所有数据加入结果集中,左侧对应字段值为空。

全外连接:在内连接的基础上,将 JOIN 关键字两侧表中不符合连接条件的数据都加入结果集中。

**例 46**:查询所有员工及其部门信息,包括不属于任何部门的员工信息。
SELECT e.empno,e.ename,d.dname,e.deptno
FROM inner_emp e LEFT JOIN inner_dept d ON e.deptno=d.deptno;

**例 47**:查询所有部门及其员工的信息,包括没有任何员工的部门信息。
SELECT e.empno,e.ename,d.dname,e.deptno
FROM inner_emp e RIGHT JOIN inner_dept d ON e.deptno=d.deptno;

**例 48**:查询所有部门、所有员工的信息。
SELECT e.empno,e.ename,d.dname,e.deptno
FROM inner_emp e FULL JOIN inner_dept d ON e.deptno=d.deptno;

(3)笛卡儿积连接

笛卡儿积连接又称交叉连接,是两个或多个表之间的无条件连接,使用关键字 CROSS JOIN。当省略连接条件、连接条件无效时,都会发生笛卡儿积连接,所有表中的所有行都相互连接。

**例 49**:将 inner_emp 表的所有行分别与 inner_dept 表的所有行进行连接。
SELECT e.*,d.* FROM inner_emp e CROSS JOIN inner_dept d;

(4)多表连接

多表连接时,可以使用多个 JOIN…ON 进行连接操作,例如:
SELECT a.val, b.val, c.val FROM a JOIN b ON (a.key = b.key1) JOIN c ON (c.key = b.key2)

多表连接的执行过程为:先启动一个 Job1,执行 A 表和 B 表的连接操作,生成一个结果

Result1；然后启动一个 Job2，执行 Result1 与 C 表的连接，生成最终结果。需要注意的是，当对 3 个或更多表进行 JOIN 操作时，如果每个 ON 子句都使用相同的连接列，则只会产生一个 MapReduce 作业。

**例 50**：查询员工及其领导的员工号、员工名及部门名。

```
SELECT e.empno ,e.ename,m.empno,m.ename, d.dname
FROM inner_emp e JOIN inner_emp m ON e.mgr=m.empno JOIN inner_dept d ON e.deptno=d.deptno;
```

该连接的执行将启动两个 MapReduce 作业，原因是两个 ON 条件采用不同的连接列。

在表连接的每个 Map/Reduce 阶段，连接序列中最后一个表被 Reducer 流式处理，而其他表被缓存。这样，可以减少 Reducer 缓存特定键值的行数据所需的内存空间大小，方法就是将数据量最大的表放在连接序列的最后。也可以通过命令提示/*+ STREAMTABLE() */指定需要流式处理的表。

**例 51**：表 A、表 B、表 C 进行连接操作，将表 A 序列化处理。

```
SELECT /*+ STREAMTABLE(a) */ a.val, b.val, c.val
FROM a JOIN b ON (a.key = b.key1) JOIN c ON (c.key = b.key1)
```

由于连接条件相同，因此将启动一个 MapReduce 作业，表 B 和表 C 中数据被缓存在 Reducer 内存中，从表 A 中返回的每条记录分别与缓存中数据进行计算。

（5）Map 连接

如果一个连接表小到足以放入内存，Hive 就可以把该表放入 Map 端的内存中执行连接操作，这称为 Map 连接。Map 连接不需要使用 Reducer，使用命令指示 /*+ MAPJOIN(b) */指定小表。

**例 52**：查询员工及其部门信息，部门表很小。

```
SELECT /*+ MAPJOIN(b) */ e.empno,e.ename,e.deptno,d.dname
FROM inner_emp e JOIN inner_dept d ON e.deptno=d.deptno;
```

需要注意的是，全外连接、右外连接需要使用 Reducer，因此不是 Map 连接。

如果要连接的表在连接上是桶，并且一个表中的桶数是另一个表中桶数的倍数，那么这些桶可以相互连接。例如，表 A 有 4 个桶，表 B 有 4 个桶，执行下面的连接：

```
SELECT /*+ MAPJOIN(b) */ a.key, a.value FROM a JOIN b ON a.key = b.key
```

执行过程中，不是为表 A 的每个 Map 端操作获取所有的表 B，而只是返回必须的桶。处理表 A 的第一个桶时，只需获取表 B 的第一个桶。这不是默认行为，需要进行下列参数设置：

```
SET hive.optimize.bucketmapjoin = true;
```

## 6.6.8 排序

在 HiveQL 中，排序包括 ORDER BY（全局排序）、SORT BY（局部排序）和 DISTRIBUTE BY…SORT BY（分区排序）3 种。

（1）ORDER BY

ORDER BY 短语对查询结果集执行一个全局排序，所有数据通过一个 Reducer 进行处理，对大规模数据集而言，执行效率非常低。ORDER BY 短语语法为：

```
SELECT expression [, expression…] FROM src
ORDER BY colName [ ASC | DESC][NULLS FIRST | NULLS LAST] [, colName …];
```

可以按升序（ASC）或降序（DESC）排序，默认为 ASC。空值可以放在最前面（NULLS FIRST），也可以放在最后面（NULLS LAST）。可以对一列或多列进行排序。

**例 53**：查询员工的员工号、员工工资信息，按工资升序排序。

```
SELECT empno,sal FROM inner_emp ORDER BY sal;
```

**例 54**：查询员工信息，按员工所在部门号升序、工资降序排序。
SELECT * FROM inner_emp ORDER BY deptno,sal DESC;

（2）SORT BY

SORT BY 短语用于对每个 Reducer 端数据进行排序，即局部排序。SORT BY 短语语法为：
SELECT expression [, expression…] FROM src SORT BY colName [ ASC | DESC] [, colName …];

**例 55**：查询员工信息，按员工入职日期升序、工资降序排序。
SELECT * FROM inner_emp SORT BY hiredate,sal DESC;

如果 Reducer 数量为 1，则 SORT BY 与 ORDER BY 结果相同。可以通过 mapreduce.job.reduces 参数设置 Reducer 的个数。例如：
SET mapreduce.job.reduces=3;

（3）DISTRIBUTE BY

在有些情况下，需要控制某个特定行应该到哪个 Reducer（通常是为了进行后续的聚集操作），可以使用 DISTRIBUTE BY 子句进行排序。DISTRIBUTE BY 可以控制在 Map 端如何拆分数据给 Reducer，根据其后的列及 Reducer 个数进行数据分发，默认采用 Hash 算法。

DISTRIBUTE BY 控制 Reducer 如何接收一行行数据，SORT BY 则控制着 Reducer 内的数据是如何进行排序的。

**例 56**：查询所有员工信息，按工资降序排序，确保同一个部门的员工数据最终都在一个 Reducer 分区中。
SELECT * FROM inner_emp DISTRIBUTE BY deptno SORT BY sal DESC;

（4）CLUSTER BY

如果在 DISTRIBUTE BY 短语和 SORT BY 短语中使用的列完全相同，并且按列值升序排序，则可以使用 CLUSTER BY 方式。

**例 57**：查询所有员工信息，员工号升序排序，确保同一个部门的员工数据最终都在一个 Reducer 分区中。以下两种写法效果是等价的：
SELECT * FROM inner_emp DISTRIBUTE BY deptno SORT BY deptno;
SELECT * FROM inner_emp CLUSTER BY deptno;

## 6.6.9 合并操作

在 Hive 中可以使用 UNION 短语将两个或多个子查询结果进行合并，语法为：
select_statement UNION [ALL | DISTINCT] select_statement UNION [ALL | DISTINCT] select_statement …

默认情况下，重复记录将被从结果集中移除，UNION ALL 将保留所有的重复记录，UNION DISTINCT 与默认情况相同。需要注意的是，每个子查询都必须具有相同的列，而且对应每个列的数据类型必须一致。

**例 58**：查询 50 号部门的员工号、工资和部门号及工资大于 8000 元的所有员工的员工号、工资和部门号，重复记录保留一次。
SELECT empno,sal,deptno FROM inner_emp WHERE deptno=50
UNION ALL
SELECT empno,sal,deptno FROM inner_emp WHERE sal>8000;

## 6.6.10 复合类型数据查询

Hive 表中复合类型数据的引用方式如表 6-2 所示，ARRAY 通过下标引用、MAP 通过键值引用、STRUCT 通过字段引用，例如，array[index]、map['key']、struct.fieldname。

**例 59**：查询员工名、员工的第一个爱好、员工的移动电话、员工所在城市信息。

SELECT name,hobby[0],phone['mobile'],address.city FROM employees;

可以通过复合类型数据的构造函数创建复合类型数据，如表 6-10 所示。

表 6-10 复合类型数据的构造函数

| 构造函数 | 操作数 | 说明 |
| --- | --- | --- |
| map | (key1, value1, key2, value2, …) | 用给定的键值对创建一个 MAP |
| named_struct | (name1, val1, name2, val2, …) | 用给定的字段名称和值创建一个 STRUCT |
| create_union | (tag, val1, val2, …) | 用 tag 参数指向的值创建 UNION |
| array | (val1, val2, …) | 用给定的元素创建一个 ARRAY |
| struct | (val1, val2, val3, …) | 用给定的字段值创建一个 STRUCT，字段名称为 col1，col2… |

**例 60**：利用复合类型数据的构造函数创建复合类型数据。

SELECT map('tel','88832280','mobile','15467856456') ,array(10,20) ,
struct('zhang',30) ,named_struct('name','zhang','sex','male') ;

执行结果如图 6-20 所示。

```
0: jdbc:hive2://master:10000> SELECT map('tel','88832280','mobile','15467856456') ,array(10,20) ,struct('zhang',30) ,named_struct('name','zhang','sex','male') ;
+------------------------------------------+---------+------------------------+-------------------------------+
|                    _c0                   |   _c1   |          _c2           |             _c3               |
+------------------------------------------+---------+------------------------+-------------------------------+
| {"tel":"88832280","mobile":"15467856456"} | [10,20] | {"col1":"zhang","col2":30} | {"name":"zhang","sex":"male"} |
```

图 6-20 复合类型数据的构造函数应用的执行结果

## 6.7 Hive 内置函数

Hive 提供了大量的内置函数用于数据查询和数据分析，包括内置函数 UDF（User-Defined Functions）、内置聚集函数 UDAF（Built-in Aggregate Functions）、内置表生成函数 UDTF（Built-in Table-Generating Functions）等，同时提供了 UDF、UDAF 和 UDTF 接口，用户可以自定义函数，以满足特定业务需求。

可以通过 SHOW FUNCTIONS 语句查看系统提供的各种内置函数，也可以通过 DESCRIBE FUNCTION 查看指定函数的具体描述信息。例如：

SHOW FUNCTIONS;
DESCRIBE FUNCTION lower;
DESCRIBE FUNCTION EXTENDED lower;

### 6.7.1 数学函数

Hive 提供了大量的数学函数，主要用于数值类型参数的处理。当参数为 NULL 时，返回值大多数也为 NULL。常用的数学函数如表 6-11 所示。

表 6-11 常用的数学函数

| 返回类型 | 函数 | 描述 |
| --- | --- | --- |
| DOUBLE | round(DOUBLE a) | 返回 DOUBLE 类型的整数值部分（遵循四舍五入） |
| DOUBLE | round(DOUBLE a, INT d) | 返回 DOUBLE 类型，保留 d 位小数（遵循四舍五入） |
| DOUBLE | bround(DOUBLE a) | 返回使用 HALF_EVEN 舍入模式的 DOUBLE 类型 |
| DOUBLE | bround(DOUBLE a, INT d) | 返回使用 HALF_EVEN 舍入模式的 DOUBLE 类型，保留 d 位小数 |

续表

| 返回类型 | 函数 | 描述 |
| --- | --- | --- |
| BIGINT | floor(DOUBLE a) | 返回小于或等于 a 的最大 BIGINT 类型值 |
| BIGINT | ceil(DOUBLE a)<br>ceiling(DOUBLE a) | 返回大于或等于 a 的最小 BIGINT 类型值 |
| DOUBLE | rand(), rand(INT seed) | 每行返回一个在 0 与 1 之间均匀分布的随机数,seed 是随机因子 |
| DOUBLE | exp(DOUBLE a)<br>exp(DECIMAL a) | 返回 e 的 a 幂次方 |
| DOUBLE | ln(DOUBLE a)<br>ln(DECIMAL a) | 返回以 e 为底的 a 的对数 |
| DOUBLE | log10(DOUBLE a)<br>log10(DECIMAL a) | 返回以 10 为底的 a 的对数 |
| DOUBLE | log2(DOUBLE a)<br>log2(DECIMAL a) | 返回以 2 为底的 a 的对数 |
| DOUBLE | log(DOUBLE base, DOUBLE a)<br>log(DECIMAL base, DECIMAL a) | 返回以 base 为底的 a 的对数 |
| DOUBLE | pow(DOUBLE a, DOUBLE p),<br>power(DOUBLE a, DOUBLE p) | 返回 a 的 p 幂次方 |
| DOUBLE | sqrt(DOUBLE a), sqrt(DECIMAL a) | 返回 a 的平方根 |
| STRING | bin(BIGINT a) | 返回 a 的二进制表示的 STRING |
| STRING | hex(BIGINT a) hex(STRING a) hex(BINARY a) | 如果 a 为 BIGINT 或 BINARY,返回以十六进制表示的 STRING;如果 a 为 STRING,将每个字符转换为十六进制并返回 STRING |
| BINARY | unhex(STRING a) | 将 a 中每个字符解析为十六进制数字并转换为数字的字节表示 |
| STRING | conv(BIGINT num, INT from_base, INT to_base)<br>conv(STRING num, INT from_base, INT to_base) | 将 num 从 from_base 进制转换为 to_base 进制,并返回 STRING 表示 |
| DOUBLE | abs(DOUBLE a) | 返回 a 的绝对值 |
| INT 或 DOUBLE | pmod(INT a, INT b),<br>pmod(DOUBLE a, DOUBLE b) | 返回 a 对 b 取模的正值 |
| DOUBLE | sin(DOUBLE a), sin(DECIMAL a) | 返回 a (a 以弧度表示) 的正弦值 |
| DOUBLE | asin(DOUBLE a), asin(DECIMAL a) | 如果-1<=a<=1,返回 a 的反正弦值,否则返回 NULL |
| DOUBLE | cos(DOUBLE a), cos(DECIMAL a) | 返回 a (a 以弧度表示) 的余弦值 |
| DOUBLE | acos(DOUBLE a), acos(DECIMAL a) | 如果-1<=a<=1,返回 a 的反余弦值,否则返回 NULL |
| DOUBLE | tan(DOUBLE a), tan(DECIMAL a) | 返回 a (a 以弧度表示) 的正切值 |
| DOUBLE | atan(DOUBLE a), atan(DECIMAL a) | 返回 a (a 以弧度表示) 的反正切值 |
| DOUBLE | degrees(DOUBLE a)<br>degrees(DECIMAL a) | 将 a 从弧度值转换为角度值 |

续表

| 返回类型 | 函数 | 描述 |
| --- | --- | --- |
| DOUBLE | radians(DOUBLE a)<br>radians(DOUBLE a) | 将 a 从角度值转换为弧度值 |
| DOUBLE 或 INT | positive(INT a)<br>positive(DOUBLE a) | 返回 a |
| DOUBLE 或 INT | negative(INT a)<br>negative(DOUBLE a) | 返回-a |
| DOUBLE 或 INT | sign(DOUBLE a)<br>sign(DECIMAL a) | 如果 a 为整数，返回 1.0；如果 a 为负数，返回-1.0；否则返回 0 |
| DOUBLE | e() | 返回常数 e 的值 |
| DOUBLE | pi() | 返回常数 pi 的值 |
| BIGINT | factorial(INT a) | 返回 a 的阶乘。a 取值范围为[0,20] |
| DOUBLE | cbrt(DOUBLE a) | 返回 a 的立方根 |
| INT<br>BIGINT | shiftleft(TINYINT\|SMALLINT\|INT a, INT b)<br>shiftleft(BIGINT a, INT b) | 按位左移，左移一个 b 位置 |
| INT<br>BIGINT | shiftright(TINYINT\|SMALLINT\|INT a, INT b)<br>shiftright(BIGINT a, INT b) | 按位右移，右移一个 b 位置 |
| INT<br>BIGINT | shiftrightunsigned(TINYINT\|SMALLINT\|INT a, INT b),<br>shiftrightunsigned(BIGINT a, INT b) | 按位无符号右移，向右移动一个 b 位置 |
| T | greatest(T v1, T v2, …) | 放回列表中最大值。当参数中包含 NULL 时，返回 NULL |
| T | least(T v1, T v2, …) | 放回列表中最小值。当参数中包含 NULL 时，返回 NULL |
| INT | width_bucket(NUMERIC expr, NUMERIC min_value, NUMERIC max_value, INT num_buckets) | 将[min_value,max_value]分成相同区域大小的桶。返回 expr 所在桶编号。如果 expr 小于 min_value，将返回 0；如果 expr 大于 max_vlaue，将返回 num_buckets+1 |

**例 61**：把 1000 元到 10000 元之间分为 10 个桶，查看每个员工工资在哪个桶内。
SELECT empno,sal,width_bucket(sal,1000,10000,10) FROM inner_emp;

## 6.7.2 集合函数

集合函数用于对复合类型数据的处理，常用的集合函数如表 6-12 所示。

表 6-12 常用的集合函数

| 返回类型 | 函数 | 描述 |
| --- | --- | --- |
| int | size(Map<K.V>) | 返回 MAP 类型中元素的数量 |
| int | size(Array<T>) | 返回 ARRAY 类型中元素的数量 |
| array<K> | map_keys(Map<K.V>) | 返回包含输入 MAP 中所有 key 值的无序数组 |
| array<V> | map_values(Map<K.V>) | 返回包含输入 MAP 中所有 value 值的无序数组 |
| boolean | array_contains(Array<T>, value) | 如果 Array 中包含 value 值，则返回 TRUE |
| array<t> | sort_array(Array<T>) | 按数组元素的自然顺序对输入数组升序排序并返回 |

**例 62**：查询每个员工的员工名、爱好个数、所有联系方式的种类和所有电话号码。

SELECT size(hobby),map_keys(phone),map_values(phone) FROM employees;

## 6.7.3 类型转换函数

类型转换函数如表 6-13 所示。

表 6-13 类型转换函数

| 返回类型 | 函数 | 描述 |
| --- | --- | --- |
| binary | binary(string\|binary) | 将参数转换为二进制数 |
| Expected "=" to follow "type" | cast(expr as <type>) | 将表达式 expr 转换为<type>类型。如果转换失败，返回 NULL |

## 6.7.4 日期函数

日期函数如表 6-14 所示。

表 6-14 日期函数

| 返回类型 | 函数 | 描述 |
| --- | --- | --- |
| string | from_unixtime(bigint unixtime[, string format]) | 将时间戳秒数转换为 UTC 时间，并用字符串表示。可以通过 format 设置输出时间格式 |
| bigint | unix_timestamp() | 以秒数方式返回当前 UNIX 时间戳 |
| bigint | unix_timestamp(string date) | 将以 yyyy-MM-dd HH:mm:ss 格式输入的 date 字符串转换为 UNIX 时间戳 |
| bigint | unix_timestamp(string date, string pattern) | 将指定格式的日期字符串转换为 UNIX 时间戳 |
| date | to_date(string timestamp) | 返回时间戳字符串中的日期部分 |
| int | year(string date) | 返回日期字符串或时间戳字符串中的年份 |
| int | quarter(date/timestamp/string) | 返回日期、时间戳或字符串中的季度，范围为 1～4 |
| int | month(string date) | 返回日期字符串或时间戳字符串中的月份，范围为 1～12 |
| int | day(string date) dayofmonth(date) | 返回日期字符串或时间戳字符串中的天数，范围为 1～31 |
| int | hour(string date) | 返回时间戳中的小时数，范围为 0～23 |
| int | minute(string date) | 返回时间戳中的分钟数，范围为 0～59 |
| int | second(string date) | 返回时间戳中的秒数，范围为 0～59 |
| int | weekofyear(string date) | 返回时间戳字符串位于一年中的第几周 |
| int | extract(field FROM source) | 从 source 中返回天数、小时等字段。source 必须是日期、时间戳、时间间隔或可以转化为日期或时间戳的字符串 |
| int | datediff(string enddate, string startdate) | 返回从开始时间 startdate 到结束时间 enddate 之间相差的天数 |
| date | date_add(date/timestamp/string startdate, tinyint/smallint/int days) | 返回开始时间 startdate 增加 days 天后的日期 |
| date | date_sub(date/timestamp/string startdate, tinyint/smallint/int days) | 返回开始日期 startdate 减去 days 天后的日期 |
| date | current_date | 返回查询开始时的日期,同一个查询中 current_date 返回相同值 |

续表

| 返回类型 | 函数 | 描述 |
|---|---|---|
| timestamp | current_timestamp | 返回查询开始时的时间戳，同一个查询中 current_timestamp 返回相同值 |
| string | add_months(string start_date, int num_months, output_date_format) | 返回开始日期 start_date 增加 num_months 个月后的日期，输出日期默认格式为'yyyy-MM-dd' |
| string | last_day(string date) | 返回 date 所属月份的最后一天，格式为 'yyyy-MM-dd HH:mm:ss' |
| string | next_day(string start_date, string day_of_week) | 返回开始日期 start_date 后由 day_of_week 指定的第一个工作日所对应的日期 |
| string | trunc(string date, string format) | 返回截取由 format 指定单元的日期字符串 |
| double | months_between(date1, date2) | 返回 date1 与 date2 之间相差的月数 |
| string | date_format(date/timestamp/string ts, string fmt) | 将日期、时间戳或字符串 ts 转换为 fmt 格式的字符串 |

**例 63**：查询员工号、入职的年、月、日和入职日期所在月份最后一天的日期。

SELECT empno,year(hiredate),month(hiredate),day(hiredate),last_day(hiredate) FROM inner_emp;

### 6.7.5 条件函数

条件函数如表 6-15 所示。

表 6-15 条件函数

| 返回类型 | 函数 | 描述 |
|---|---|---|
| T | if(boolean testCondition, T valueTrue, T valueFalseOrNull) | 如果 testCondition 为 true，返回 valueTrue；否则返回 valueFalseOrNull |
| boolean | isnull( a ) | 如果 a 为 NULL，返回 true；否则返回 false |
| boolean | isnotnull( a ) | 如果 a 不为 NULL，返回 true；否则返回 false |
| T | nvl(T value, T default_value) | 如果 value 为 NULL，返回 defalut_value；否则返回 value |
| T | COALESCE(T v1, T v2, …) | 返回参数列表中第一个不是 NULL 的值 |
| T | CASE a WHEN b THEN c [WHEN d THEN e]* [ELSE f] END | 当 a=b，返回 c；a=d，返回 e；否则返回 f |
| T | CASE WHEN a THEN b [WHEN c THEN d]* [ELSE e] END | 当 a 为 true，返回 b；c 为 true，返回 d；否则返回 e |
| T | nullif( a, b ) | 如果 a=b，则返回 NULL；否则返回 a |
| void | assert_true(boolean condition) | 如果 condition 不成立，则抛出异常；否则返回 NULL |

**例 64**：查询员工号、部门号，如果部门号为 10，输出"10 号部门"，如果为 20，输出"20 号部门"，否则输出"其他部门"。

SELECT empno,deptno,(CASE deptno WHEN 10 THEN '10号部门' WHEN 20 THEN '20号部门' ELSE '其他部门' END) dname  FROM inner_emp;

### 6.7.6 字符串函数

字符串函数如表 6-16 所示。

表 6-16 字符串函数

| 返回类型 | 函数 | 描述 |
| --- | --- | --- |
| int | ascii(string str) | 返回字符串 str 首字符的 ASCII 码值 |
| string | base64(binary bin) | 将二进制数 bin 转换为基于 64 位的字符串 |
| int | character_length(string str) | 返回字符串 str 中包含的 UTF-8 字符个数 |
| string | chr(bigint\|double A) | 返回二进制数等于 A 的 ASCII 字符 |
| string | concat(string\|binary A, string\|binary B…) | 将字符串或字节数组按序拼接成一个字符串返回 |
| array<struct<string,double>> | context_ngrams(array<array<string>>, array<string>, int K, int pf) | 在给定 context 字符串情况下,返回句子分词后的前 K 个词语组合 |
| string | concat_ws(string SEP, string A, string B…) | 类似 concat(),但字符串 A 与 B 拼接时使用分隔符 SEP |
| string | concat_ws(string SEP, array<string>) | 类似 concat_ws(),将字符数组元素进行拼接 |
| string | decode(binary bin, string charset) | 使用指定的字符集将二进制数 bin 解码成字符串 |
| string | elt(N int,str1 string,str2 string,str3 string,…) | 返回索引 N 指定位置的字符串 |
| binary | encode(string src, string charset) | 使用指定的字符集将字符串 src 编码成二进制值 |
| int | field(val T,val1 T,val2 T,val3 T,…) | 返回 val 在 val1、val2、val3…中的索引位置,不存在返回 0 |
| int | find_in_set(string str, string strList) | 返回 str 在 strList 中第一次出现的问题,strList 以逗号分隔 |
| string | format_number(number x, int d) | 将数值 x 格式化为'#,###,###.##',四舍五入保留 d 位小数 |
| string | get_json_object(string json_string, string path) | 从指定的 json 路径 path 的 json_string 中提取 json 对象,并返回提取的 json 对象的 json 字符串 |
| boolean | in_file(string str, string filename) | 如果 filename 中有完整一行数据与 str 完全匹配,返回 true |
| int | instr(string str, string substr) | 返回字符串 substr 在字符串 str 中第一次出现的位置 |
| int | length(string A) | 返回字符串 A 的长度 |
| int | locate(string substr, string str[, int pos]) | 返回字符串 substr 在 str 的 pos 位置后第一次出现的位置 |
| string | lower(string A) , lcase(string A) | 将字符串 A 中的所有字母转换为小写字母 |
| string | lpad(string str, int len, string pad) | 在字符串 str 左侧填充字符串 pad,使其长度达到 len |
| string | ltrim(string A) | 将字符串 A 左侧的空格全部去掉 |
| array<struct<string,double>> | ngrams(array<array<string>>, int N, int K, int pf) | 返回句子分词后的前 K 个由 N 个字符构成的字符串 |
| int | octet_length(string str) | 返回以 UTF-8 编码保存字符串 str 所需的 8 位字节数 |
| string | parse_url(string urlString, string partToExtract [, string keyToExtract]) | 从指定 URL 中抽取指定部分的内容,urlString 表示 URL 字符串,partToExtract 表示要抽取的部分 |
| string | printf(String format, Obj… args) | 按照 printf 格式化输出字符串 |
| string | quote(String text) | 返回用引号引起来的字符串 |
| string | regexp_extract(string subject, string pattern, int index) | 抽取字符串 subject 中符合正则表达式模式 pattern 的第 index 部分的子字符串 |
| string | regexp_replace(string INITIAL_STRING, string PATTERN, string REPLACEMENT) | 将字符串 INITIAL_STRING 中符合正则表达式模式 PATTERN 的部分用字符串 REPLACEMENT 替换 |

续表

| 返回类型 | 函数 | 描述 |
|---|---|---|
| string | repeat(string str, int n) | 将字符串 str 重复 n 次 |
| string | replace(string A, string OLD, string NEW) | 将字符串 A 中的 OLD 用 NEW 替换 |
| string | reverse(string A) | 返回将 A 反转后的字符串 |
| string | rpad(string str, int len, string pad) | 在字符串 str 右侧填充字符串 pad，使其长度达到 len |
| string | rtrim(string A) | 将字符串 A 右侧的空格全部去掉 |
| array<array<string>> | sentences(string str, string lang, string locale) | 将字符串 str 转换为句子数组，每个句子由单词数组构成 |
| string | space(int n) | 返回一个由 n 个空格组成的字符串 |
| array | split(string str, string pat) | 按正则表达式 pat 分隔字符串 str，返回分隔后的字符串数组 |
| map<string,string> | str_to_map(text[, delimiter1, delimiter2]) | 将 text 分隔成 Map 元素。其中，delimiter1 将 text 分隔成键值对，delimiter2 将键值对分隔为键和值 |
| string | substr(string\|binary A, int start) substring(string\|binary A, int start) | 返回从 A 的 start 位置到 A 结尾的子字符串 |
| string | substr(string\|binary A, int start, int len) substring(string\|binary A, int start, int len) | 返回从 A 的 start 位置开始，长度为 len 的子字符串 |
| string | substring_index(string A, string delim, int count) | 使用分隔符 delim 将字符串 A 分隔，去掉 count 部分。count 为整数；从左向右截取；count 为负数，从右向左截取 |
| string | translate(string\|char\|varchar input, string\|char\|varchar from, string\|char\|varchar to) | 使用 to 字符串中的相应字符替换 from 字符串中存在的字符来转换输入字符串 |
| string | trim(string A) | 将字符串 A 左、右两侧的空格去掉 |
| binary | unbase64(string str) | 将基于 64 位的字符串转换为二进制字符串 |
| string | upper(string A) ucase(string A) | 将字符串 A 中的所有字母转换为大写字母 |
| string | initcap(string A) | 将字符串 A 中每个单词的第一个字母大写，其他字母小写 |
| int | levenshtein(string A, string B) | 返回字符串 A 与 B 之间的 levenshtein 距离 |
| string | soundex(string A) | 返回字符串的 soundex 代码 |

**例 65**：查找员工的员工号、员工名、职位长度，要求职位从第四位起匹配'ACCOUNT'。
SELECT empno,job, length(ename) length FROM inner_emp WHERE substr(job, 4)='ACCOUNT';

### 6.7.7 内置聚合函数

内置聚合函数如表 6-17 所示。

表 6-17 内置聚合函数

| 返回类型 | 函数 | 描述 |
|---|---|---|
| BIGINT | count(*) | 返回检索行的总数，包括包含 NULL 值的行 |
| | count(expr) | 返回表达式 expr 不为 NULL 的行数 |
| | count(DISTINCT expr[, expr…]) | 返回表达式 expr 不为 NULL，且去重后的行数 |

续表

| 返回类型 | 函数 | 描述 |
| --- | --- | --- |
| DOUBLE | sum(col) | 返回组中指定列的所有非 NULL 值的总和 |
| | sum(DISTINCT col) | 返回组中指定列去重后的所有非 NULL 值的总和 |
| DOUBLE | avg(col) | 返回组中指定列的所有非 NULL 值的平均值 |
| | avg(DISTINCT col) | 返回组中指定列去重后的所有非 NULL 值的平均值 |
| DOUBLE | min(col) | 返回组中指定列的所有非 NULL 值的最小值 |
| DOUBLE | max(col) | 返回组中指定列的所有非 NULL 值的最大值 |
| DOUBLE | variance(col), var_pop(col) | 返回组中指定数字列对应值的方差 |
| DOUBLE | var_samp(col) | 返回组中指定数字列对应值的无偏样本方差 |
| DOUBLE | stddev_pop(col) | 返回组中指定数字列对应值的标准差 |
| DOUBLE | stddev_samp(col) | 返回组中指定数字列对应值的无偏样本标准差 |
| DOUBLE | covar_pop(col1, col2) | 返回组中指定的一对数字列的总体协方差 |
| DOUBLE | covar_samp(col1, col2) | 返回组中指定的一对数字列的样本协方差 |
| DOUBLE | corr(col1, col2) | 返回组中指定的一对数字列的 Pearson 协方差 |
| DOUBLE | percentile(BIGINT col, p) | 返回 col 在 p（p 介于 0 与 1 之间）处对应的百分比 |
| array<double> | percentile(BIGINT col, array(p1 [, p2]…)) | 返回 col 在 pi（pi 介于 0 与 1 之间）处对应的百分比 |
| DOUBLE | percentile_approx(DOUBLE col, p [, B]) | 返回 col 在 p 处对应的百分比 |
| DOUBLE | regr_avgx(independent, dependent) | 类似 avg(deptentdent) |
| DOUBLE | regr_avgy(independent, dependent) | 类似 avg(inpeptdent) |
| DOUBLE | regr_count(independent, dependent) | 返回用于拟合线性回归的非空对数 |
| DOUBLE | regr_intercept(independent, dependent) | 返回线性回归直线的 y 轴截距 |
| DOUBLE | regr_r2(independent, dependent) | 返回回归的判定系数 |
| DOUBLE | regr_slope(independent, dependent) | 返回线性回归直线的斜率 |
| DOUBLE | regr_sxx(independent, dependent) | 相当于 regr_count(independent, dependent) * var_pop(dependent) |
| DOUBLE | regr_sxy(independent, dependent) | 相当于 regr_count(independent, dependent) * covar_pop(independent, dependent) |
| DOUBLE | regr_syy(independent, dependent) | 相当于 regr_count(independent, dependent) * var_pop(independent) |
| array<struct {'x','y'}> | histogram_numeric(col, b) | 用 b 个非均匀间隔计算组内数字列的直方图，输出的数组大小为 b，(x,y) 分别表示直方图的中心和高度 |
| array | collect_set(col) | 返回 col 列元素去重后的对象集合 |
| array | collect_list(col) | 返回 col 列元素包含的重复对象列表 |
| INTEGER | ntile(INTEGER x) | 将有序分区拆分为 x 个称为桶的组，为分区中每一行分配一个桶编号，可以方便地计算三分位数、四分位数、十分位数和其他常见的统计 |

## 6.7.8 内置表生成函数

内置表生成函数如表 6-18 所示。

表 6-18 内置表生成函数

| 返回类型 | 函数 | 描述 |
|---|---|---|
| T | explode(ARRAY<T> a) | 将一个数组展开为多行，每行一个数组元素 |
| Tkey,Tvalue | explode(MAP<Tkey,Tvalue> m) | 将一个 MAP 类型的数据展开为多行，每行一个键值对 |
| int,T | posexplode(ARRAY<T> a) | 将一个数组展开为多行，每行包含一个位置索引和一个数组元素 |
| T1,…,Tn | inline(ARRAY<STRUCT<f1:T1,…,fn:Tn>> a) | 将结构体数组展开为多行，每行为数组中的一个结构体元素 |
| T1,…,Tn/r | stack(int r,T1 V1,…,Tn/r Vn) | 将 n 个值 V1,…,Vn 分成 r 行，每行 n/r 列，r 为常数 |
| string1,…,stringn | json_tuple(string jsonStr,string k1,…,string kn) | 接收 json 字符串和 n 个键，返回 n 个值的元组 |
| string1,…,stringn | parse_url_tuple(string urlStr,string p1,…,string pn) | 接收 URL 字符串和一组 n 个 URL 部分，返回 n 个值的元组 |

**例 66**：查询所有员工的联系方法，每种联系方式及电话单独为一行输出。

SELECT explode(phone) FROM employees limit 5;

执行结果如图 6-21 所示。

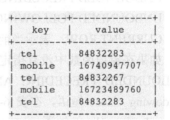

图 6-21 例 66 执行结果

### 6.7.9 窗口函数

在进行复杂的数据分析时可以使用窗口函数，窗口函数包括窗口聚合函数、窗口排序函数和窗口分析函数，如表 6-19 所示。

窗口函数应用的基本形式为：

function (arg1,arg2…argn)
OVER([PARTITION BY expr1,expr2…] [ORDER BY expr3 [ASC|DESC NULLS FIRST|LAST],expr4…]
[windowing_clause])

说明：

① function 可以是窗口聚合函数、窗口排序函数和窗口分析函数。

表 6-19 窗口函数

| 函数 | 说明 |
|---|---|
| * COUNT([DISTINCT|ALL]*) OVER () | 返回记录个数 |
| * COUNT (DISTINCT|ALL expr ) OVER () | 返回结果集中列值非空记录个数 |
| * MIN ( DISTINCT |ALL expr ) OVER () | 返回最小值 |
| * MAX ( DISTINCT |ALL expr ) OVER () | 返回最大值 |
| * AVG (DISTINCT|ALL expr )OVER () | 返回平均值 |
| RANK () OVER () | 返回记录的排序号（如果有同序号存在，序号跳跃） |
| ROW_NUMBER()OVER() | 返回记录的排序号，从 1 开始不重复地持续下去 |
| DENSE_RANK()OVER () | 返回记录的排序号（序号不跳跃） |
| NTILE ( expr ) OVER () | 根据 expr 将有序数据集分成若干片段，返回每条记录对应的片段编号 |
| LEAD(expr[, offset [, default]])OVER () | 返回当前记录后偏移量为 offset 的记录在指定列上的值 |
| LAG(expr[ , offset [, default]])OVER () | 返回当前记录前偏移量为 offset 的记录在指定列上的值 |
| *FIRST_VALUE (expr) OVER () | 返回排序结果集中的第一条记录 |
| * LAST_VALUE ( expr )OVER () | 返回排序结果集中的最后一条记录 |

② PARTITION BY：对每个分区进行统计，如果没有指定 PARTITOIN BY 短语，则对整个窗口进行统计。

③ ORDER BY：进行累计统计。如果没有指定 ORDER BY 短语，则对整个分区或窗口进行统计。

④ windowing_clause：定义一个变化或固定的数据窗口，结构如图 6-22 所示。

```
(ROWS | RANGE) BETWEEN (UNBOUNDED | [num]) PRECEDING AND ([num] PRECEDING | CURRENT ROW | (UNBOUNDED | [num]) FOLLOWING)
(ROWS | RANGE) BETWEEN CURRENT ROW AND (CURRENT ROW | (UNBOUNDED | [num]) FOLLOWING)
(ROWS | RANGE) BETWEEN [num] FOLLOWING AND (UNBOUNDED | [num]) FOLLOWING
```

图 6-22　windowing_clause 短语结构

ROWS：以物理行指定窗口中的数据，必须在 ORDER BY 短语之后使用。

RANGE：以逻辑偏移量指定窗口中的数据，必须在 ORDER BY 短语之后使用。

BETWEEN…AND…：指定窗口的起始记录和结尾记录。如果忽略 BETWEEN…AND…，则表示从分区的第一条记录开始到当前记录结束。

UNBOUNDED PRECEDING：指明以分区的第一条记录作为窗口的第一条记录。

UNBOUNDED FOLLOWING：指明以分区的最后一条记录作为窗口的最后一条记录。

CURRENT ROW：表示当前记录作为窗口的起始记录或结尾记录。

如果在窗口函数中没有指定 windowing_clause 短语，则默认设置为 RANGE BETWEEN UNBOUNDED PRECEDING AND CURRENT ROW。并不是所有的窗口函数都可以使用 windowing_clause 短语，在表 6-19 所示的窗口函数中，只有带星号（*）的窗口函数才可以使用 windowing_clause 短语。

**例 67**：统计分析 50 号部门员工的总人数、累计人数、工资总和、累计工资和、平均工资、累计平均工资。

```
SELECT empno,COUNT(*)OVER() AS totalCount,
COUNT(*)OVER(ORDER BY empno) AS cumulativeCount ,
SUM(sal)OVER() AS totalSum ,SUM(sal)OVER(ORDER BY empno) AS cumulativeSum,
AVG(sal)OVER() AS totalAvg ,AVG(sal)OVER(ORDER BY empno) AS cumulativeAvg
FROM inner_empWHERE deptno=20 ORDER BY empno;
```

执行结果如图 6-23 所示。

```
+-------+------------+-----------------+----------+----------------+----------+---------------+
| empno | totalcount | cumulativecount | totalsum | cumulativesum  | totalavg | cumulativeavg |
+-------+------------+-----------------+----------+----------------+----------+---------------+
| 201   | 2          | 1               | 19000.0  | 13000.0        | 9500.0   | 13000.0       |
| 202   | 2          | 2               | 19000.0  | 19000.0        | 9500.0   | 9500.0        |
+-------+------------+-----------------+----------+----------------+----------+---------------+
```

图 6-23　窗口聚合函数不分区应用示例

**例 68**：以部门为单位，分析统计 20、30、40 号部门员工的人数、平均工资、工资总和。

```
SELECT deptno,empno,COUNT(*)OVER(PARTITION BY deptno) AS totalCount,
COUNT(*)OVER(PARTITION BY deptno ORDER BY empno) AS cumulativeCount,
SUM(sal)OVER(PARTITION BY deptno) AS deptavg,
SUM(sal)OVER(PARTITION BY deptno ORDER BY empno) AS cumulativeSum,
AVG(sal)OVER(PARTITION BY deptno) AS deptAvg,
AVG(sal)OVER(PARTITION BY deptno ORDER BY empno) AS cumulativeAvg
FROM inner_emp WHERE deptno IN (20,30,40) ORDER BY deptno;
```

执行结果如图 6-24 所示。

**例 69**：查询 30、60 号部门中员工工资的排序信息。

```
SELECT deptno,empno,sal,
```

```
+--------+-------+------------+----------------+---------+---------------+--------+---------------+
| deptno | empno | totalcount | cumulativecount| deptavg | cumulativesum | deptavg| cumulativeavg |
+--------+-------+------------+----------------+---------+---------------+--------+---------------+
| 20     | 202   | 2          | 2              | 19000.0 | 19000.0       | 9500.0 | 9500.0        |
| 20     | 201   | 2          | 1              | 19000.0 | 13000.0       | 9500.0 | 13000.0       |
| 30     | 119   | 5          | 5              | 13900.0 | 13900.0       | 2780.0 | 2780.0        |
| 30     | 118   | 5          | 4              | 13900.0 | 11400.0       | 2780.0 | 2850.0        |
| 30     | 117   | 5          | 3              | 13900.0 | 8800.0        | 2780.0 | 2933.3333333333335 |
| 30     | 116   | 5          | 2              | 13900.0 | 6000.0        | 2780.0 | 3000.0        |
| 30     | 115   | 5          | 1              | 13900.0 | 3100.0        | 2780.0 | 3100.0        |
| 40     | 203   | 1          | 1              | 6500.0  | 6500.0        | 6500.0 | 6500.0        |
+--------+-------+------------+----------------+---------+---------------+--------+---------------+
```

图 6-24　窗口聚合函数分区应用示例

```
RANK() OVER (PARTITION BY deptno ORDER BY sal) AS RANK_ORDER,
DENSE_RANK() OVER (PARTITION BY deptno ORDER BY sal) AS DENSE_RANK_ORDER,
ROW_NUMBER()OVER(PARTITION BY deptno ORDER BY sal) AS ROW_NUMBER_ORDER
FROM inner_emp WHERE deptno IN (30,60);
```

执行结果如图 6-25 所示。

```
+--------+-------+--------+------------+------------------+------------------+
| deptno | empno | sal    | rank_order | dense_rank_order | row_number_order |
+--------+-------+--------+------------+------------------+------------------+
| 30     | 119   | 2500.0 | 1          | 1                | 1                |
| 30     | 118   | 2600.0 | 2          | 2                | 2                |
| 30     | 117   | 2800.0 | 3          | 3                | 3                |
| 30     | 116   | 2900.0 | 4          | 4                | 4                |
| 30     | 115   | 3100.0 | 5          | 5                | 5                |
| 60     | 107   | 4200.0 | 1          | 1                | 1                |
| 60     | 106   | 4800.0 | 2          | 2                | 2                |
| 60     | 104   | 6000.0 | 3          | 3                | 3                |
+--------+-------+--------+------------+------------------+------------------+
```

图 6-25　窗口排序函数应用示例

**例 70**：查询部门号、员工号、员工工资、员工所在部门最低工资的员工号、最高工资的员工号。

```
SELECT deptno, empno, sal,
FIRST_VALUE(empno)OVER (PARTITION BY deptnoORDER BY sal ASC ROWS UNBOUNDED PRECEDING) AS lowest_sal,
LAST_VALUE(empno)OVER(PARTITION BY deptno ORDER BY sal ROWS BETWEEN UNBOUNDED PRECEDING AND UNBOUNDED FOLLOWING) AS lowest_sal
FROM inner_emp ORDER BY deptno;
```

执行结果如图 6-26 所示。

```
+--------+-------+---------+------------+------------+
| deptno | empno | sal     | lowest_sal | lowest_sal |
+--------+-------+---------+------------+------------+
| NULL   | 198   | 2600.0  | 198        | 102        |
| NULL   | 199   | 2600.0  | 198        | 102        |
| NULL   | 105   | 4800.0  | 198        | 102        |
| NULL   | 178   | 7000.0  | 198        | 102        |
| NULL   | 111   | 7700.0  | 198        | 102        |
| NULL   | 112   | 7800.0  | 198        | 102        |
| NULL   | 103   | 9000.0  | 198        | 102        |
| NULL   | 114   | 11000.0 | 198        | 102        |
| NULL   | 102   | 17000.0 | 198        | 102        |
| 10     | 200   | 4400.0  | 200        | 200        |
| 20     | 202   | 6000.0  | 202        | 201        |
| 20     | 201   | 13000.0 | 202        | 201        |
+--------+-------+---------+------------+------------+
```

图 6-26　窗口分析函数应用示例

### 6.7.10　其他函数

其他函数如表 6-20 所示。

表 6-20　其他函数

| 返回类型 | 函数 | 描述 |
| --- | --- | --- |
| int | hash(a1[, a2…]) | 返回参数的 Hash 值 |
| string | current_user() | 返回当前用户名 |
| string | logged_in_user() | 从会话中返回当前登录用户名 |
| string | current_database() | 返回当前数据库名称 |
| string | md5(string/binary) | 返回对一个 string 或 binary 进行 MD5 加密后的字符串 |
| string | version() | 返回 Hive 版本号 |

**例 71**：查询当前用户及当前数据库名称。

SELECT current_user(),current_database();

### 6.7.11　词频统计实例

在 4.4 节中介绍了利用 MapReduce 程序设计计算文件中单词出现的次数，利用 Hive 同样可以实现词频统计，而且非常简单。

（1）首先在 Hive 中创建一个表。

CREATE TABLE docs(line string);

（2）将包含单词的文件加载到 docs 表中。

LOAD DATA LOCAL INPATH '/home/hadoop/data/word*.txt' INTO TABLE docs;

（3）统计词频，并将统计结果保存到一个新表中。

CREATE TABLE word_count
AS
SELECT word,count(*) AS quantity FROM (SELECT explode(split(line,' ')) AS word FROM docs) words GROUP BY word ORDER BY word;

（4）查询词频统计结果：

SELECT * FROM word_count;

执行结果如图 6-27 所示。

```
+-----------------+-------------------+
| word_count.word | word_count.quantity |
+-----------------+-------------------+
| a               | 4                 |
| count           | 1                 |
| data            | 1                 |
| example         | 4                 |
| hadoop          | 1                 |
| is              | 4                 |
| mapreduce       | 1                 |
| sort            | 1                 |
| that            | 2                 |
| this            | 2                 |
| word            | 1                 |
+-----------------+-------------------+
```

图 6-27　词频统计结果

## 6.8　Hive 高级应用

### 6.8.1　用户自定义函数

在 Hive 中，除可以使用内置的 UDF、UDAF、UDTF 函数外，用户还可以根据业务需要自定义函数。其方法是创建一个继承 UDF 的类，并重写 evaluate 方法，经过编译后加载到 Hive 函数库中，然后就可以直接在 HiveQL 语句中调用用户自定义的函数了。

（1）在 Eclipse 中创建一个名为 hive 的 Maven 工程。

（2）编辑 pom.xml 文件，引入 Hadoop 和 Hive 依赖。

```xml
<project xmlns="http://maven.apache.org/POM/4.0.0"
    xmlns:xsi="http://www.w3.org/2001/XMLSchema-instance"
    xsi:schemaLocation="http://maven.apache.org/POM/4.0.0
        https://maven.apache.org/xsd/maven-4.0.0.xsd">
    <modelVersion>4.0.0</modelVersion>
    <groupId>com.bigdata</groupId>
    <artifactId>hive</artifactId>
    <version>0.0.1-SNAPSHOT</version>
    <properties>
        <project.build.sourceEncoding>UTF-8</project.build.sourceEncoding>
        <hadoop.version>3.1.2</hadoop.version>
    </properties>
    <dependencies>
        <dependency>
            <groupId>org.apache.hadoop</groupId>
            <artifactId>hadoop-client</artifactId>
            <version>3.1.2</version>
        </dependency>
        <dependency>
            <groupId>junit</groupId>
            <artifactId>junit</artifactId>
            <version>3.8.2</version>
            <scope>test</scope>
        </dependency>
        <dependency>
            <groupId>org.apache.hive</groupId>
            <artifactId>hive-exec</artifactId>
            <version>3.1.2</version>
        </dependency>
        <dependency>
            <groupId>org.apache.hive</groupId>
            <artifactId>hive-jdbc</artifactId>
            <version>3.1.2</version>
        </dependency>
    </dependencies>
</project>
```

（3）定义 DateUDF 类，继承于 UDF 类，实现自定义函数。

```java
package com.bigdata.example;
import org.apache.hadoop.hive.ql.exec.UDF;
import org.apache.hadoop.io.Text;
public class DateUDF extends UDF {
    public Text evaluate(Text date) {
        String str = date.toString();
        String year = str.substring(0, 4);
```

```
        int i=str.indexOf("-",5);
        String month = str.substring(5, i);
        String day = str.substring(i+1);
        return new Text("year: " + year + " month: " + month + " day: " + day);
    }
    // 测试
    public static void main() {
        DateUDF udf = new DateUDF();
        System.out.println(udf.evaluate(new Text("2021-3-15")));
    }
}
```

需要注意的是，UDF 必须要有返回值，可以是 null，但不能为 void。推荐使用 Text/LongWritabe 等 Hadoop 序列化类型。

（4）测试完成后，导出 JAR 文件。JAR 文件保存到/home/hadoop/dateUDF.jar。

（5）在 Hive 中执行下列语句，将 JAR 文件添加到 Hive 环境中。

add jar /home/hadoop/dateUDF.jar;

（6）在 Hive 环境中定义临时函数。

CREATE TEMPORARY FUNCTION dateformat AS 'com.bigdata.example.DateUDF';

（7）在 SELECT 语句中调用自定义的 UDF 函数。

SELECT empno,dateformat(hiredate) FROM Inner_emp LIMIT 5;

执行结果如图 6-28 所示。

```
+--------+------------------------------+
| empno  |             _c1              |
+--------+------------------------------+
| 198    | year: 2007 month: 6 day: 21  |
| 199    | year: 2008 month: 1 day: 13  |
| 200    | year: 2003 month: 9 day: 17  |
| 201    | year: 2004 month: 2 day: 17  |
| 202    | year: 2005 month: 8 day: 17  |
+--------+------------------------------+
```

图 6-28　自定义函数的调用执行结果

需要注意的是，在 Hive 环境中定义的临时函数只在当前会话中有效，一旦退出 Hive，重新启动进入 Hive 命令行，上述临时函数就会失效。如果要保持临时函数不失效，需要将导出的 JAR 文件放入 HIVE_HONE/lib 目录下。

### 6.8.2　Hive 与 HBase 整合

Hive 提供的 HiveQL 语言简化了 MapReduce 的使用，而 HBase 提供了低延迟的数据库访问。如果两者结合，可以充分利用 MapReduce 的优势针对 HBase 存储的海量数据进行离线的计算和分析，充分发挥 Hive 数据分析的优势，避免了繁杂的 MapReduce 程序设计。

Hive 与 HBase 整合后，可以达到以下目标：

① 在 Hive 中创建的表能直接创建保存到 HBase 中。

② 向 Hive 表中插入数据，数据会同步更新到 HBase 对应的表中。

③ HBase 表的列族值变更，也会在 Hive 对应表中变更。

在 Hive 中创建与 HBase 表关联表的语法为：

CREATE [EXTERNAL] TABLE [IF NOT EXISTS][db_name.]table_name (

```
(col_name data_type [COMMENT col_comment], … )]
STORED BY 'org.apache.hadoop.hive.hbase.HBaseStorageHandler'
WITH SERDEPROPERTIES ("hbase.columns.mapping" = …)
TBLPROPERTIES ("hbase.table.name" = …);
```

其中，HBaseStorageHandler 是 Hive 表与 HBase 进行通信的类，hbase.columns.mapping 指定 Hive 表列与 HBase 列族或列族限定符的对应关系，hbase.table.name 指定与新建的 Hive 表对应的 HBase 表。如果 HBase 中表已经存在，在 Hive 中需要创建一个外部表与 HBase 表关联；如果 HBase 中表不存在，则需要在 Hive 中创建内部表与 HBase 表关联。

Hive 与 HBase 整合主要是依靠 HBaseStorageHandler 进行通信的。利用 HbaseStorageHandler，Hive 可以获取到 Hive 表对应的 HBase 表名、列族及列族限定符、InputFormat 和 OutputFormat 类，可以创建和删除 HBase 表等。Hive 访问 HBase 中表数据，实质上是通过 MapReduce 读取 HBase 表数据，其实现是在 MapReduce 中使用 HiveHBaseTableInputFormat 完成对 HBase 表的切分，获取 RecordReader 对象来读取数据的。对 HBase 表的切分原则是一个 Region 切分成一个 Split，即表中有多少个 Region，MapReduce 中就启动多少个 Map 任务。读取 HBase 表数据都通过构建 Scanner 对象，对表进行全表扫描，如果有过滤条件，则转化为 Filter。当过滤条件为行键时，则转化为对行键的过滤；Scanner 通过 RPC 调用 RegionServer 的 next() 来获取数据。

进行 Hive 与 HBase 整合之前，需要启动下列服务。
- 启动 Hadoop 集群服务：start-all.sh。
- 启动 HBase 服务：start-hbase.sh。
- 启动 HiveServer2 服务：hiveserver2 &。

（1）在 Hive 中创建表

```
USE dbhive;
CREATE TABLE IF NOT EXISTS hive_hbase_dept(deptno int, dname string,manager_id string)
STORED BY "org.apache.hadoop.hive.hbase.HBaseStorageHandler"
WITH SERDEPROPERTIES ("hbase.columns.mapping" = ":key,info:name,info:manager")
TBLPROPERTIES ("hbase.table.name" ="dept");
```

Hive 创建的 hive_hbase_dept 表与 HBase 的 dept 表关联。其中，:key 表示 dept 表中的行键与 hive_hbase_dept 表中的第一列 deptno 对应，info:id 与 dname 对应，info:manager 与 manager_id 对应。

表创建成功后，执行 SHOW TABLES 命令，可以看到新建的 hive_hbase_dept 表。

（2）在 HBase 中执行 list 命令，可以看到新建的 dept 表。

（3）向 Hive 表中导入数据。向 Hive 表中导入数据时需要注意，只能使用 INSERT 语句导入子查询，而不能使用 LOAD 语句导入数据，因为数据最终存储在 HBase 表中，而非 Hive 表中。

```
INSERT OVERWRITE TABLE hive_hbase_dept SELECT * FROM inner_dept WHERE deptno=20;
```

此时，系统启动 MapReduce 作业，将数据插入 hive_hbase_dept 后，在 Hive 的该表的存储位置看不到数据，数据被插入与该关联的 HBase 的 dept 表中了。

（4）查询数据

可以在 Hive 中查询 hive_hbase_dept 表，也可以在 HBase 中查询 dept 表，结果相同。

```
0: jdbc:hive2://master:10000> SELECT * FROM hive_hbase_dept;
hbase(main):005:0> scan 'dept'
```

（5）在 HBase 中向 dpet 表中插入数据后，在 Hive 中也可以看到更新后的数据。

```
hbase(main):006:0> put 'dept',300,'info:name','new dept'
```

```
hbase(main):007:0> scan 'dept'
0: jdbc:hive2://master:10000> SELECT * FROM hive_hbase_dept;
```

## 6.9 Hive 程序设计

除使用 Shell 命令进行 Hive 操作外，还可以通过 JDBC 连接 Hive，使用 Hive Java API 对 Hive 进行操作。与访问关系数据库类似，对 Hive 进行操作也需要注册驱动、创建连接，不同之处在于不能对 Hive 中的数据进行删除或更新。

下面通过一个案例介绍如何利用 Java API 操作 Hive。

（1）在 Eclipse 中创建一个名为 hive 的 Maven 工程。

（2）编辑 pom.xml 文件，引入 Hadoop 和 Hive 依赖。代码详见 6.8.1 节中的介绍。

（3）创建一个类，实现对 Hive 的操作，包括创建数据库、切换数据库、创建表、加载数据、查询表等。

```java
package com.bigdata.example;
import java.sql.Connection;
import java.sql.DriverManager;
import java.sql.ResultSet;
import java.sql.Statement;
public class HiveJDBC {
    public static void main(String[] args) {
        try {
            // 1.加载驱动
            Class.forName("org.apache.hive.jdbc.HiveDriver");
            // 2.获得连接
            Connection conn = DriverManager.getConnection("jdbc:hive2://master:10000/default","hadoop","hadoop");
            // 3.获得数据库操作对象
            Statement statement = conn.createStatement();
            // 4.操作数据库
            String sql="";
            // (1)新建数据库
            sql = "CREATE DATABASE IF NOT EXISTS empdb";
            statement.execute(sql);
            // (2)切换数据库
            sql = "use empdb";
            statement.execute(sql);
            // (3)新建数据表
            sql="CREATE TABLE IF NOT EXISTS emp(empno int ,ename char(20),job string,mgr int, hiredate string,sal float,comm float,deptno int) row format delimited fields terminated by ','";
            statement.execute(sql);
            // (4)导入数据
            sql = "LOAD DATA LOCAL INPATH '/home/hadoop/hivedata/emp.txt' OVERWRITE INTO TABLE emp";
            statement.execute(sql);
            // (5)数据查询分析
            sql = "SELECT deptno,count(*) AS num,avg(sal) AS avgsal FROM emp GROUP BY deptno";
            statement.execute(sql);
            // 5.接收返回值结果
```

```
            ResultSet resultSet = statement.executeQuery(sql);
            while(resultSet.next()){
                System.out.println(resultSet.getString("deptno") + "\t" + resultSet.getInt("num") + "\t" + resultSet.getFloat("avgsal"));
            }
            // 6.关闭连接，释放资源
            resultSet.close();
            statement.close();
            conn.close();
        } catch (Exception e) {
            e.printStackTrace();
        }
    }
}
```

（4）编译调试。

在 Eclipse 中运行该工程，在控制台可以看到输出的结果，如图 6-29 所示。

| 10  | 1  | 4400.0   |
|-----|----|----------|
| 20  | 2  | 9500.0   |
| 30  | 5  | 2780.0   |
| 40  | 1  | 6500.0   |
| 50  | 43 | 3516.279 |
| 60  | 3  | 5000.0   |
| 70  | 1  | 10000.0  |
| 80  | 34 | 8955.883 |
| 90  | 2  | 20500.0  |
| 100 | 4  | 9027.0   |
| 110 | 2  | 10154.0  |

图 6-29　Hive JDBC 数据查询结果

（5）将工程导出为 JAR 文件（/home/hadoop/hivejdbc.jar），并部署执行。

Hadoop jar /home/hadoop/hivejdbc.jar com.bigdata.example.HiveJDBC

# 本 章 小 结

本章首先系统介绍了 Hive 系统架构、工作原理、数据存储模型、数据类型及数据存储格式等理论知识，然后详细介绍了 Hive 安装与配置、Beeline 操作、Hive 操作（DDL 操作、DML 操作、数据查询）、Hive 内置函数与自定义函数、Hive 与 HBase 整合和基于 Hive Java API 的程序设计。通过本章学习，读者可以了解 Hive 的体系结构与工作机制，掌握通过 Shell 命令对 Hive 进行对象定义、数据导入、数据查询与数据分析，掌握利用 Java API 定义函数并进行 Hive 程序设计。

# 思考题与习题

**1. 简答题**

（1）简述 Hive 适合用于数据仓库的原因。

（2）简述 Hive 架构组成及各个组件的作用。

（3）简述 Hive 表的类型及其特点。

（4）简述 Hive 中数据导入、导出的方法。

（5）Hive 与 HBase 整合的目标是什么？

## 2. 选择题

（1）Hive 是由哪家公司开源的大数据处理组件？（　　）

A．Google　　　B．Apache　　　C．Facebook　　　D．Baidu

（2）Hive 的计算引擎是什么？（　　）

A．Spark　　　B．MapReduce　　　C．HDFS　　　D．Flink

（3）下列哪项不是 Hive 使用的场景？（　　）

A．实时的在线数据分析

B．数据挖掘（用户行为分析、兴趣分区、区域展示）

C．数据汇总（每天/每周用户点击数、点击排行）

D．非实时分析（日志分析、统计分析）

（4）关于 Hive 与传统关系数据库比较，错误的是（　　）。

A．Hive 基于 HDFS 存储，理论上存储量可以无限扩展，而传统关系数据库的存储会有上限

B．由于 Hive 基于大数据平台，所以查询效率比传统关系数据库快

C．传统关系数据库索引发展非常成熟，而 Hive 的索引机制还很低效

D．由于 Hive 数据存储在 HDFS 中，因此可以保证数据的高容错性、高可靠性

（5）关于 Hive 与 Hadoop 其他组件的关系，描述错误的是（　　）。

A．Hive 最终将数据存储在 HDFS 中

B．HiveQL 的基本思想是执行 MapReduce 任务

C．Hive 是 Hadoop 平台的数据仓库工具

D．Hive 对 HBase 有强依赖

（6）针对 Hive 中的分区概念，如下描述错误的是（　　）。

A．分区字段要在创建表时定义

B．分区字段只能有一个，不可以创建多级分区

C．使用分区，可以减少某些查询的数据扫描范围，进而提高查询效率

D．分区字段可以作为 WHERE 子句条件

（7）加载数据到 Hive 表，哪种方式不正确？（　　）

A．直接将本地路径的文件加载到 Hive 表中

B．将 HDFS 上的文件加载到 Hive 表中

C．Hive 只支持单条记录插入的方法，因此可以直接在命令行插入单条记录

D．将其他表的结果集插入 Hive 表

（8）下列关于 HiveQL 语句描述，不正确的是（　　）。

A．支持表连接操作　　　B．支持分组操作

C．支持各种子查询操作　　　D．支持复杂的有条件查询

（9）关于 Hive 内部表描述不正确的是（　　）。

A．删除内部表时，不会删除表对应的文件

B．内部表的创建与数据加载是两个独立的过程

C．向内部表中加载数据时，数据文件将移动到表对应的存储目录中

D．内部表中数据默认存储在与表同名的目录中

（10）关于 Hive 架构描述不正确的是（　　）。

A．MetaStore 是用来存放元数据的，可以是关系数据库，也可以是非关系数据库

B．ThriftServer 提供 thrift 接口，作为 JDBC 和 ODBC 的服务端，并将 Hive 和其他应用程序集成起来

C．Driver 用户管理 HiveQL 执行的生命周期并贯穿 Hive 任务的整个执行期间

D．Compiler 用于编译 HiveQL 并将其转化为一系列相互依赖的 Map 任务和 Reduce 任务

3．实训题

（1）利用 Shell 操作 Hive

① 查看系统中所有数据库。

② 如果不存在名为 test 的数据库，则创建 test 数据库。

③ 切换到 test 数据库，查看该数据库中有哪些表。

④ 在 test 中创建一个 student1 的内部表，表中包括 sno（学号）、sname（姓名）、age（年龄）、classid（班级）、dept（系别），并向表中插入 5 条记录。

⑤ 在 test 中创建一个 student2 的内部表，表中包括 sno（学号）、sname（姓名）、age（年龄）、classid（班级）、dept（系别），将本地文件 localstudents 中的数据导入表中（本地文件 localstudents 需要提前创建并写入 10 条记录，第一条为学生自己的信息），并查询表中信息。

⑥ 将上述文件 localstudents 传送到 HDFS 的当前目录，然后以追加方式导入 student2 表中，并查询表中信息。

⑦ 创建一个包含学生姓名、课程名、成绩的数据文件 scfile，并插入 10 条记录，上传到 HDFS，然后创建一个外部表 sc，对应 scfile 文件，查询表中数据信息。

（2）编写一个 Java 程序，查询 studnet2 表中所有数据，并输出显示。

# 第 7 章 数据迁移工具 Sqoop

Sqoop 是关系数据库与 Hadoop 集群之间的数据迁移工具，可以方便地将关系数据库的源数据导入 HDFS、Hive 或 HBase 中，也可以将 HDFS、Hive 数据分析、处理后的结果导出到关系数据库中。本章将介绍 Sqoop 体系结构、常用命令和数据导入与导出方法。

- Sqoop 概述：Sqoop 功能、架构、Sqoop 版本。
- Sqoop 安装与配置：Sqoop 安装、配置、测试。
- Sqoop 常用命令：Sqoop 常用命令及其功能介绍。
- Sqoop 数据导入：从 MySQL 导入数据到 HDFS、Hive、HBase。
- Sqoop 数据导出：从 HDFS、Hive 导出数据到 MySQL。

## 7.1 Sqoop 概述

Sqoop 的含义是"SQL 到 Hadoop 和 Hadoop 到 SQL"，是 Apache 的顶级项目，是 Hadoop 和关系数据库（RDBMS）服务器之间数据迁移的工具。可以使用 Sqoop 将数据从关系数据库（比如 MySQL、Oracle 等）导入 Hadoop 的 HDFS、Hive、HBase 等，或者将数据从 HDFS、Hive、HBase 导出到 MySQL、Oracle 等关系数据库中，如图 7-1 所示。

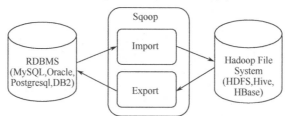

图 7-1 Sqoop 导入与导出

通常，在运营式系统中，有价值的数据都保存在关系数据库中。在进行数据分析时，关系数据库是重要的数据来源，可以利用 Sqoop 将数据导入 Hadoop 集群的存储组件中，充分利用 Hadoop 集群海量数据存储能力和分析计算能力，进行数据分析处理。由于 Hadoop 集群延迟性比较大，通常需要将数据分析结果导出到关系数据库，以便进行可视化或其他应用开发，此时就可以利用 Sqoop 实现从 Hadoop 集群到关系数据库的数据导出。

Sqoop 命令的本质是将导入或导出命令转换成 MapReduce 程序来实现，Sqoop 架构如图 7-2 所示。

到目前为止，Sqoop 主要存在两个版本：Sqoop 1 和 Sqoop 2，其中 Sqoop 1.4.4 之前的所有版本称为 Sqoop 1，之后的所有版本称为 Sqoop 2，当前最新版本为 Sqoop 1.99.1，最新稳定版本为 Sqoop 1.4.7。Sqoop 1 与 Sqoop 2 在架构和使用上差别较大，互不兼容。本书采用的版本是 Sqoop 1.4.7。

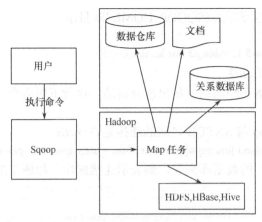

图 7-2 Sqoop 架构

## 7.2 Sqoop 安装与配置

在 Sqoop 安装配置过程中，需要根据关系数据库的不同，添加不同的驱动程序，本书采用的数据库是 MySQL，因此添加 MySQL 驱动程序。

（1）下载 Sqoop 安装包

到 Sqoop 官网（https://downloads.apache.org）下载 Sqoop 安装文件 sqoop-1.4.7.bin__hadoop-2.6.0.tar.gz 到~/Downloads 目录。

（2）解压 Sqoop 安装包

```
$ cd ~/Downloads
$ sudo tar -zxvf sqoop-1.4.7.bin__hadoop-2.6.0.tar.gz -C /usr/local
```

（3）重命名解压后的目录并修改目录所有者

```
$ cd /usr/local
$ sudo mv sqoop-1.4.7.bin__hadoop-2.6.0-alpha sqoop
$ sudo chown -R hadoop:hadoop sqoop
```

（4）配置环境变量

打开当前用户的环境变量配置文件：

```
$ gedit ~/.bashrc
```

在配置文件第一行输入如下信息：

```
export SQOOP_HOME=/usr/local/sqoop
export PATH=$PATH:$SQOOP_HOME/bin
```

保存该文件后，执行下面命令让配置文件立即生效：

```
$ source ~/.bashrc
```

（5）创建、编辑配置文件 sqoop-env.sh

```
$ cd /usr/local/sqoop/conf
$ cp sqoop-env-template.sh    sqoop-env.sh
$ gedit sqoop-env.sh
```

在 sqoop-env.sh 中写入下列信息：

```
export HADOOP_COMMON_HOME=/usr/local/hadoop
export HADOOP_MAPRED_HOME=/usr/local/hadoop
export HBASE_HOME=/usr/local/hbase
export HIVE_HOME=/usr/local/hive
```

（6）将 MySQL 驱动包复制到$SQOOP_HOME/lib 目录中

$ cd ~/Downloads
$ cp ./mysql-connector-java-5.1.39-bin.jar /usr/local/sqoop/lib

（7）测试与 MySQL 的连接

首先确保 MySQL 服务已经启动，如果没有启动，执行下面命令启动：

$ sudo service mysql start

然后就可以测试 Sqoop 与 MySQL 之间的连接是否成功：

$ sqoop list-databases --connect jdbc:mysql://master:3306/ --username root --password root

如果获取到了 MySQL 的数据库列表，则表示连接成功，如图 7-3 所示。

```
2021-04-05 09:11:07,511 INFO sqoop.Sqoop: Running Sqoop version: 1.4.7
2021-04-05 09:11:07,688 WARN tool.BaseSqoopTool: Setting your password on the command-line is insecure. Consider using -P instead.
2021-04-05 09:11:08,541 INFO manager.MySQLManager: Preparing to use a MySQL streaming resultset.
Mon Apr 05 09:11:09 CST 2021 WARN: Establishing SSL connection without server's identity verification is not recommended. According to MySQL 5.5.45+, 5.6.26+ and 5.7.6+ requirements SSL connection must be established by default if explicit option isn't set. For compliance with existing applications not using SSL the verifyServerCertificate property is set to 'false'. You need either to explicitly disable SSL by setting useSSL=false, or set useSSL=true and provide truststore for server certificate verification.
information_schema
hive
mysql
performance_schema
sys
test
testcode
```

图 7-3　使用 list-databases 测试 Sqoop 与 MySQL 的连接

## 7.3　Sqoop 常用命令

Sqoop 提供了一系列命令用于数据导入、数据导出等，可以执行 sqoop help 查看 Sqoop 提供的所有命令，如图 7-4 所示。

```
hadoop@master:~$ sqoop help
2021-04-05 09:08:35,988 INFO sqoop.Sqoop: Running Sqoop version: 1.4.
usage: sqoop COMMAND [ARGS]

Available commands:
  codegen            Generate code to interact with database records
  create-hive-table  Import a table definition into Hive
  eval               Evaluate a SQL statement and display the results
  export             Export an HDFS directory to a database table
  help               List available commands
  import             Import a table from a database to HDFS
  import-all-tables  Import tables from a database to HDFS
  import-mainframe   Import datasets from a mainframe server to HDFS
  job                Work with saved jobs
  list-databases     List available databases on a server
  list-tables        List available tables in a database
  merge              Merge results of incremental imports
  metastore          Run a standalone Sqoop metastore
  version            Display version information

See 'sqoop help COMMAND' for information on a specific command.
```

图 7-4　Sqoop 提供的命令

Sqoop 命令使用的基本形式为 sqoop COMMAND [arguments]，例如 sqoop version。可以使用 sqoop help COMMAND 查看命令的帮助信息，例如 sqoop help import。

（1）import

import 命令用于从 RDBMS 中导入数据到 HDFS，数据库中的每行数据表示为 HDFS 中的一条记录，以 HDFS 文件和文本文件或二进制文件格式存储。

关于 import 命令的使用详见 7.4 节中的介绍。

（2）import-all-tables

import-all-tables 命令将 RDBMS 中指定库中的所有表导入 HDFS，每个表在 HDFS 中对应一个独立的存储目录。例如：

```
$ sqoop import-all-tables --connect jdbc:mysql://master:3306/test
```

（3）import-mainframe

import-mainframe 命令将大型机上分区数据集（PDS）中的所有顺序数据集导入 HDFS。PDS 类似于开放系统上的目录，数据集中的记录只能包含字符数据。在 HDFS 中整个记录存储为单个文本字段。例如：

```
$ sqoop import-mainframe --connect z390 --dataset emps --username hadoop -password hadoop
```

（4）export

export 命令将一组文件从 HDFS 导出到 RDBMS 的表中。根据用户指定的分隔符，将输入文件读取并解析为一组记录插入表中。

关于 export 命令的使用详见 7.5 节中的介绍。

（5）job

job 命令用于创建作业和使用保存的作业。保存的作业会记住作业使用的参数，因此可以通过其句柄调用作业来重新执行作业。如果保存的作业配置为执行增量导入，则保存的作业中有关最近导入的行的状态将更新，以允许作业仅导入最新行。例如：

```
$ sqoop job --create myjob -- import --connect jdbc:mysql://master:3306/test --username root --password root --table employees
```

（6）metastore

metastore 命令将 Sqoop 配置到一个共享元数据存储库。多个本地用户或远程用户可以定义作业和执行保存在元数据中的作业（使用 sqoopjob 创建）。客户端必须配置为连接到 sqoop-site.xml 文件中指定的元数据库或者使用--meta connect 参数指定元数据库。

（7）merge

merge 命令允许将两个数据集合并，其中一个数据集中的条目覆盖旧数据集的条目。例如，last-modify 模式运行的增量导入将在 HDFS 中生成多个数据集，其中每个数据集中依次出现较新的数据。merge 命令可以把两个数据集"展平"为一个，为每个最新记录获取一个主键。例如：

```
$ sqoop merge --new-data newer --onto older --target-dir merged --jar-file datatypes.jar --class-name Emp --merge-key id
```

（8）codegen

codegen 命令生成 Java 类，这些类封装和解释导入的记录。表中记录转化为 Java 类是导入过程的一部分，但也可以单独执行。例如，如果 Java 源代码丢失，则可以重新创建 Java 类，新版本可以在字段间使用不同分隔符。例如：

```
$ sqoop codegen --connect jdbc:mysql://master:3306/test --username root --password root --table employees
```

（9）create-hive-table

create-hive-table 命令基于先前导入 HDFS 的数据库表或计划导入的数据库表的定义构建 Hive 元数据，这将高效执行 sqoop import 导入中的 hive-import 步骤，而不需要运行前面的导入。如果数据已经加载到 HDFS，那么可以使用此命令完成将数据导入 Hive 的管道。还可以使用该命令创建 Hive 表，然后在用户运行预处理步骤后将数据导入并填充到表中。例如：

```
$ sqoop create-hive-table --connect jdbc:mysql://master:3306/test --username root --password root --table employees   --hive-table hive_emp
```

（10）eval

eval 命令允许用户快速运行简单的 SQL 查询，并结果将打印到控制台。用户可以预览导入查询，以确保导入所需的数据。例如：

$ sqoop eval --connect jdbc:mysql://localhost:3306/test --username root--password root --query "SELECT * FROM employees LIMIT 10"

（11）list-databases

list-databases 命令可以列出 RDBMS 中所有数据库的名称。例如：

sqoop list-databases --connect jdbc:mysql://localhost:3306/ --username root --password root

（12）sqoop list-tables

list-tables 命令可以列出指定的 RDBMS 中特定数据库中的所有表。例如：

sqoop list-tables --connect jdbc:mysql://localhost:3306/test --username root --password root

## 7.4　Sqoop 数据导入

Sqoop 数据导入使用 sqoop import 命令实现，可以将 RDBMS 表中数据导入 HDFS、Hive 及 HBase。

在介绍 Sqoop 数据导入之前，首先在 MySQL 中创建一个 test 数据库，在 test 数据库中创建一个 employees 表，并导入数据。

$ mysql -uroot -proot
mysql>CREATE DATABASE IF NOT EXISTS test;
mysql>USE test;
mysql>CREATE TABLE emp(
empno int ,ename char(20),job char(20),mgr int,hiredate datetime,sal float,comm float,deptno int);
mysql>LOAD DATA LOCAL INFILE '/home/hadoop/data/emp.txt' INTO TABLE emp FIELDS TERMINATED BY ',';

### 7.4.1　Sqoop 命令参数

Sqoop 命令的使用都是通过参数控制的，所有的 Sqoop 命令都有一些公共参数和各个命令的专用参数。

#### 1．公共参数

Sqoop 命令的公共参数如表 7-1 所示。

表 7-1　Sqoop 命令的公共参数

| 参数 | 描述 |
| --- | --- |
| --connect <jdbc-uri> | 指定 JDBC 连接的 URL |
| --connection-manager <class-name> | 指定连接使用的管理类 |
| --driver <class-name> | 手动指定 JDBC 驱动类 |
| --hadoop-mapred-home <dir> | 覆盖$HADOOP_MAPRED_HOME 设置 |
| --help | 打印命令使用说明 |
| --password-file <password-file> | 指定包含授权密码的文件 |
| -P | 通过控制台手动输入密码 |
| --password <password> | 在命名行中直接指定授权密码 |
| --username <username> | 指定授权用户名 |
| --verbose | 打印 Sqoop 任务的更多信息 |
| --connection-param-file <filename> | 可选参数，用于提供连接参数 |
| --relaxed-isolation | 设置每个 Mapper 连接事务隔离 |

## 2. import 命令常用参数

在使用 import 命令向 Hadoop 导入数据时常用的参数如表 7-2 所示。

表 7-2　import 命令的常用参数

| 参数 | 描述 |
| --- | --- |
| --append | 导入的数据追加到 HDFS 已经存在的数据集中 |
| --as-avrodatafile | 导入数据到 avro 文件 |
| --as-sequenceFile | 导入数据到 sequence 文件 |
| --as-textfile | 导入数据到纯文本文件（默认） |
| --as-parquetfile | 导入数据到 parquet 文件 |
| --boundary-query \<statement\> | 指定创建分片时的边界 |
| --columns \<col,col,col…\> | 指定从表中导入的字段 |
| --delete-target-dir | 如果导入的路径存在，则删除 |
| --direct | 使用 direct 连接 |
| --fetch-size \<n\> | 从数据库一次性读入的记录数 |
| --inline-lob-limit \<n\> | 设定内置 LOB 数据类型的最大值 |
| -m, --num-mappers \<n\> | 指定并行导入数据的 Map 任务个数，默认为 4 个 |
| -e, --query \<statement\> | 导入 SQL 语句的查询结果 |
| --split-by \<column-name\> | 一般与-m 参数一起使用，指定分隔 split 的字段 |
| --autoreset-to-one-mapper | 如果导入的表没有主键，也没有提供 split-by 字段，将自动使用一个 Mapper |
| --table \<table-name\> | 指定要导入的名 |
| --target-dir \<dir\> | 指定目标 HDFS 路径 |
| --warehouse-dir \<dir\> | 指定目标路径的根目录 |
| --where \<where clause\> | SQL 查询语句中的查询条件 |
| -z,--compress | 打开压缩功能 |
| --compression-codec \<c\> | 使用 Hadoop 的压缩方式，默认为 gzip |
| --null-string \<null-string\> | 源表中的空字符串导入后转换为 null-string，默认为"null" |
| --null-non-string \<null-string\> | 源表中的非字符串空值导入后转换为 null-string |
| --enclosed-by \<char\> | 给字段值前后加上的指定字符 |
| --escaped-by \<char\> | 设置转意字符 |
| --fields-terminated-by \<char\> | 设置字段分隔符 |
| --lines-terminated-by \<char\> | 设置行结束符 |
| --mysql-delimiters | 使用 MySQL 默认分隔符。字段分隔符为逗号","、行分隔符为"\n"、转意字符为"\"、可选必填字符为"：" |

## 7.4.2　数据从 MySQL 导入 HDFS

将数据从 MySQL 导入 HDFS 时，可以将表中的所有数据、部分数据、指定字段的数据导入 HDFS 文件，也可以通过 SQL 语句将一个或多个表中数据的查询结果导入 HDFS。可以进行数据的全量导入，也可以进行数据的增量导入。

## 1. 全量导入

将关系数据库中的数据按特定条件一次性导入 Hadoop 的 HDFS 中，称为全量导入。

在进行数据导入时，通常需要指定下列参数。

--connect：指定关系数据库的 URL 及数据库名称。

--username：指定连接关系数据库的用户名。

--password：指定连接关系数据库的用户密码。

--table：指定要导入 HDFS 的表。

--target-dir：指定导入数据存储位置。如果不指定，默认存储集群的当前目录中；如果指定的目录已经存在，将产生错误。

--delete-target-dir：如果参数 target-dir 指定的目录已经存在，则删除该目录。

-m：指定启动 Map 任务的数量，即并行度，默认值为 4。

--split-by：如果-m 不为 1，需要指定分片字段。

--fields-terminated-by：指定 HDFS 文件中字段分隔符，默认为逗号","。

**例 1**：将 MySQL 中 test 数据库中的 employees 表导入 HDFS 中。

```
$ sqoop import --connect jdbc:mysql://master:3306/test --username root --password root --table emp --target-dir /emps/emp1 --delete-target-dir -m 2 --split-by empno
```

执行完成后，可以使用 hdfs dfs -ls 命令查看 target-dir 参数指定的目录下新创建的文件，文件名以 part-m 为前缀，文件数量为参数 m 指定的值。可以使用 hadoop dfs -cat 命令查看文件的内容，可以看到 HDFS 文件中每行对应表中的一条记录，字段之间默认使用逗号隔开。如图 7-5 所示。

```
hadoop@master:~$ hdfs dfs -ls /emps/emp1
Found 3 items
-rw-r--r--   1 hadoop supergroup          0 2021-03-31 22:28 /emps/emp1/_SUCCESS
-rw-r--r--   1 hadoop supergroup       3308 2021-03-31 22:28 /emps/emp1/part-m-00000
-rw-r--r--   1 hadoop supergroup       3317 2021-03-31 22:28 /emps/emp1/part-m-00001
hadoop@master:~$ hdfs dfs -cat /emps/emp1/part*
100,King,AD_PRES,0,2003-06-17 00:00:00.0,24000.0,7200.0,90
101,Kochhar,AD_VP,100,2005-09-21 00:00:00.0,17000.0,5100.0,90
102,De Haan,AD_VP,100,2001-01-13 00:00:00.0,17000.0,5100.0,0
```

图 7-5 全表导入 HDFS 后的结果

在数据导入时，可以通过 columns 参数指定导入的字段、通过 where 参数设置数据选择条件。

**例 2**：将 emp 表中 sal 值大于 5000 元的员工的 empno、ename、hiredate 和 sal 列的数据导入 HDFS 文件中，字段间使用\t 分隔。

```
$ sqoop import --connect jdbc:mysql://master:3306/test --username root --password root --table emp --columns empno,ename,hiredate,sal --where 'sal>5000' --target-dir /emps/emp2 --delete-target-dir --fields-terminated-by '\t'  -m 2 --split-by empno
```

除通过指定 table、columns、where 等参数进行数据导入外，Sqoop 还支持指定 query 或者 e 参数传入一个 SQL 语句，把该 SQL 语句的执行结果导入 HDFS 中。使用 query 参数或 e 参数时，需要指定 target-dir 参数，并需要在 SQL 语句中增加一个$CONDITIONS 字符串。

**例 3**：将 emp 表中各个部门的部门号、人数和平均工资导入 HDFS 中。

```
$ sqoop import --connect jdbc:mysql://master:3306/test --username root --password root --query 'SELECT deptno,count(*) num,avg(sal) avgsal FROM emp WHERE $CONDITIONS GROUP BY deptno' --target-dir /emps/emp3 --delete-target-dir -m 1
```

如果 SQL 语句外层使用单引号，内层使用双引号，则$CONDITIONS 的$符号不需要转义；

如果 SQL 语句外层使用双引号，内层使用单引号，那么$CONDITIONS 的$符号需要使用转义符号\进行转义。

**例 4**：将 emp 表中 sal 大于 2000 元，job 为 ST_CLERK 的员工信息导入 HDFS 中。

```
$ sqoop import --connect jdbc:mysql://master:3306/test --username root --password root --query 'SELECT * FROM emp WHERE sal>2000 AND job="ST_CLERK" AND $CONDITIONS' --target-dir /emps/emp4 --delete-target-dir -m 1
```

或者

```
$ sqoop import --connect jdbc:mysql://master:3306/test --username root --password root --query "SELECT * FROM emp WHERE sal>2000 AND job='ST_CLERK' AND \$CONDITIONS" --target-dir /emps/emp4 --delete-target-dir -m 1
```

**2. 增量导入**

在生产环境中，系统可能会定期从与业务相关的关系数据库向 Hadoop 导入数据，导入后进行后续离线分析。此时不可能再将所有数据重新导入一遍，而是采用另一种导入模式，即增量导入。增量导入分两种，一种是基于递增列的 append 方式的增量导入，第二种是基于时间列的 lastmodified 方式的增量导入。

增量导入时，需要指定下列 3 个参数。

--check-column (col)：指定检查列，根据该列的值决定哪些行需要导入。

--incremental (mode)：指定增量导入方式。append 表示导入 check-column 列值大于 last-value 值的行；lastmodified 表示导入 check-column 列值为 last-value 指定日期之后的行。

--last-value：指定导入 check-column 列的值大于 last-value 值的行。

（1）append 方式的增量导入

append 方式的增量导入需要在关系数据库表中有一个数值递增列，基于该列值进行增量导入。进行增量导入之前，需要进行一次完全导入，以后在此基础上进行增量导入。

**例 5**：利用 append 增量导入方式将 emp 表中数据导入 HDFS 中。

① 首先对 emp 表进行一次全量导入：

```
$ sqoop import --connect jdbc:mysql://master:3306/test --username root --password root --table emp --target-dir /emps/emp5 --delete-target-dir --m 2 --split-by empno
```

数据导入完成后，可以看到在 HDFS 中 empno 最大值为 206。

② 在 MySQL 数据库的 emp 表中插入两条 empno 大于 206 的记录。

```
mysql>INSERT INTO emp VALUES(207,'zhangsan','SH_CLERK',110,'2020-10-1',5000,500,20);
mysql>INSERT INTO emp VALUES(208,'lisi','SH_CLERK',110,'2020-5-10',3500,200,30);
```

③ 采用 append 增量导入方式将 emp 表中新增的 empno 值大于 206 的数据导入 HDFS 中。

```
$ sqoop import --connect jdbc:mysql://master:3306/test --username root --password root --table emp --incremental append --check-column empno --last-value 206 --target-dir /emps/emp5 --m 2 --split-by empno
```

需要注意的是，在增量导入中，不能使用--delete-target-dir 参数。增量导入完成后，查看 HDFS 文件，可以看到在原来数据的基础上，将 empno 大于 206 的数据导入进来了。

（2）lastmodified 方式的增量导入

lastmodified 方式增量导入需要关系数据库表中存在一个时间戳字段，基于该时间戳进行增量导入。

**例 6**：利用 lastmodified 增量导入方式将 emp 表中数据导入 HDFS 中。

① 首先对 emp 表进行一次全量导入：

```
$ sqoop import --connect jdbc:mysql://master:3306/test --username root --password root --table emp --target-dir /emps/emp6 --delete-target-dir --m 2 --split-by empno
```

数据导入完成后，可以看到在 HDFS 中日期字段 hiredate 最大值为"2020-5-10"。

② 在 MySQL 数据库的 emp 表中插入两条 hiredate 日期在"2020-5-10"之后的记录。

mysql> INSERT INTO emp VALUES(300,'wangwu','SH_CLERK',110,'2021-3-1',5000,500,20);
mysql> INSERT INTO emp VALUES(301,'laoli','SH_CLERK',120,'2021-5-1',3500,200,30);

③ 采用 append 增量导入方式将 emp 表中新增的 hiredate 值在"2020-10-01"之后的数据导入 HDFS 中。

$ sqoop import --connect jdbc:mysql://master:3306/test --username root   --password root   --table emp --incremental lastmodified --check-column hiredate --last-value "2020-10-1" --merge-key empno --target-dir /emps/emp6 --m 2 --split-by empno

需要注意的是，lastmodified 增量导入时需要使用 -merge-key 参数指定数据合并列（主键），即合并键值相同的记录。完成数据导入后，查看 HDFS 文件，可以看到在原来数据的基础上，将 hiredate 值在"2020-10-01"之后的数据导入进来了。

### 7.4.3 数据从 MySQL 导入 Hive

将数据导入 Hive 的过程本质上是先将数据导入临时 HDFS 文件中，然后加载到 Hive 中，同时将临时 HDFS 文件移动到 Hive 默认存储位置，并删除临时 HDFS 文件。

Sqoop import 命令中与 Hive 数据导入相关的参数如表 7-3 所示。

表 7-3 与 Hive 数据导入相关的参数

| 参数 | 说明 |
| --- | --- |
| --hive-home <dir> | Hive 的安装目录，可以通过该参数覆盖之前默认配置的目录 |
| --hive-import | 将数据从关系数据库中导入 Hive 表中 |
| --hive-overwrite | 覆盖掉在 Hive 表中已经存在的数据 |
| --create-hive-table | 默认为 false，如果目标表已经存在了，作业执行失败 |
| --hive-table <table-name> | 指定导入数据的 Hive 表名称，默认使用 MySQL 的表名 |
| --hive-drop-import-delims | 导入数据到 Hive 时，去掉字符串类型字段中的\r、\n、\01 字符 |
| --hive-delims-replacement <arg> | 用自定义的字符串替换字符串类型字段中的\r、\n、\01 字符 |
| --hive-partition-key <key> | 指定 Hive 分区字段名 |
| --hive-partition-value <value> | 指定导入 hive-partition-key <key>分区的值 |
| --hive-database <database-name> | 指定 Hive 中的库 |
| --map-column-hive <arg> | 覆盖 SQL 数据类型到 Hive 数据类型的默认映射 |

向 Hive 中导入数据时，可以在 Hive 中自动创建表，也可以预先在 Hive 中创建表，为了防止数据类型的不兼容，建议在 Hive 中预先创建表。如果没有指定 Hive 表名，则默认创建与关系数据库中表同名的表。如果指定的 Hive 表已存在，数据可以采用追加方式或覆盖方式导入。如果没有指定 Hive 中的库名，则默认为 default 库；如果指定的库不存在，将产生错误。

**例 7**：将 MySQL 数据库中的 emp 表中的数据导入 Hive 中。

$ sqoop import --connect jdbc:mysql://master:3306/test --username root --password root --table emp --hive-import --hive-database empdb   --hive-table hiveemp -m1

导入完成后，可以查看 hiveemp 表中的数据，如图 7-6 所示。

向 Hive 中导入数据时，可以指定数据导入过程中的临时 HDFS 目录、字段分隔符、行分隔符等。

```
0: jdbc:hive2://master:10000> SELECT * FROM hiveemp LIMIT 5;
+--------------+--------------+------------+-------------+----------------------+------------+-------------+---------------+
| hiveemp.empno| hiveemp.ename| hiveemp.job| hiveemp.mgr | hiveemp.hiredate     | hiveemp.sal| hiveemp.comm| hiveemp.deptno|
+--------------+--------------+------------+-------------+----------------------+------------+-------------+---------------+
| 198          | Connell      | SH_CLERK   | 124         | 2007-06-21 00:00:00.0| 2600.0     | 780.0       | 0             |
| 199          | Grant        | SH_CLERK   | 124         | 2008-01-13 00:00:00.0| 2600.0     | 780.0       | 0             |
| 200          | Whalen       | AD_ASST    | 101         | 2003-09-17 00:00:00.0| 4400.0     | 1320.0      | 10            |
| 201          | Hartstein    | MK_MAN     | 100         | 2004-02-17 00:00:00.0| 13000.0    | 3900.0      | 20            |
| 202          | Fay          | MK_REP     | 201         | 2005-08-17 00:00:00.0| 6000.0     | 1800.0      | 20            |
+--------------+--------------+------------+-------------+----------------------+------------+-------------+---------------+
```

图 7-6 数据导入 Hive 表

**例 8**：将 MySQL 数据库中的 emp 表导入 Hive 的 empdb 库中，表名为 emp。如果 emp 表已经存在，数据采用覆盖方式导入。

```
$ sqoop import --connect jdbc:mysql://master:3306/test --username root --password root --table emp
--hive-import --hive-database empdb --hive-overwrite --target-dir /sqoop/mysql/hive --delete-target-dir
--fields-terminated-by '\t' --lines-terminated-by '\n' -m 1
```

导入成功后，进入 Hive 的 empdb 库，可以查看新创建的 emp 表及其数据。与此同时，临时目录 /sqoop/mysql/hive 已经被删除，数据文件移动到 /user/hive/warehouse/empdb.db/emp 目录中，文件中字段分隔符为 "\t"，如图 7-7 所示。

```
hadoop@master:~$ hdfs dfs -ls /sqoop/mysql/hive
ls: `/sqoop/mysql/hive': No such file or directory
hadoop@master:~$ hdfs dfs -ls /user/hive/warehouse/empdb.db/emp
Found 3 items
-rw-r--r--   1 hadoop supergroup          0 2021-04-01 21:46 /user/hive/warehouse/empdb.db/emp/_SUCCESS
-rw-r--r--   1 hadoop supergroup       6872 2021-04-01 21:46 /user/hive/warehouse/empdb.db/emp/part-m-00000
-rw-r--r--   1 hadoop supergroup       6872 2021-04-01 21:48 /user/hive/warehouse/empdb.db/emp/part-m-00000_copy_1
hadoop@master:~$ hdfs dfs -cat /user/hive/warehouse/empdb.db/emp/part-m-00000
198    Connell    SH_CLERK    124    2007-06-21 00:00:00.0    2600.0    780.0    0
199    Grant      SH_CLERK    124    2008-01-13 00:00:00.0    2600.0    780.0    0
200    Whalen     AD_ASST 101    2003-09-17 00:00:00.0    4400.0    1320.0    10
```

图 7-7 数据导入 Hive 后的结果

## 7.4.4 数据从 MySQL 导入 HBase

与 Hive 数据导入不同，向 HBase 中导入数据之前必须先在 HBase 中创建相应的表，然后才可以进行数据导入。

Sqoop import 命令中与 HBase 数据导入相关的参数如下。

column-family：指定 HBase 表的列族。

hbase-bulkload：是否允许批量导入。

hbase-table：指定导入目标的 HBase 表。

hbase-row-key：指定的 HBase 表的行键的列。

如果没有指定 hbase-row-key 参数，Sqoop 默认使用 split-by 参数指定列作为行键列。如果没有指定 split-by 参数，将尝试使用关系数据库表中的主键列作为行键列。如果关系数据库表具有复合主键，那么 hbase-row-key 参数值必须是以逗号为分隔符的复合属性列表，相应行键值为使用下画线为分隔符的相应行中的复合属性对应的值的组合。如果指定的 HBase 表或列族不存在，则结束作业的运行。为了减少 HBase 负载，可以通过设置 hbase-bulkload 参数启动批量加载，而不是直接写入。

**例 9**：将 MySQL 数据库中的 emp 表中 empno、ename、sal、deptno 列的数据导入 HBase 的 hbaseemp 表中。

① 启动 Hbase 服务，在 HBase 中创建一个 hbaseemp 表，指定列族为 info。

$ start-hbase.sh
$ hbase shell
hbase(main):002:0> create 'hbaseemp','info'

② 使用 sqoop import 命令将 emp 表中指定列的数据导入 HBase 的 hbaseemp 表中。

$ sqoop import --connect jdbc:mysql://master:3306/test --username root --password root --table emp --columns empno,ename,sal,deptno --hbase-table hbaseemp --column-family info --hbase-row-key empno    -m 1

③ 导入成功后，可以在 HBase 中查看 hbaseemp 表中的数据。

hbase(main):003:0>scan 'hbaseemp',{LIMIT=>10}

## 7.5 Sqoop 数据导出

Sqoop 数据导出使用 sqoop export 命令实现，将一组文件从 HDFS 导出到关系数据库已存在的表中。根据用户指定的分隔符，读取输入文件并解析为一组记录。默认操作是将解析后的记录转换为一组 INSERT 语句，然后将记录插入数据库表中。在更新模式（Update Mode）中，Sqoop 根据记录生成 UPDATE 语句，更新数据库表中已经存在的记录。在调用模式（Call Mode）中，Sqoop 将为每一条记录执行一个存储过程调用。

### 7.5.1 Sqoop export 命令参数

除 7.4.1 节介绍的 Sqoop 命令公共参数外，export 命令还有一些常用的参数，如表 7-4 所示。

表 7-4  export 命令常用的参数

| 参数 | 说明 |
| --- | --- |
| --columns < col,col,col… > | 指定导出到关系数据库表的字段 |
| --direct | 指定使用 direct 模式 |
| --export-dir < dir > | 指定要导出 HDFS 源文件路径 |
| -m, --num-mappers < n > | 指定并行导出操作的 Map 任务个数，默认为 4 个 |
| --table < table-name > | 指定关系数据库中的表名 |
| --call < stored-proc-name > | 指定调用的存储过程名称 |
| --update-key <col-name> | 指定用于更新的字段，多个字段可以用逗号分隔 |
| --update-mode < mode > | 指定当某记录在数据库中找不到对应记录时的操作。取值为 updateonly（默认）或 allowinsert |
| --input-null-string < null-string > | 将 string 字段中的 null 值解析为特定符号 |
| --input-null-non-string < null-string > | 将非 string 字段中的 null 值解析为特定符号 |
| --staging-table < staging-table-name > | 在最终插入目标表之前临时存储数据的表 |
| --clear-staging-table | 表示可以删除 staging-table 表中的任何数据 |
| --batch | 对底层数据操作使用 batch 模式 |
| --input-enclosed-by <char> | 设置字段值前后加上的指定字符 |
| --input-escaped-by <char> | 指定转义字符 |
| --input-fields-terminated-by <char> | 指定源文件中字段之间的分隔符 |
| --input-lines-terminated-by <char> | 指定源文件中行之间的分隔符 |
| --input-optionally-enclosed-by <char> | 给带有双引号或单引号的字段前后加上指定字符 |

在 Sqoop export 命令中，必须使用--export-dir 参数指定 HDFS 源文件的路径，使用--table 或--call 参数指定数据导出的表或调用的存储过程。默认情况下，向表中所有列导出数据，可以通过--columns 参数指定部分列，但其他列需要有默认值或者可以为 NULL，否则导出作业将失败。

## 7.5.2 从 HDFS 导出数据到 MySQL

从 HDFS 导出数据到 MySQL 的操作包括数据插入和数据更新两种模式。

### 1. 数据插入模式

默认情况下，export 命令将每一条导出记录转化为一条 INSERT 语句，以追加方式将记录插入目标数据库表中。如果目标数据库表有约束（如主键约束），需要注意避免导出记录违反这些约束条件的限制。数据插入模式主要应用于将数据导出到一张新表或者空表中。

**例 10**：将 HDFS 上的 /emps/emp1 目录中的文件导出到 MySQL 的 test 数据库的 exportemp 表中。

① 在 MySQL 的 test 数据库中创建一个新表 exportemp，注意表结构与文件结构匹配。

```
$ mysql -uroot -proot
mysql> USE test;
mysql> CREATE TABLE exportemp(
empno int ,ename char(20),job char(20),mgr int,hiredate datetime,sal float,comm float,deptno int);
```

② 使用 sqoop export 命令将数据导出到 exportemp 表中。

```
$ sqoop export --connect jdbc:mysql://master:3306/test --username root --password root --table exportemp --export-dir /emps/emp1 --input-fields-terminated-by ',' --m 1
```

③ 导出完成后，可以查看 exportemp 表中的数据。

```
mysql> SELECT * FROM exportemp LIMIT 5;
mysql> SELECT count(*) FROM exportemp;
```

结果如图 7-8 所示。

```
mysql> SELECT * FROM exportemp LIMIT 5;
+-------+--------+---------+------+---------------------+-------+------+--------+
| empno | ename  | job     | mgr  | hiredate            | sal   | comm | deptno |
+-------+--------+---------+------+---------------------+-------+------+--------+
|   100 | King   | AD_PRES |    0 | 2003-06-17 00:00:00 | 24000 | 7200 |     90 |
|   101 | Kochhar| AD_VP   |  100 | 2005-09-21 00:00:00 | 17000 | 5100 |     90 |
|   102 | De Haan| AD_VP   |  100 | 2001-01-13 00:00:00 | 17000 | 5100 |      0 |
|   103 | Hunold | IT_PROG |  102 | 2006-01-03 00:00:00 |  9000 | 2700 |      0 |
|   104 | Ernst  | IT_PROG |  103 | 2007-05-21 00:00:00 |  6000 | 1800 |     60 |
+-------+--------+---------+------+---------------------+-------+------+--------+
5 rows in set (0.00 sec)

mysql> SELECT count(*) FROM exportemp;
+----------+
| count(*) |
+----------+
|      107 |
+----------+
1 row in set (0.00 sec)
```

图 7-8 以数据插入模式导出数据到 MySQL

### 2. 数据更新模式

如果在 sqoop export 命令中指定了 update-key 参数，Sqoop 将每一条导出记录转化为一条 UPDATE 语句，然后根据 --update-key 指定的列值匹配更新关系数据库表中已经存在的记录。

如果一条 UPDATE 语句匹配不到对应的记录或者匹配到多条记录，不会发生任何的错误，导出过程会正常继续执行后面的 SQL 语句，新的记录不会插入目标表中。如果在使用 update-key 参数的同时指定 update-mode 参数为 allowinsert，UPDATE 语句匹配不到记录时就会以 INSERT 的形式插入目标表中。update-key 参数可以设置多个字段，多个字段之间用逗号隔开。

**例 11**：将 HDFS 上的 /emps/emp2 目录中的文件以数据更新模式导出到 MySQL 数据库的 test 库的 exportemp 表中。如果导出的数据不存在，则插入数据。

```
$ sqoop export --connect jdbc:mysql://master:3306/test --username root --password root --table exportemp --columns empno,ename,hiredate,sal --update-key empno --update-mode allowinsert --input-fields-terminated-by '\t' --export-dir /emps/emp2 --m 1
```

导出完成后，查询 exportemp 表可以发现新增了 58 行数据，如图 7-9 所示。

```
mysql> SELECT count(*) FROM exportemp;
+----------+
| count(*) |
+----------+
|      165 |
+----------+
```

图 7-9 以数据更新模式导出数据到 MySQL

### 3．批量导出

数据导出时，默认情况下是读取一行 HDFS 文件的数据，就插入或更新到关系数据库中的表，这样做容易造成性能低下。可以使用参数-Dsqoop.export.records.pre.statement 指定批量导出，即一次性导出指定的数据到关系数据库的表中。

**例 12**：将 HDFS 上的 /emps/emp 目录中的文件批量导出到 MySQL 数据库的 test 库的 exportemp 表中。

```
$ sqoop export -Dsqoop.export.records.pre.statement --connect jdbc:mysql://master:3306/test --username root --password root --table exportemp   --export-dir   /emps/emp1 --input-fields-terminated-by ',' --m 1
```

## 7.5.3 从 Hive 导出数据到 MySQL

将 Hive 表中数据导出到关系数据库的方法与导出 HDFS 数据的方法相同，只需要将 --export-dir 参数设置为 Hive 表对应的目录即可。

**例 13**：统计 Hive 中 empdb 库中的 emp 表中不同部门的部门号、人数、平均工资、最高工资和最低工资，并将结果导出到 MySQL 的 test 数据库的 deptstat 表中。

① 在 Hive 中进行数据统计分析，并将结果保存到一个名为 deptresult 的表中。

```
use empdb;
CREATE TABLE deptresult
AS
SELECT deptno,count(*) num,round(avg(sal),2) avgsal,round(max(sal),2) maxsal,round(min(sal),2) minsal
FROM emp GROUP BY deptno ORDER BY deptno;
```

② 在 MySQL 的 test 数据库中创建 deptstat 表，用于接收导出的数据。

```
mysql> use test;
mysql> CREATE TABLE deptstat (empno int,num int,avgsal float,maxsal float,minsal float);
```

③ 利用 sqoop export 命令将 Hive 中的 deptresult 数据导出到 MySQL 的 deptstat 表中。

```
$ sqoop export --connect jdbc:mysql://localhost/test --username root --password root --table deptstat --export-dir /user/hive/warehouse/empdb.db/deptresult --input-fields-terminated-by '\01' -m 1
```

④ 在 MySQL 中查看导出结果，如图 7-10 所示。

```
mysql> SELECT * FROM deptstat;
+-------+-----+---------+--------+--------+
| empno | num | avgsal  | maxsal | minsal |
+-------+-----+---------+--------+--------+
|     0 |  18 | 7722.22 |  17000 |   2600 |
|    10 |   2 |    4400 |   4400 |   4400 |
|    20 |   8 |    7250 |  13000 |   5000 |
|    30 |  14 | 2985.71 |   3500 |   2500 |
|    40 |   2 |    6500 |   6500 |   6500 |
|    50 |  86 | 3516.28 |   8200 |   2100 |
|    60 |   6 |    5000 |   6000 |   4200 |
|    70 |   2 |   10000 |  10000 |  10000 |
|    80 |  68 | 8955.88 |  14000 |   6100 |
|    90 |   4 |   20500 |  24000 |  17000 |
|   100 |   8 |    9027 |  12008 |   6900 |
|   110 |   4 |   10154 |  12008 |   8300 |
+-------+-----+---------+--------+--------+
```

图 7-10 将 Hive 数据分析结果导出到 MySQL

## 7.5.4 中文乱码问题

由于 Hadoop 平台与 MySQL 数据库中数据编码方式不同，利用 Sqoop 进行数据导入、导出中文数据时会产生乱码的问题，解决方法是设置统一的 UTF-8 编码。具体方法为：

① 在 MySQL 中创建数据库时设定数据库编码方式为 UTF-8，指定 DEFAULT CHARACTER SET utf8 COLLATE utf8_general_ci。

② 在 MySQL 中创建表时设定数据编码方式为 UTF-8，指定 CHARSET utf8 COLLATE utf8_general_ci。

③ 在 Sqoop import 或 export 命令中，--connect 参数指定的 URL 中设定 UTF-8 编码，指定 useUnicode=true&characterEncoding=utf-8"。

**例 14**：在 MySQL 数据库中创建一个包含中文信息的表 student，并插入数据，然后将该表中数据导入 HDFS，再从 HDFS 导出到 student 表中，以验证中文是否正常显示。

① 在 MySQL 中创建一个 testcode 库，并在库中创建一个 student 表，插入两条记录。

```
mysql> CREATE DATABASE testcode DEFAULT CHARACTER SET UTF8 COLLATE utf8_general_ci;
mysql> USE testcode;
mysql> CREATE TABLE student (sno char(10),sname char(10),sage int,sex char(10),info varchar(1000))
        CHARSET UTF8 COLLATE utf8_general_ci;
mysql> INSERT INTO student VALUES ( 100,"张三",25,"male","good"),(101,"李四",30,"female","hello");
```

② 将 student 表中数据导入 HDFS，并查看导入结果。

```
$ sqoop import --connect "jdbc:mysql://localhost:3306/testcode?useUnicode=true&characterEncoding=utf-8"
--username root --password root --table student --target-dir ./studentdata --delete-target-dir --m 1
```

数据导入结果如图 7-11 所示，中文正常显示。

```
hadoop@master:~$ hdfs dfs -ls ./studentdata
Found 2 items
-rw-r--r--   1 hadoop supergroup          0 2021-04-03 10:15 studentdata/_SUCCESS
-rw-r--r--   1 hadoop supergroup         51 2021-04-03 10:15 studentdata/part-m-00000
hadoop@master:~$ hdfs dfs -cat ./studentdata/part-m-00000
100,张三,25,male,good
101,李四,30,female,hello
```

图 7-11 包含中文信息的数据导入结果

③ 将 HDFS 中包含中文信息的文件导出到 MySQL 表中，并查看导出数据。

```
$ sqoop export --connect "jdbc:mysql://localhost:3306/testcode?useUnicode=true&characterEncoding=utf-8"
--username root--password root --table student --export-dir ./studentdata-m 1
```

数据导出结果如图 7-12 所示，中文正常显示。

```
mysql> SELECT * FROM student;
+------+-------+------+--------+-------+
| sno  | sname | sage | sex    | info  |
+------+-------+------+--------+-------+
| 100  | 张三  |   25 | male   | good  |
| 101  | 李四  |   30 | female | hello |
| 100  | 张三  |   25 | male   | good  |
| 101  | 李四  |   30 | female | hello |
+------+-------+------+--------+-------+
```

图 7-11 包含中文信息的数据导出结果

## 本 章 小 结

本章首先介绍了 Sqoop 的基本功能、架构、版本等基础知识，然后介绍了 Sqoop 的安装与配置及 Sqoop 常用命令，最后详细介绍了 Sqoop import 数据导入和 Sqoop export 数据导出的实现。通过本章的学习，读者可以掌握如何将关系数据库的数据导入 Hadoop 集群，包括导入 HDFS、Hive 和 HBase，可以将 Hadoop 集群中分析处理后的结果导出到关系数据库。

## 思考题与习题

### 1. 简答题
（1）简述在 Hadoop 生态系统中 Sqoop 组件的作用。
（2）Sqoop 的主要命令有哪些？各有什么功能？
（3）简述 Sqoop import 命令常用参数及其作用。
（4）Sqoop import 命令的操作模式有哪些？有何异同？
（5）简述 Sqoop export 命令常用参数及其作用。
（6）简述存在中文信息时，如何解决数据导入、导出时出现乱码的问题。

### 2. 选择题
（1）关于 Sqoop 的描述错误的是（    ）。
A. Sqoop 是 Hadoop 与关系数据库之间数据传递的工具
B. Sqoop 是 SQL 到 Hadoop 和 Hadoop 到 SQL
C. Sqoop 可以实现 Hadoop 内部数据之间的传递
D. Sqoop 执行是通过 MapReduce 实现的

（2）关于 Sqoop 的命令描述错误的是（    ）。
A. codegen 用于将关系数据库表映射为一个 java 文件、java class 类及相关的 JAR 文件
B. create-hive-table 用于生成与关系数据库表的表结构对应的 Hive 表
C. eval 用于执行 SQL 语句，并能将结果显示在控制台
D. job 用于生成一个 Sqoop 的任务，该任务并不执行，除非使用命令执行该任务

（3）下列不是 Sqoop import 导入数据的格式的是（    ）。
A. textfile　　　　　　B. josnFile　　　　　　C. avro　　　　　　D. sequenceFile

（4）下列不是有效的 Sqoop 命令的是（    ）。
A. sqoop import　　　B. sqoop list-database　　C. sqoop create-table　　D. sqoop list-tables

（5）使用 Sqoop import 导入数据时，若使用 query 参数，下列描述不正确的是（    ）。
A. 必须设置 split-by 参数　　　　　　　　B. 必须设置 target-dir 参数
C. 必须指定$CONDITIONS 字符串　　　　D. 不需要使用 table 参数指定源表

（6）关于 Sqoop import 的增量导入描述不正确的是（    ）。
A. 可以基于 append 方式增量导入
B. 可以基于 lastmodified 方式增量导入
C. 增量导入前需要先进行一次全量导入
D. lastmodified 增量导入方式不需要表中存储时间戳列

（7）使用 Sqoop import 向 Hive 中导入数据时，必须指定的参数为（    ）。

A. hive-import　　　B. target-dir　　　C. split-by　　　D. hive-table

（8）使用 Sqoop import 向 Hbase 中导入数据时，不需要指定的参数为（　　）。

A. hbase-table　　　B. hbase-row-key　　　C. column-family　　　D. target-dir

（9）使用 Sqoop export 向 MySQL 中导出数据时，下列描述不正确的是（　　）。

A. 需要先在 MySQL 中创建好表，用于存储数据

B. 通过参数 fields-terminated-by 设置文件中字段分隔符

C. 通过 export-dir 指定导出文件目录

D. 通过 update-mode 设置数据更新模式

（10）利用 Sqoop 进行数据导入、导出操作时，如果存在中文，容易出现乱码问题，下列哪一项无助于乱码的解决？（　　）

A. 在 MySQL 中创建数据库时指定编码方式为 UTF-8

B. 在 MySQL 中创建表时指定数据编码方式为 UTF-8

C. 在 sqoop export 或 sqoop import 命令中指定 UTF 编码方式

D. 在 Hadoop 中指定 UTF 编码方式

### 3. 实训题

（1）在 MySQL 数据库中创建一个名为 Sqoop 的库，编码方式采用 UTF-8。

（2）在 MySQL 的 Sqoop 库中，创建一个名为 buyer 的表，包含 buyer_id、reg_date、reg_ip、buyer_status 四个字段，buyer_id 为主键，字符类型均为 varchar(100)。向 buyer 表中插入下列数据：

| buyer_id | reg_date | reg_ip | buyer_status |
|---|---|---|---|
| 10001 | 2020-2-1 | 192.168.102.11 | Active |
| 10002 | 2020-2-2 | 192.168.102.12 | Active |

（3）将 buyer 表数据导入 HDFS 的/mysqoop 目录中。

（4）将 buyer 表数据导入 Hive 的 sqoop 库的 hivebuyer 表中，并查看导入结果。

（5）将 buyer 表数据导入 HBase 的 hbasebuyer 表中，并查看导入结果。

（6）向 MySQL 的 buyer 表中添加一条数据：

| buyer_id | reg_date | reg_ip | buyer_status |
|---|---|---|---|
| 10008 | 2020-2-5 | 192.168.102.33 | Active |

采用增量导入的方法将 buyer 表中 buyer_id 等于 10007 之后的数据导入 HDFS 的/mysqoop 的目录中，并查看导入后的结果。

（7）将 HDFS 的/mysqoop 目录中数据导出到 MySQL 的 exportbuyer 表中，并查看结果。

（8）将 Hive 中的 hivebuyer 表中数据导出到 MySQL 的 exportbuyer 表中，并查看结果。

# 第 8 章　数据采集工具 Flume

Flume 是一个分布式数据采集工具，可以将不同来源的数据采集到不同的目的地，特别在海量网站日志采集中得到广泛的应用，可以将不同 Web 服务中的日志采集、传输、聚合后保存到 Hadoop 集群。本章将介绍 Flume 体系结构、Flume 组件及数据采集案例等。
- Flume 概述：Flume 功能特性、Flume 版本、Flume 架构。
- Flume 安装与配置：Flume 安装、Flume 配置、Flume 测试。
- Flume 组件：Source、Channel、Sink、Interceptor、Selector、Sink Processor。
- Flume 数据采集案例与实施：实时采集本地文件到 HDFS、多源与多目的地数据采集。

## 8.1　Flume 概述

### 8.1.1　Flume 简介

Flume 是一个分布式的、高可靠的、高可用的日志采集系统，它将海量日志数据从不同的数据源进行采集、聚合并传输到一个集中的数据中心进行存储。Flume 不仅仅局限于实现日志数据的聚合，由于数据源是可以定制的，还可以用于传输大量数据，包括但不限于网络流量数据、社交媒体生成的数据、电子邮件信息等所有可能的数据源。Flume 最初由 Clouderra 提供，当前是 Apache 软件基金会（ASF）的顶级项目，称为 Apache Flume。

Flume 采用 Java 语言实现，采用系统框架设计，内置 Source、Channel、Sink 等多种插件，模块分明，既易于开发，也可以轻松与其他系统集成。事务机制保证了信息传递的可靠性。

Flume 有两个版本，Flume 0.9x 版本统称 Flume-og，Flume1.x 版本统称 Flume-ng。由于 Flume-ng 经过重大重构，与 Flume-og 有很大不同，使用时需要注意。本书采用的是 Flume 的最新版本 Flume1.9。

### 8.1.2　Flume 架构

event 是 Flume 数据流的基本单位，Agent 是 Flume 中最小的独立运行单位。Agent 是一个 JVM 进程，运行在日志采集节点上，包含 Source（源）、Channel（通道）和 Sink（目的地）3 个组件，将外部数据源产生的数据以 event 的形式传输到目的地。Flume 的基本架构如图 8-1 所示。

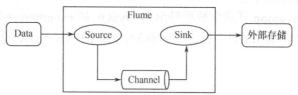

图 8-1　Flume 的基本架构

Source 组件从外部数据源读入 event，并写入 Channel。Channel 组件暂存由 Source 写入的 event，直到被 Sink 成功消费，event 才会被删除。Sink 组件从 Channel 读取 event，并写入目的地。

Flume 数据流由 event 组成，针对文本文件，每行数据对应一个 event。每个 event 包括 event header 和 eventbody 两部分。event header 是一些列键值对，用于传输标识类信息。event body 是一个字节数组，用于存储实际要传输的数据。

根据运行 Flume Agent 数量不同，Flume 架构分为单 Agent 数据流模型架构和多 Agent 数据流模型架构。根据 Source、Channel、Sink 组件数量不同，Flume 架构还可以分为多路数据流模型架构、Sink 组数据流模型架构等。

（1）单 Agent 数据流模型架构

单 Agent 数据流模型架构如图 8-2 所示，Agent 由单个 Source、Channel 和 Sink 组成，整个数据流向为 Web Server→Source→Channel→Sink→HDFS。

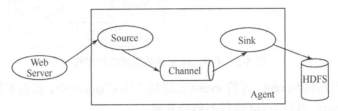

图 8-2　单 Agent 数据流模型架构

（2）多 Agent 串行传输数据流架构

多 Agent 串行传输数据流架构如图 8-3 所示，多个 Agent 串在一起，将数据从 Agent foo 传输到 Agent bar，再传输到目的地。

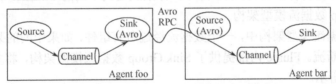

图 8-3　多 Agent 串行传输数据流架构

（3）多 Agent 聚合数据流模型架构

多 Agent 聚合数据流模型架构如图 8-4 所示，将位于不同服务器的 Agent1、Agent2 和 Agent3 收集到的数据汇集到一个中心节点 Agent4 上，再由 Agent4 汇聚将数据写入 HDFS 中。通常称 Agent1、Agent2、Agent3 为 Flume Agent，称 Agent4 为 Flume Collector。

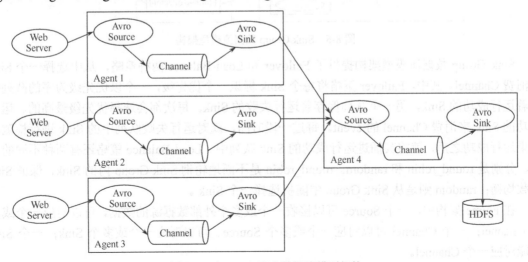

图 8-4　多 Agent 聚合数据流模型架构

（4）单 Agent 多路数据流模型架构

单 Agent 多路数据流模型架构如图 8-5 所示，一个 Agent 可由一个 Source、多个 Channel、多个 Sink 组成多路数据流。

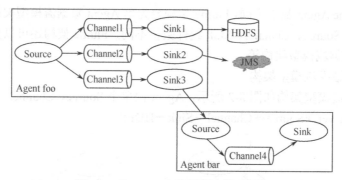

图 8-5　单 Agent 多路数据流模型架构

一个 Source 接收外部 event，并将 event 发送到 3 路 Channel 中，然后不同的 Sink 消费不同的 Channel 内的 event，再将 event 进行不同的处理。

Source 将 event 发送到 Channel 有两种不同的策略，分别为 replicating 和 multiplexing。其中，replicating 是 Source 将每个 event 都发送到所有的 Channel 中，即将 event 复制成 3 份发送到不同的 Channel 中。multiplexing 是 Source 根据一些映射关系，将不同种类的 event 发送到不同的 Channel 中，即将所有 event 分成 3 部分，分别发送到不同的 Channel 中。

（5）Sink Group 数据流模型架构

前面介绍的数据流模型架构中，一个数据流经过所有组件，如果中间的某个组件出现故障，会导致整个数据流断流。Flume 内部提供了 Sink Group 数据流模型架构，将多个 Sink 组件组合在一起，如图 8-6 所示。

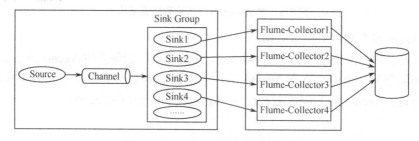

图 8-6　Sink Group 数据流模型架构

Sink Group 数据流模型架构提供了 Failover 和 Load-Balance 两种策略，从中选择一个 Sink 来消费 Channel。其中，Failover 策略将每个 Sink 标识一个优先级，一个以优先级为序的队列保存着运行成功的 Sink，另一个队列保存着运行失败的 Sink。每次都会选择优先级最高的、运行成功的 Sink 来消费 Channel 的 event。每过一段时间，就对运行失败队列中的 Sink 进行检测，如果运行成功之后，就将其插进运行成功的 Sink 队列中。Load-Balance 策略还有两种不同的机制，分别是 round_robin 和 random。round_robin 是不断地轮询 Sink Group 内的 Sink，保证 Sink 负载均衡；random 则是从 Sink Group 中随机选择一个 Sink。

在 Flume 架构中，一个 Source 可以接收一个或多个外部数据源的数据，可以对应一个或多个 Channel；一个 Channel 可以对应一个或多个 Source、可以对应一个或多个 Sink；一个 Sink 只能对应一个 Channel。

## 8.2　Flume 安装与配置

Flume 依赖于 Java 运行环境，因此在安装 Flume 之前需要先安装好 JDK，JDK 版本在 1.8 或之后。

### 1．安装 Flume

（1）下载 Flume 安装包

到 Flume 官网（http://flume.apache.org/download.html）下载 Flume 安装文件 apache-flume-1.9.0-bin.tar.gz 到 ~/Downloads 目录。

（2）解压 Flume 安装包

```
$ cd ~/Downloads
$ sudo tar -zxvf apache-flume-1.9.0-bin.tar.gz -C /usr/local
```

（3）重命名解压后的目录并修改目录所有者

```
$ cd /usr/local
$ sudo mv apache-flume-1.9.0-bin flume
$ sudo chown -R hadoop:hadoop flume
```

（4）配置环境变量

打开当前用户的环境变量配置文件：

```
$ gedit ~/.bashrc
```

在配置文件第一行输入如下信息：

```
export FLUME_HOME=/usr/local/flume
export PATH=$PATH:$FLUME_HOME/bin
```

保存该文件后，执行下面命令让配置文件立即生效。

```
$ source ~/.bashrc
```

（5）修改 flume-env.sh 配置文件

```
$ cd /usr/local/flume/conf
$ cp ./flume-env.sh.template ./flume-env.sh
$ gedit ./flume-env.sh
```

打开 flume-env.sh 文件以后，在文件中增加一行内容，用于设置 JAVA_HOME 变量：

```
export JAVA_HOME=/usr/lib/jvm/jdk1.8.0_121
```

### 2．配置 Flume

配置 Flume 的主要任务是编辑 Flume 配置文件。Flume 配置文件是一个 Java 属性文件，里面放置一系列键值对，用于描述 Agent 的 Source、Channel 和 Sink 组件的配置信息。Flume 配置文件的位置和名称没有固定要求，在启动 Flume 时指定，但一般都放置在 Flume 安装目录的 conf 文件夹中。

在 Flume 安装目录下的 conf 目录（/usr/local/flume/conf）中创建一个名为 example.conf 的配置文件，描述一个数据来源为 netcat 终端、目的地为日志控制台的单节点 Flume。

```
#单节点Flume配置
#为名为a1的Agent的Source、Channel和Sink组件命名
a1.sources = r1
a1.sinks = k1
a1.channels = c1
# Source组件属性配置
a1.sources.r1.type = netcat
a1.sources.r1.bind = localhost
```

```
a1.sources.r1.port = 44444
# Sink组件属性配置
a1.sinks.k1.type = logger
# Channel组件属性配置
a1.channels.c1.type = memory
a1.channels.c1.capacity = 1000
a1.channels.c1.transactionCapacity = 100
# Source组件和Sink组件与Channel组件绑定
a1.sources.r1.channels = c1
a1.sinks.k1.channel = c1
```

该配置文件定义了一个名为 a1 的 Agent，a1 的 Source 组件监听 localhost 主机的 44444 端口上的数据源，并将接收到的 event 发送给 Channel 组件。a1 的 Channel 组件将接收到的 event 缓存到内存，最大存储容量为 1000 个 event，每次从 Source 接收或发送给 Sink 的 event 以事务为单位，每个事务中包含 event 的最大数量为 100。a1 的 Sink 组件读取 Channel 组件中的 event 并随日志信息一起输出到控制台。

一个 Flume 配置文件中可以定义一个或多个 Agent，在启动 Flume 时指定启动哪个 Agent。

### 3. 启动 Flume

① 打开一个终端，执行下列命令，启动 Flume。

```
$ cd /usr/local/flume
$ flume-ng agent --name a1 --conf ./conf --conf-file ./conf/example.conf -Dflume.root.logger=INFO,console
```

参数说明：

--name、-n <name>：指定要启动的 Agent 名称。

--conf、-c <conf>：指定配置文件所在的目录。

--conf-file、-f<file>：指定配置文件。

Flume 启动后，日志信息如图 8-7 所示。

```
2021-04-05 11:00:21,808 (conf-file-poller-0) [INFO - org.apache.flume.node.Application.startAllComponents(Application.java:196)] Starting Sin
k k1
2021-04-05 11:00:21,826 (conf-file-poller-0) [INFO - org.apache.flume.node.Application.startAllComponents(Application.java:207)] Starting Sou
rce r1
2021-04-05 11:00:21,827 (lifecycleSupervisor-1-0) [INFO - org.apache.flume.source.NetcatSource.start(NetcatSource.java:155)] Source starting
2021-04-05 11:00:21,896 (lifecycleSupervisor-1-0) [INFO - org.apache.flume.source.NetcatSource.start(NetcatSource.java:166)] Created serverSo
cket:sun.nio.ch.ServerSocketChannelImpl[/127.0.0.1:44444]
```

图 8-7 Flume 启动后的日志信息

② 打开另外一个终端，执行 telnet 命令，连接 44444 端口。

```
$ telnet localhost 44444
```

③ 在 telnet 终端中，输入以换行符分隔的文本，每行文本将被转换为一个 event。例如，输入“hello world”“hello everyone”两行文本，如图 8-8 所示。

```
hadoop@master:~$ telnet localhost 44444
Trying 127.0.0.1...
Connected to localhost.localdomain.
Escape character is '^]'.
hello world
OK
hello everyone
OK
```

图 8-8 Source 组件接收的源数据

④ 启动 Flume 的控制台查看接收的数据，如图 8-9 所示。

```
cket:sun.nio.ch.ServerSocketChannelImpl[/127.0.0.1:44444]
2021-04-05 11:08:19,961 (SinkRunner-PollingRunner-DefaultSinkProcessor) [INFO - org.apache.flume.sink.LoggerSink.process(LoggerSink.java:95)]
 Event: { headers:{} body: 68 65 6C 6C 6F 20 77 6F 72 6C 64 0D          hello world. }
2021-04-05 11:08:22,716 (SinkRunner-PollingRunner-DefaultSinkProcessor) [INFO - org.apache.flume.sink.LoggerSink.process(LoggerSink.java:95)]
 Event: { headers:{} body: 68 65 6C 6C 6F 20 65 76 65 72 79 6F 6E 65 0D    hello everyone. }
```

图 8-9 Flume 接收并传输到控制台的数据

## 8.3 Flume 组件

在 Flume 中数据流是以 event 为单元组成的,一个 event 在一个 Agent 中传输处理的完整流程如图 8-10 所示。在 Source 组件将 event 传输给 Channel 组件之前,可以通过 Interceptor(拦截器)组件、Selector(选择器)组件进行预处理。在 Channel 组件将数据传输给 Sink 组件之前,可以进行 SinkProcessor 预处理。

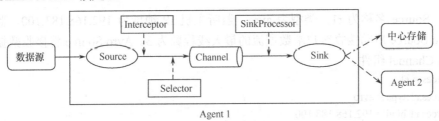

图 8-10 event 完整数据传输处理流程

### 8.3.1 Source 组件

Source 组件用于接收各种外部数据源,并将收集到的数据以 event 为单位发送到 Channel 组件。一个 Source 组件可以接收多个数据源的数据,可以向多个 Channel 组件发送 event。Flume 内置了多种类型的 Source 组件,如表 8-1 所示。

表 8-1 Flume 内置的 Source 组件

| 类型 | 描述 |
| --- | --- |
| Avro Source | 监听 Avro 端口并从外部 Avro 客户端接收 event |
| Exec Source | 启动时运行给定的 UNIX 命令,在标准输出设备上连续生成数据 |
| JMS Source | 从 JMS 目的地(如队列或主题)读取信息 |
| Kafka Source | 作为 Apache Kafka 的一个消费者,从 Kafka 主题中读取信息 |
| Netcat Source | 打开指定端口监听数据,数据是以行分隔符分隔的文本,每行文本被转换为一个 event 发送给 Channel |
| HTTP Source | 接收 HTTP 的 GET 或 POST 请求的数据作为 event |
| Spooling Directory Source | 通过将要接收的文件放入磁盘上的"假脱机"目录来接收数据 |
| Taildir Source | 实时接收指定的文件中追加写入的新行 |

**1. Avro Source**

Avro Source 支持 Avro 协议,接收 RPC 事件请求。Avro Source 通过监听 Avro 端口接收外部 Avro 客户端的 event,在 Flume 的多层架构中经常被使用以接收上游 Avro Sink 发送的 event。

Avro Source 常用的属性配置如表 8-2 所示,其中带*的为必须配置的属性。

表 8-2 Avro Source 常用的属性配置

| 属性名称 | 描述 |
| --- | --- |
| * type | 指定 Source 组件类型,此处应为 avro |
| * bind | 监听的主机名或 IP 地址 |
| * port | 监听的端口号 |
| * channels | Source 对应的 Channel 组件名称列表,用空格分隔 |

续表

| 属性名称 | 描述 |
|---|---|
| compression-type | 是否使用压缩，如果使用压缩，则值为 deflate。默认值为 none，即不压缩 |
| interceptors | Source 组件包含的拦截器名称列表，用空格分隔 |
| threads | 要生成的最大线程数。当需要接收多个 Avro 客户端的数据流时要设置合适的线程数，否则会造成 Avro 客户端数据流的积压 |

例如，Source 名称为 r1，类型为 avro，监听主机 IP 地址为 192.168.183.100，监听端口为 8888。Avro Source 启动接收客户端数据流的最大线程数为 3。Avro Source 将接收数据发送到名称为 c1 的 Channel 组件。

```
a1.sources = r1
a1.sources.r1.type = avro
a1.sources.r1.bind = 192.168.183.100
a1.sources.r1.port = 8888
a1.sources.r1.threads= 3
a1.sources.r1.channels = c1
```

2. Exec Source

Exec Source 支持 Linux 命令，收集标准输出数据或者通过 tail -F file 的方式监听指定文件。Exec Source 可以实现实时的消息传输，但是它并不记录已经读取文件的位置，不支持断点续传，当 Exec Source 重启或者挂掉都会造成后续增加的消息丢失，一般在测试环境中使用。

Exec Source 常用的属性配置如表 8-3 所示，其中带*的为必须配置的属性。

表 8-3　Exec Source 常用的属性配置

| 属性名称 | 描述 |
|---|---|
| * type | 指定 Source 组件类型，此处应为 exec |
| * command | 要执行的 Linux 命令 |
| * channels | Source 对应的 Channel 组件名称列表，用空格分隔 |
| interceptors | Source 组件包含的拦截器名称列表，用空格分隔 |
| batchSize | 一次读取并发送给 Sink 组件的最大行数 |

例如，Source 名称为 r1，类型为 exec，r1 通过 tail -F 命令监听 exectest.log 文件，将 event 发送到名为 c1 的 Channel 组件。

```
a1.sources = r1
a1.sources.r1.type = exec
a1.sources.r1.command = tail -F /home/hadoop/flume/execsource/exectest.log
a1.sources.r1.channels = c1
```

3. Kafka Source

Kafka Source 对应分布式消息队列 Kafka，作为 Kafka 的消费者持续从 Kafka 中消费数据，如果多个 Kafka Source 同时消费 Kafka 中的同一个主题（topic），则 Kafka Source 的 kafka.consumer.group.id 应设置成相同的组 id，多个 Kafka Source 之间不会消费重复的数据，每一个 Source 都会消费 topic 下的不同数据。

Kafka Source 常用的属性配置如表 8-4 所示，其中带*的为必须配置的属性。

表 8-4  Kafka Source 常用的属性配置

| 属性名称 | 描述 |
| --- | --- |
| * type | 指定 Source 组件类型，此处应为 org.apache.flume.source.kafka.KafkaSource |
| * channels | Source 对应的 Channel 组件名称列表，用空格分隔 |
| * kafka.bootstrap.servers | Kafka broker 列表，格式为 ip1:port1, ip2:port2…，多个值之间用逗号隔开 |
| * kafka.topics | 消费的 topic 名称列表，用逗号分隔 |
| * kafka.topics.regex | 用正则表达式表示 Source 消费的 topic 列表，优先级高于 kafka.topics |
| kafka.consumer.group.id | Kafka Source 所属消费组的 id，默认值为 flume |
| batchSize | 批量写入 Channel 的最大 event 数量，默认值为 1000 |
| batchDurationMillis | 等待批量写入 Channel 的最长时间。该参数和 batchSize 参数只要有一个先满足，都会触发批量写入 Channel 操作，默认值为 1000ms |

例如，Source 名称为 r1，Channel 名称为 c1，Source 类型为 Kafka Source，r1 批量写入 c1 的最大 event 数量为 1000，r1 等待批量写入 c1 的最长时间为 2s，r1 消费的主题名称为 flumetopic，r1 所属的 Kafka Source 消费组 id 为 flumegroupid。

```
a1.sources = r1
a1.channels = c1
a1.sources.r1.type = org.apache.flume.source.kafka.KafkaSource
a1.sources.r1.channels = c1
a1t.sources.r1.kafka.bootstrap.servers=192.168.1.1:9092,192.168.1.2:9092
a1.sources.r1.kafka.topics = flumetopic
a1.sources.r1.batchSize = 1000
a1.sources.r1.batchDurationMillis = 2000
a1.sources.r1.kafka.consumer.group.id =flumegroupid
```

### 4．Spooling Directory Source

Spooling Directory Source 允许通过将要接收的文件放入磁盘上的"假脱机"目录来接收数据，监听指定目录，在新文件出现时解析其中的 event。将新文件完全读入 Channel 后，默认情况下，通过重命名该文件来指示完成，或者可以删除该文件，或者使用 trackerDir 跟踪已处理的文件。

与 Exec Source 不同，Spooling Directory Source 是可靠的，即使 Flume 重启或终止，也不会丢失数据。Spooling Directory Source 要求"假脱机"目录中的文件名称必须是唯一的，且文件是不可变的。如果文件放入"假脱机"目录后被写入，或者文件名被重用，Flume 将在日志文件中打印错误并停止处理。为了避免上述问题的出现，可以在将文件移动到"假脱机"目录时，向文件名中添加唯一标识符，例如时间戳。

Spooling Directory Source 常用的属性配置如表 8-5 所示，其中带*的为必须配置的属性。

表 8-5  Spooling Directory Source 常用的属性配置

| 属性名称 | 描述 |
| --- | --- |
| * type | 指定 Source 组件类型，此处应为 spoolDir |
| * channels | Source 对应的 Channel 组件名称列表，用空格分隔 |
| * spoolDir | Source 监听的"假脱机"目录 |
| fileSuffix | 采集完新文件数据后，给文件添加的后缀名称，默认值为.COMPLETED |
| fileHeader | 是否添加文件的绝对路径到 event header 中，默认值为 false |
| fileHeaderKey | 添加到 event header 中文件绝对路径的键，默认值为 file |
| deletePolicy | 是否删除采集完后的文件，取值为 never（默认）或 immediate |
| interceptors | Source 组件包含的拦截器名称列表，用空格分隔 |

例如，Source 名称为 r1，Channel 名称为 c1，Source 类型为 Spooling Directory Source，r1 监听的文件夹路径为 /home/hadoop/apps/flume/spoolDir，r1 将 event 发送到 c1，并在 event header 中添加文件绝对路径信息。

```
a1.sources = r1
a1.channels = c1
a1.sources.r1.type = spooldir
a1.sources.r1.channels = c1
a1.sources.r1.spoolDir = /home/hadoop/flume/spoolDir
a1.sources.r1.fileHeader = true
```

### 5. Taildir Source

Taildir Source 监听一个或多个指定文件，一旦检测到文件中有新行写入时便进行实时跟踪。如果新行正在写入，Taildir Source 将等待写入完成，并重新读取新行数据。

Taildir Source 是可靠的，定期以 JSON 格式将每个文件的最后一次读取位置写入指定的位置文件中，从而可以实现断点续传，并且能够保证没有重复数据的读取。

文件将按修改时间的顺序被采集，修改时间最长的文件将首先被采集。采集完数据后，不会重命名或删除正在跟踪的文件，也不会对其进行任何修改。当前 Taildir Source 不支持跟踪二进制文件，只支持文本文件进行逐行读取。

Taildir Source 常用的属性配置如表 8-6 所示，其中带 * 的为必须配置的属性。

**表 8-6 Taildir Source 常用的属性配置**

| 属性名称 | 描述 |
| --- | --- |
| * type | 指定 Source 组件类型，此处应为 TAILDIR |
| * channels | Source 对应的 Channel 组件名称列表，用空格分隔 |
| * filegroups | 逗号分隔的文件组列表，每个文件组包含一系列监听的文件 |
| * filegroups.<filegroupName> | 文件组绝对路径，正则表达式只能用于文件名 |
| fileHeader | 是否添加文件的绝对路径到 event header 中，默认值为 false |
| fileHeaderKey | 添加到 event header 中文件绝对路径的键值，默认值为 file |
| headers.<filegroupName>.<headerKey> | fileHeaderKey 指定的键（key）对应的值（value） |

例如，配置一个 Flume，Agent 名称为 taildiragent，Source 名称为 r1，Sink 名称为 k1，Channel 名称为 c1，Source 类型为 TAILDIR，r1 对接的 Channel 名称为 c1，设置保存监听文件读取位置信息的文件路径，监听文件列表包含两个监听文件组 f1、f2，f1 监听指定的 example.log 文件，f2 通过正则表达式匹配指定路径下包含 log 关键字的所有文件。k1 类型为 logger，将消费的 event 输出到控制台。

```
a1.sources = r1
a1.channels = c1
a1.sources.r1.type = TAILDIR
a1.sources.r1.positionFile=/home/hadoop/flume/taildir/position/taildir_position.json
a1.sources.r1.filegroups = f1 f2
a1.sources.r1.filegroups.f1=/home/hadoop/flume/taildir/test1/example.log
a1.sources.r1.filegroups.f2 =/home/hadoop/flume/taildir/test2/.*log.*
a1.sources.r1.channels = c1
```

## 8.3.2 Channel 组件

Channel 组件被设计为 event 中转暂存区，存储 Source 收集且没有被 Sink 消费的 event。为了平衡 Source 收集和 Sink 读取数据的速度，Channel 可视为 Flume 内部的消息队列。Channel 组件是线程安全的并且具有事务性，支持 Source 写失败重复写和 Sink 读失败重复读等操作。

常用的 Channel 类型有 Memory Channel、File Channel、Kafka Channel 和 JDBC Channel 等。

### 1. Memory Channel

Memory Channel 使用内存作为 Channel，读写速度快，但是存储数据量小，Flume 进程挂掉、服务器停机或者重启都会导致数据丢失。如果部署 Flume Agent 的服务器的内存资源充足，并且不关心数据丢失，这样场景下可以使用 Memory Channel。

Memory Channel 常用的属性配置如表 8-7 所示，其中带*的为必须配置的属性。

表 8-7 Memory Channel 常用的属性配置

| 属性名称 | 描述 |
| --- | --- |
| * type | 指定 Channel 组件类型，此处应为 memory |
| capacity | Channel 中存储的最大 event 数量，默认值为 100 |
| transactionCapacity | 一次事务中写入和读取的 event 最大数量，默认值为 100 |
| keep-alive | 在 Channel 中写入或读取 event 等待完成的超时时间，默认为 3s |
| byteCapacityBufferPercentage | 缓冲空间占 Channel 容量（byteCapacity）的百分比，为 event header 保留了空间，默认值为 20（单位百分比） |
| byteCapacity | Channel 占用内存的最大容量，默认值为 JVM 最大内存的 80% |

例如：

```
a1.channels = c1
a1.channels.c1.type = memory
a1.channels.c1.capacity = 10000
a1.channels.c1.transactionCapacity = 10000
a1.channels.c1.byteCapacityBufferPercentage = 20
a1.channels.c1.byteCapacity = 800000
```

### 2. File Channel

File Channel 将 event 写入磁盘文件中，与 Memory Channel 相比，其存储容量大，无数据丢失风险。File Channel 的数据存储路径可以配置多个磁盘文件路径，提高写入文件性能。Flume 将 event 顺序写入 File Channel 文件的末尾,在配置文件中通过设置 maxFileSize 参数设置数据文件大小上限。

当一个已关闭的只读数据文件中的 event 被完全读取完成，并且 Sink 已经提交读取完成的事务时，Flume 将删除该数据文件。通过设置检查点和备份检查点，在 Agent 重启之后快速将 File Channel 中的数据按顺序回放到内存中。

File Channel 常用的属性配置如表 8-8 所示，其中带*的为必须配置的属性。

表 8-8 File Channel 常用的属性配置

| 属性名称 | 描述 |
| --- | --- |
| * type | 指定 Channel 组件类型，此处应为 file |
| checkpointDir | 检查点文件存储目录，默认为~/.flume/file-channel/checkpoint |
| useDualCheckpoints | 是否开启备份检查点，默认为 false。备份检查点的作用是当 Agent 意外出错导致写入检查点文件异常时，重新启动 File Channel 时通过备份检查点将数据回放到内存中。如果不开启备份检查点，在数据回放的过程中发现检查点文件异常，会对所有数据进行全回放，全回放的过程相当耗时 |

续表

| 属性名称 | 描述 |
|---|---|
| backupCheckpointDir | 备份检查点目录，不能与检查点目录或数据目录相同 |
| dataDirs | 以逗号分隔的存储日志文件的目录列表。可以使用不同磁盘上的多个目录，通过磁盘的并行写入来提高 File Channel 性能 |
| checkpointInterval | 每次写检查点的时间间隔，默认值为 30000ms |
| transactionCapacity | 一次事务中写入和读取的最大 event 数量，默认值为 10000 |
| maxFileSize | 每个日志文件的大小上限，默认值为 2146435071 字节 |
| minimumRequiredSpace | 最小空闲空间。当空闲空间低于该值时，File Channel 停止数据读写操作 |
| capacity | File Channel 可容纳的最大 event 数量，默认值为 1000000 |
| keep-alive | 在 Channel 中写入或读取 event 等待完成的超时时间，默认值为 3s |

例如，Channel 名称为 c1，Channel 类型为 file，检查点路径为/home/hadoop/flume/file_channnel_checkpoint，数据存放路径为/home/hadoop/flume/file_channnel_data，开启备份检查点，备份检查点路径为/home/hadoop/flume/file_channnel_backup。

a1.channels = c1
a1.channels.c1.type = file
a1.channels.c1.dataDirs = /home/hadoop/flume/filechannel/data
a1.channels.c1.checkpointDir = /home/hadoop/flume/filechannel/checkpoint
a1.channels.c1.useDualCheckpoints = true
a1.channels.c1.backupCheckpointDir = /home/hadoop/flume/filechannel/backup

### 3．Kafka Channel

Kafka Channel 将 event 写入 Kafka 集群中。Kafka 提供了高可用性和复制特性，在 Agent 或个别 Kafkabroker 崩溃的情况下，event 仍然可以被其他 Sink 读取。相对于 Memory Channel 和 File Channel，Kafka Channel 的存储容量更大、容错能力更强，弥补了它们的短板，如果合理利用 Kafka 的性能，能够达到事半功倍的效果。

Kafka Channel 常用的属性配置如表 8-9 所示，其中带*的为必须配置的属性。

表 8-9　Kafka Channel 常用的属性配置

| 属性名称 | 描述 |
|---|---|
| * type | Channel 组件类型，应为 org.apache.flume.channel.kafka.KafkaChannel |
| kafka.bootstrap.servers | 以逗号分隔的 Kafka broker 列表，格式为 ip1:port1, ip2:port2… |
| kafka.topic | Channel 使用的 topic 名称，默认值为 flume-channel |
| kafka.consumer.group.id | 在 Kafka 中注册的 Channel 使用的 Consumer 组 Id，默认值为 flume |
| parseAsFlumeEvent | 是否以 Avro FlumeEvent 模式写入 Kafka Channel 中，默认值为 true |
| pollTimeout | 轮询超时时间，默认值为 500ms |
| kafka.consumer.auto.offset.reset | 如果没有初始偏移量或当前偏移量不存在时该如何处理。earliest 表示从最早的偏移量开始消费数据，latest 表示从最新的偏移量开始消费数据，none 表示如果没有指定的偏移量则抛出异常。默认值为 latest |

例如，Channel 名称为 c1，类型为 KafkaChannel，使用的 topic 名称为 flumechannel，Consumer 组 ID 为 flumecg。

a1.channels = c1

```
a1.channels.c1.type = org.apache.flume.channel.kafka.KafkaChannel
a1.channels.c1.kafka.bootstrap.servers = 192.168.183.102:9092,192.168.183.103:9092
a1.channels.c1.kafka.topic = flumechannel
a1.channels.c1.kafka.consumer.group.id = flumecg
```

### 8.3.3 Sink 组件

Sink 组件从 Channel 组件消费 event，输出到外部存储，或者输出到下一个阶段的 Agent。

一个 Sink 组件只能从一个 Channel 组件中消费 event。当 Sink 读取 event 成功后，会向 Channel 提交事务，如果事务提交成功，读取完的 event 将会被 Channel 删除，否则 Channel 会等待 Sink 重新消费处理失败的 event。

Flume 提供了丰富类型的内置 Sink 组件，如表 8-10 所示。

表 8-10 Flume 内置的 Sink 组件

| 属性名称 | 描述 |
| --- | --- |
| HDFS Sink | 将 event 写入 Hadoop 的 HDFS 文件系统中 |
| Hive Sink | 将包含分隔符的文本和 JSON 数据的 event 直接写入 Hive 表或分区中 |
| Logger Sink | 在 INFO 级别记录 event，主要用于测试或调试 |
| Avro Sink | 将接收的 event 转换为 Avro event，发送到指定的主机/端口 |
| Kafka Sink | 将 event 写入 Kafka 主题中 |
| File Roll Sink | 在本地文件系统中存储 event |
| HTTP Sink | 利用 HTTP POST 请求将 event 发送到一个远程服务 |
| HBase Sink | 将 event 写入 HBase 表中 |

#### 1. Avro Sink

Avro Sink 常用于对接下一层的 Avro Source，通过发送 RPC 请求将 Event 发送到下一层的 Avro Source。为了减少 event 传输占用大量的网络资源，Avro Sink 提供了端到端的批量压缩数据传输。

Avro Sink 常用的属性配置如表 8-11 所示，其中带*的为必须配置的属性。

表 8-11 Avro Sink 常用的属性配置

| 属性名称 | 描述 |
| --- | --- |
| * type | Sink 组件类型，应为 avro |
| * channel | 指定的 Channel 组件名称，一个 Sink 只能对应一个 Channel 组件 |
| * hostname | 绑定的目标 Avro Source 主机名称或者 IP 地址 |
| * port | 绑定的目标 Avro Source 端口号 |
| batch-size | 批量发送 event，默认值为 100 |
| compression-type | 取值为 deflate 或 none，默认值为 none。压缩类型与 Avro Source 压缩类型一致 |
| compression-level | 压缩级别，0 表示不压缩，从 1 到 9 数字越大，压缩效果越好，默认值为 6 |

例如，Sink 名称为 k1，类型为 avro，对应 Channel 为 c1。

```
a1.channels = c1
a1.sinks = k1
a1.sinks.k1.type = avro
a1.sinks.k1.channel = c1
```

```
a1.sinks.k1.hostname = 10.10.10.10
a1.sinks.k1.port = 4545
```

### 2. HDFS Sink

HDFS Sink 是将 event 写入 HDFS 文件系统中，目前支持文本文件和序列化文件，支持文件的压缩。可以根据运行时间、数据量大小或 events 数量定期滚动（关闭当前文件并创建新文件）。可以根据时间戳、event 来源机器名或 IP 地址等属性来进行数据分区存储。HDFS 存储路径中可能包含格式化转义序列，这些转义序列将被 HDFSSink 替换，以生成用于存储 event 的目录或文件名。

HDFS Sink 的存储路径中支持使用的转义序列如表 8-12 所示。

表 8-12　HDFS Sink 的存储路径中支持使用的转义序列

| 别名 | 描述 | 别名 | 描述 |
| --- | --- | --- | --- |
| %{host} | 替换 event 头部名为 "host" 的值 | %t | 以毫秒表示的 UNIX 时间 |
| %a | 工作日名称缩写（Mon,Tue） | %A | 工作日全称（Monday,Tuesday） |
| %b | 月份名称缩写（Jan,Feb） | %B | 月份全称（January,February） |
| %c | 日期时间（Thu Mar 3 23:05:25 2021） | %d | 月中的某一天（01） |
| %e | 月中某一天，不填充（1） | %D | 日期，与 %m/%d/%y 相同 |
| %H | 小时（00..23） | %I | 小时（01..12） |
| %j | 一年中的某天（001..366） | %k | 小时（0..23） |
| %m | 月份（01..12） | %n | 月份，不填充（1..12） |
| %M | 分钟（00..59） | %p | 等价于 am 或 pm |
| %s | 从 1970-01-01 00:00:00 UTC 开始的秒 | %S | 秒（00..59） |
| %y | 年的最后两位（00..99） | %Y | 年（2021） |
| %[IP] | 替换 Agent 运行的主机 IP | %[localhost] | 替换 Agent 运行的主机名 |

HDFS Sink 常用的属性配置如表 8-13 所示，其中带*的为必须配置的属性。

表 8-13　HDFS Sink 常用的属性配置

| 属性名称 | 描述 |
| --- | --- |
| * type | Sink 组件类型，应为 hdfs |
| * channel | 指定的 Channel 组件名称，一个 Sink 只能对应一个 Channel 组件 |
| * hdfs.path | HDFS 存储目录的路径，支持按日期时间分区 |
| hdfs.filePrefix | event 输出到 HDFS 的文件名前缀，默认前缀为 FlumeData |
| hdfs.fileSuffix | event 输出到 HDFS 的文件名后缀 |
| hdfs.inUsePrefix | 正在写入的临时文件名前缀 |
| hdfs.inUseSuffix | 正在写入的临时文件名后缀，默认值为 .tmp |
| hdfs.rollInterval | HDFS 文件滚动生成时间间隔，默认值为 30s。0 表示文件不根据时间滚动生成 |
| hdfs.rollSize | 触发文件滚动的大小，默认值为 1024B。0 表示文件不根据文件大小滚动生成 |
| hdfs.rollCount | 文件滚动之前写入的 event 数量，默认值为 10。0 表示文件不根据 event 数量滚动生成 |
| hdfs.idleTimeout | 文件不活跃超时，将自动关闭。默认值为 0，表示禁用此功能，不自动关闭文件 |
| hdfs.batchSize | 批量写入 HDFS 的 event 数量，默认值为 100 |
| hdfs.codeC | 文件压缩，包括 gzip、bzip2、lzop、snappy，默认不采用压缩 |

续表

| 属性名称 | 描述 |
|---|---|
| hdfs.fileType | 文件格式，DataStream 输出的文件不压缩，CompressedStream 对输出的文件进行压缩，需要设置 hdfs.codeC 指定压缩格式。默认值为 SequenceFile |
| hdfs.maxOpenFiles | 允许打开文件的数量，超过该值，最旧文件将关闭，默认为 5000 |
| hdfs.writeFormat | 序列文件中记录格式，取值 Text 或 Writable，默认值为 Writable |
| hdfs.round | 时间戳向下取整，用于 HDFS 文件按照时间分区，默认值为 false |
| hdfs.roundValue | 当 round 设置为 true，配合 roundUnit 时间单位一起使用，例如 roundUnit 值为 minute，该值设置为 1 表示 1 分钟之内的数据写到一个文件中。默认值为 1 |
| hdfs.roundUnit | 时间取整单位，可以取值 second、minute、hour，默认值为 second |
| hdfs.timeZone | 写入 HDFS 文件使用的时区，默认值为 Local Time（本地时间） |
| hdfs.useLocalTimeStamp | 是否使用本地时间戳替换 event header 中的时间戳，默认值为 false |

例如，Source r1 使用 timestamp 拦截器，在 event header 中添加 timestamp 时间戳信息。使用 HDFS Sink 将 event 写入 HDFS 的/data/flume 路径下，并且按照年、月、日分区，写入 HDFS 的 Text 文件以 hdfssink-开头，每分钟生成一个文件，时间向下取整，HDFS 操作超时时间为 1 分钟。

```
a1.sources = r1
a1.channels = c1
a1.sinks = k1
a1.sources.r1.interceptors = i1
a1.sources.r1.interceptors.i1.type = timestamp
a1.sources.r1.interceptors.i1.preserveExisting = false
a1.sinks.k1.channel = c1
a1.sinks.k1.hdfs.path = /data/flume/%y-%m-%d
a1.sinks.k1.hdfs.fileprefix = hdfssink-
a1.sinks.k1.hdfs.fileType = Datastream
a1.sinks.k1 .hdfs.writeFormat = Texta1.sinks.k1.hdfs.round = true
a1.sinks.k1.hdfs.roundvalue = 1
a1.sinks.k1.hdfs.roundunit = minute
a1.sinks.k1.hdfs.callTimeout = 60000
```

### 3. Kafka Sink

Flume 可以通过 Kafka Sink 将 event 发布到 Kafka 指定的主题中，将 Flume 与 Kafka 集成，以便基于推送处理系统能够处理来自各种 Flume 源的数据。

Kafka Sink 常用的属性配置如表 8-14 所示，其中带*的为必须配置的属性。

表 8-14  Kafka Sink 常用的属性配置

| 属性名称 | 描述 |
|---|---|
| * type | Sink 组件类型，应为 org.apache.flume.sink.kafka.KafkaSink |
| * kafka.bootstrap.servers | Kafka broker 列表，格式为 host:port。建议配置至少两个 Kafka broker，之间用逗号隔开 |
| kafka.topic | Kafka 中主题名称，默认值为 default-flume-topic |
| flumeBatchSize | 每次批量发送的消息条数，默认值为 100 |
| kafka.producer.acks | 在成功写入消息之前，必须确认多少个副本。0 表示不需要确认，1 表示只需 leader 确认，-1 表示所有副本都需确认 |
| useFlumeEventFormat | 默认值为 false，只将 event body 发送到 Kafka Topic 中。如果为 true，将以 flume avro 二进制格式保存 event |

例如，使用 Kafka Sink 向"FlumeKafkaSinkTopic"主题批量发送消息，批量发送的消息数量为 100。

```
a1.sinks = k1
a1.sinks.k1.channel = c1
a1.sinks.k1.type = org.apache.flume.sink.kafka. Kafkasink
a1.sinks.k1.kafka.topic = FlumeKafkasinkTopic1
a1.sinks. k1.kafka .bootstrap.servers = 192.168.183.102:9092,192.168.183.103:9092
a1.sinks.k1. kafka. flumeBatchsize = 100
a1.sinks.k1.kafka.producer.acks = 1
```

### 8.3.4 Interceptor 组件

Source 组件将 event 写入 Channel 组件之前可以调用拦截器。拦截器是实现 org.apache.flume. interceptor.Interceptor 接口的类，可以修改或者删除正在传送中的 event。Flume 拦截器在配置文件的 Source 组件中进行设置，一个 Source 组件可以设置多个拦截器，使用空格连接起来，按次序执行。一个拦截器将处理后的 event 传给下一个拦截器，如果一个拦截器删除某个 event，那么该 event 不会再返回给下一个拦截器。

下面是一个拦截器配置示例，配置两个名为 i1 和 i2 的拦截器，i1 拦截器的类型为 HostInterceptor，i2 拦截器的类型为 TimestampInterceptor，拦截器类型可以指定完整的类名，也可以指定拦截器的别名。event 经过 HostInterceptor 处理后，传送给 TimestampInterceptor。

```
a1.sources = r1
a1.sinks = k1
a1.channels = c1
a1.sources.r1.interceptors = i1 i2
a1.sources.r1.interceptors.i1.type = org.apache.flume.interceptor.HostInterceptor$Builder
a1.sources.r1.interceptors.i1.preserveExisting = false
a1.sources.r1.interceptors.i1.hostHeader = hostname
a1.sources.r1.interceptors.i2.type = org.apache.flume.interceptor.TimestampInterceptor$Builder
a1.sinks.k1.filePrefix = FlumeData.%{CollectorHost}.%Y-%m-%d
a1.sinks.k1.channel = c1
```

Flume 内置了多种拦截器，如 Timestamp Interceptor、Host Interceptor、Static Interceptor、Remove Header Interceptor、UUID Interceptor、Search and Replace Interceptor、Regex Filtering Interceptor、Regex Extractor Interceptor 等，也可以自定义拦截器。

**1. Timestamp Interceptor（时间戳拦截器）**

Flume 可以使用时间戳拦截器在 eventheader 中添加当前的时间戳信息（ms），加入的数据为一个键值对，其中 key 值为"timestamp"，value 值为拦截器拦截 event 时的时间戳。

时间戳拦截器常用的属性配置如表 8-15 所示，其中带*的为必须配置的属性。

表 8-15 时间戳拦截器常用的属性配置

| 属性名称 | 描述 |
| --- | --- |
| * type | 拦截器类型，可以为 timestamp 或 org.apache.flume.interceptor.TimestampInterceptor |
| headName | 存放时间戳的 event header 中的键值，默认为 timestamp |
| preserveExisting | 如果 event header 中存在时间戳，是否保留原来的时间戳信息，默认值为 false |

时间戳拦截器常用于对文件或目录按照日期进行命名。例如，在使用 HDFS Sink 向 HDFS 中写入文件时，可以根据 event header 中的时间戳对文件或文件所在目录进行命名。

例如，将文件放入按年月分类的文件夹中，使用时间戳作为文件名前缀。

```
a1.sinks.k1.hdfs.path=hdfs://master:9000/flume/%Y-%m
a1.sinks.k1.hdfs.filePrefix=%Y-%m-%d-%H
```

下面是一个时间戳拦截器配置示例：

```
a1.sources = r1
a1.channels = c1
a1.sources.r1.channels =   c1
a1.sources.r1.type = seq
a1.sources.r1.interceptors = i1
a1.sources.r1.interceptors.i1.type = timestamp
```

### 2. Host Interceptor（主机拦截器）

Flume 可以使用主机拦截器在 eventheader 中添加运行 Flume Agent 的服务器主机名称或者 IP 地址，以键值对形式添加到 eventheader 中，key 值为"host"，value 值为主机名或 IP 地址。利用主机拦截器，可以将 event 按照主机名或 IP 地址写入不同的 Channel 组件中，便于后续的 Sink 组件对不同 Channel 组件中的数据进行分开处理。

主机拦截器常用的属性配置如表 8-16 所示，其中带*的为必须配置的属性。

表 8-16 主机拦截器常用的属性配置

| 属性名称 | 描述 |
| --- | --- |
| * type | 拦截器类型，应为 host |
| preserveExisting | 如果 eventheader 中已经存在 host，是否保留，默认值为 false |
| useIP | 是否使用 IP 地址，默认值为 true，否则使用主机名 |
| hostHeader | 主机名或 IP 地址在 event header 中使用的 key 值，默认值为 host |

下面是一个主机拦截器的配置示例：

```
a1.sources = r1
a1.sources.r1.interceptors = i1
a1.sources.r1.interceptors.i1.type = host
a1.sources.r1.interceptors.i1.useIP = false
```

### 3. Static Interceptor（静态拦截器）

Flume 可以使用静态拦截器在 event header 中添加一个静态信息的键值对。目前 Flume 还不支持在静态拦截器中一次加入多个静态键值对，但可以一次使用多个静态拦截器形成一个静态拦截器链，在每个静态拦截器都定义一个静态的键值对。

静态拦截器常用的属性配置如表 8-17 所示，其中带*的为必须配置的属性。

表 8-17 静态拦截器常用的属性配置

| 属性名称 | 描述 |
| --- | --- |
| * type | 拦截器类型，应为 static |
| preserveExisting | 如果 event header 中 static 已经存在，是否保留，默认值为 true |
| key | 静态数据的 key 值 |
| value | 静态数据的 value 值 |

下面是一个静态拦截器的配置示例：

```
a1.sources = r1
a1.sources.r1.interceptors = i1
a1.sources.r1.interceptors.i1.type = static
a1.sources.r1.interceptors.i1.key = datacenter
```

a1.sources.r1.interceptors.i1.value = NEW_YORK

### 4. Remove Header Interceptor（移除头信息拦截器）

Flume 可以使用移除头信息拦截器删除 eventheaders 中指定的一个或多个 header。可以删除静态定义的 header、基于正则表达式的 header 或在一个列表中的 header。如果没有这些定义，或者没有匹配到 header，event 不会被修改。如果只有一个 header 需要移除，通过名字指定可以提供更好的性能。

移除头信息拦截器常用的属性配置如表 8-18 所示，其中带*的为必须配置的属性。

表 8-18 移除头信息拦截器常用的属性配置

| 属性名称 | 描述 |
| --- | --- |
| * type | 拦截器类型，应为 remove_header |
| withName | 要移除的 event header 中的 key 值 |
| fromList | 要移除的 header 中 key 值列表，使用 fromListSeparator 分隔 |
| fromListSeparator | 使用正则表达式分隔 fromList 中的 key 值列表，默认值为\s*,\s*，即由任意数量的空格字符包围的逗号 |

### 5. UUID Interceptor（UUID 拦截器）

UUID 拦截器用于在每个被拦截的 event header 中生成一个 UUID 字符串，例如：b5755073-77a9-43c1-8fad-b7a586fc1b97，生成的 UUID 可以在 Sink 中读取并使用。在高可用性和高性能的 Flume 网络中，利用 UUID 可以在复制和重分发 event 时对其进行后续重复数据消除。

UUID 拦截器常用的属性配置如表 8-19 所示，其中带*的为必须配置的属性。

表 8-19 UUID 拦截器常用的属性配置

| 属性名称 | 描述 |
| --- | --- |
| * type | 拦截器类型，应为 org.apache.flume.sink.solr.morphline.UUIDInterceptor$Builder |
| preserveExisting | 如果 event header 中 UUID 已经存在，是否保留，默认值为 true |
| headerName | UUID 在 event header 中使用的 key 值，默认值为 id |
| prefix | 每个生成的 UUID 的前缀字符串常量 |

下面是一个 UUID 拦截器配置的示例：

a1.sources = r1
a1.sources. r1.interceptors = i1
a1.sources. r1.interceptors.i1.type= org.apache.flume.sink.solr.morphline.UUIDInterceptor$Builder
a1.sources. r1.interceptors.i1.headerName = uuid
a1.sources. r1.interceptors.i1.preserveExisting = true
a1.sources. r1.interceptors.i1.prefix = UUID_

### 6. Search and Replace Interceptor（搜索替换拦截器）

搜索替换拦截器提供了简单的基于 Java 正则表达式的字符串搜索和替换功能，可以对 event body 中匹配内容进行替换，并且使用与 Java Matcher.replaceAll()方法相同的规则。

搜索替换拦截器常用的属性配置如表 8-20 所示，其中带*的为必须配置的属性。

表 8-20 搜索替换拦截器常用的属性配置

| 属性名称 | 描述 |
| --- | --- |
| * type | 拦截器类型，应为 search_replace |
| searchPattern | 搜索替换的匹配规则 |
| replaceString | 替换的字符串 |
| charset | event body 的字符编码，默认为 UTF-8 |

下面是一个搜索替换拦截器的配置示例：

```
a1.sources = r1
a1.sources.r1.interceptors = search-replace
a1.sources.r1.interceptors.search-replace.type = search_replace
#移除event body中所有的字母、数字和下画线
a1.sources.r1.interceptors.search-replace.searchPattern = ^[A-Za-z0-9_]+
a1.sources.r1.interceptors.search-replace.replaceString =
```

### 7. Regex Filtering Interceptor（正则表达式过滤拦截器）

正则表达式过滤拦截器通过将 event body 解析为文本并将文本与正则表达式匹配，从而有选择地过滤 event，可以包含 event 或排除 event。

正则表达式过滤拦截器常用的属性配置如表 8-21 所示，其中带*的为必须配置的属性。

表 8-21 正则表达式过滤拦截器常用的属性配置

| 属性名称 | 描述 |
| --- | --- |
| * type | 拦截器类型，应为 regex_filter |
| regex | 用于匹配 event 的正则表达式，默认值为".*" |
| excludeEvents | 如果为 true，排除正则表达式匹配的 event，否则包含匹配的 event。默认值为 false |

下面是一个搜索替换拦截器的配置示例：

```
a1.sources = r1
a1.sources.r1.interceptors = i1
a1.sources.r1.interceptors.i1.type = regex_filter
#过滤掉不是以bigdata开头的event
a1.sources.r1.interceptors.i1.regex = ^bigdata.*
a1.sources.r1.interceptors.i1.excludeEvents = false
```

### 8. Regex Extractor Interceptor（正则表达式抽取过滤器）

正则表达式抽取过滤器使用正则表达式抽取 event body 中的内容，并将该内容加入 event header 中。支持在将匹配内容添加到 event header 之前进行系列化设置。

正则表达式抽取过滤器常用的属性配置如表 8-22 所示，其中带*的为必须配置的属性。

表 8-22 正则表达式抽取过滤器常用的属性配置

| 属性名称 | 描述 |
| --- | --- |
| * type | 拦截器类型，应为 regex_extractor |
| regex | 用于匹配 event 的正则表达式 |
| serializers | 以空格分隔的名称列表，构建 event header 中的 key 与 value 的映射 |
| serializers.<s1>.type | 默认值为 default |

下面是一个正则表达式抽取过滤器的配置示例：

```
a1.sources = r1
a1.sources.r1.interceptors = i1
a1.sources.r1.interceptors.i1.type = regex_extractor
#该配置从原始events中抽取出cookieid和ip，加入events header中
a1.sources.r1.interceptors.regex = cookieid is (.*?) and ip is (.*?)
a1.sources.r1.interceptors.i1.serializers = s1 s2
a1.sources.r1.interceptors.i1.serializers.s1.type = default
a1.sources.r1.interceptors.i1.serializers.s1.name = cookieid
```

```
a1.sources.r1.interceptors.i1.serializers.s2.type = default
a1.sources.r1.interceptors.i1.serializers.s2.name = ip
```

### 8.3.5 Selector 组件

Flume Source 可以将 event 写入多个 Channel，可以通过 Channel Selector（Channel 选择器）决定写入 Channel 的方式。

Flume 内置两种 Channel 选择器，分别为 Replicating Channel Selector（复制 Channel 选择器）和 Multiplexing Channel Selector（多路 Channel 选择器）。如果 Channel 选择器没有指定，默认是复制 Channel 选择器，即 Source 以复制的方式将一个 event 同时写入多个 Channel 中，不同的 Sink 可以从不同的 Channel 中获取相同的 event。多路 Channel 选择器根据 event header 中不同键值数据来判断 event 应被写入哪个 Channel 中。

#### 1. 复制 Channel 选择器

复制 Channel 选择器常用的属性配置如表 8-23 所示。

表 8-23 复制 Channel 选择器常用的属性配置

| 属性名称 | 描述 |
| --- | --- |
| selector.type | Channel 选择器类型，取值为 replicating |
| selector.optional | 标记为可选的一组 Channel，当 event 写入失败时，不会抛出异常 |

Channel 选择器在 Source 组件中配置。下面是一个复制 Channel 选择器的配置示例，声明 c1、c2、c3 三个 Channel，其中 c3 是可选 Channel，即 c3 写入 event 失败时不影响系统运行，但 c1 或 c2 写入 event 失败时将抛出异常，事务执行失败。

```
a1.sources = r1
a1.channels = c1 c2 c3
a1.sources.r1.selector.type = replicating
a1.sources.r1.channels = c1 c2 c3
a1.sources.r1.selector.optional = c3
```

#### 2. 多路 Channel 选择器

多路 Channel 选择器常用的属性配置如表 8-24 所示。

表 8-24 多路 Channel 选择器常用的属性配置

| 属性名称 | 描述 |
| --- | --- |
| selector.type | Channel 选择器类型，取值为 multiplexing |
| selector.header | 指定 event header 中的 key 值，默认为 flume.selector.header |
| selector.default | 默认传送的 Channel 名称，没有被匹配的 event 将传送到该 Channel，默认值为 "-" |
| selector.mapping.* | 指定传送的 Channel 名称。*为 selector.header 属性设置的 key 值对应的 value 值，从而根据 value 值选择匹配的 event 发送到指定的 Channel |

下面是一个多路 Channel 选择器配置示例，声明了 c1、c2、c3、c4 四个 Channel。event header 中 key 值为 state，value 值为 CZ 的 event 发送给 c1，value 值为 US 的 event 发送给 c2 和 c3，其他 event 发送给 c4。

```
a1.sources = r1
a1.channels = c1 c2 c3 c4
a1.sources.r1.selector.type = multiplexing
```

```
a1.sources.r1.selector.header = state
a1.sources.r1.selector.mapping.CZ = c1
a1.sources.r1.selector.mapping.US = c2 c3
a1.sources.r1.selector.default = c4
```

### 8.3.6 Sink Processor

Sink Group 允许将多个 Sink 放到一个组中，组内只有一个 Sink 消费 Channel。Sink Processor 可以实现组内所有 Sink 组件之间的负载均衡或者在暂时故障的情况下实现从一个 Sink 到另一个 Sink 的故障转移。

Sink Processor 共有 3 种类型，分别是 Default Sink Processor、Load Balancing Sink Processor 和 Failover Sink Processor。Default Sink Processor 对应的是单个的 Sink，Load Balancing Sink Processor 和 Failover Sink Processor 对应的是 Sink Group，Load Balancing Sink Processor 可以实现负载均衡的功能，Failover Sink Processor 可以实现故障转移的功能。

Sink Processor 常用的属性配置如表 8-25 所示，其中带*的为必须配置的属性。

表 8-25 Sink Processor 常用的属性配置

| 属性名称 | 描述 |
| --- | --- |
| * sinks | 在一个 Sink Group 中使用空格分隔的 Sink 列表 |
| * processor.type | Sink Processor 类型，取值为 default、failover 或 load_balance，默认为 default |

例如：

```
a1.sinkgroups = g1
a1.sinkgroups.g1.sinks = k1 k2
a1.sinkgroups.g1.processor.type = load_balance
```

#### 1. Default Sink Processor

Default Sink Processor 只有一个 Sink，无须创建 Sink Processor。

#### 2. Failover Sink Processor

Failover Sink Processor 通过配置维护了一个优先级列表，保证每一个有效的 event 都会被处理。故障转移的工作原理是：将连续失败的 Sink 分配到一个冷却池中，在池中为 Sink 分配一个冷冻期，在重试之前，冷冻期随着故障的增加而增加。一旦 Sink 成功发送一个 event，Sink 将被恢复到活动池中。Sink 具有一个与其关联的优先级，数字越大，优先级越高。如果一个 Sink 发送 event 时失败，则下一个具有最高优先级的 Sink 将尝试发送 event。例如，优先级为 100 的 Sink 在优先级为 80 的 Sink 之前被激活。如果未指定优先级，则根据在配置中指定的 Sink 顺序来确定优先级。

Failover Sink Processor 常用的属性配置如表 8-26 所示，其中带*的为必须配置的属性。

表 8-26 Failover Sink Processor 常用的属性配置

| 属性名称 | 描述 |
| --- | --- |
| * sinks | 在一个 Sink Group 中使用空格分隔的 Sink 列表 |
| * processor.type | Sink Processor 类型，取值为 failover |
| * processor.priority.<sinkName> | 为 Sink 分配优先级，优先级数字必须唯一，数字越大优先级越高 |
| processor.maxpenalty | 故障 Sink 的最大回退周期，单位为 ms，默认值为 30000 |

下面是一个 Failover Sink Processor 配置示例，首先声明一个 sinkgroups，然后设置 2 个 Sink，分别为 k1 与 k2，优先级分别为 5 和 10，而 processor.maxpenalty 被设置为 10s。

```
a1.sinkgroups = g1
a1.sinkgroups.g1.sinks = k1 k2
a1.sinkgroups.g1.processor.type = failover
a1.sinkgroups.g1.processor.priority.k1 = 5
a1.sinkgroups.g1.processor.priority.k2 = 10
a1.sinkgroups.g1.processor.maxpenalty = 10000
```

#### 3. Load Balancing Sink Processor

Load Balancing Sink Processor 提供在多个 Sink 之间负载均衡的功能，维护一个活动 Sink 列表，负载必须分布在这些 Sink 上。通过循环（round_robin）机制或者随机选择（random）机制实现负载分配，默认选择机制为循环机制，但可以通过配置覆盖默认设置。还可以通过继承 AbstractSinkSelector 类来实现用户自定义的选择机制。

Load Balancing Sink Processor 常用的属性配置如表 8-27 所示，其中带*的为必须配置的属性。

表 8-27 Load Balancing Sink Processor 常用的属性配置

| 属性名称 | 描述 |
| --- | --- |
| * sinks | 在一个 Sink Group 中使用空格分隔的 sinks 列表 |
| * processor.type | Sink Processor 类型，取值为 load_balance |
| processor.backoff | 取值为 true 时，会将失败的 Sink 放进一个冷却池中。默认值为 false |
| processor.selector | 选择机制，取值为 round_robin 或 random，默认值为 round_robin |
| processor.selector.maxTimeOut | 失败的 Sink 在冷却池中的最大驻留时间，默认值为 30000ms |

下面是一个 Load Balancing Sink Processor 配置示例，首先声明一个 sinkgroups，然后设置 2 个 Sink，分别为 k1 和 k2，随机选择 k1、k2 实现负载均衡。

```
a1.sinkgroups = g1
a1.sinkgroups.g1.sinks = k1 k2
a1.sinkgroups.g1.processor.type = load_balance
a1.sinkgroups.g1.processor.backoff = true
a1.sinkgroups.g1.processor.selector = random
a1.sinkgroups.sg1.processor.selector.maxTimeOut = 10000
```

## 8.4 Flume 数据采集案例与实施

### 8.4.1 实时采集本地文件到 HDFS

#### 1. 案例说明

利用 Flume 数据采集技术，可以方便地将本地文件内容实时传输到 HDFS，如图 8-11 所示。需要在 Flume 运行节点上安装并配置好 Hadoop 集群环境。

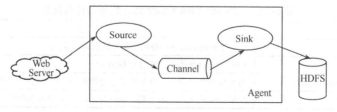

图 8-11 采集本地文件到 HDFS

对文件中数据进行采集，Source 组件可以采用 Exec Source、Spooling Directory Source 或

Taildir Source，三者的区别如下。

Exec Source：适用于监控一个实时追加的文件，但不能保证数据不丢失。

Spooling Directory Source：能够保证数据不丢失，且能够实现断点续传，但延迟较高，不能实时监控。

Taildir Source：能够实现断点续传，保证数据不丢失，还能够进行实时监控。

本例将采用 Taildir Source，将本地/home/hadoop/logs/access.log 中的日志信息实时上传到 HDFS 中，并根据本地日期进行分区存储。

2．操作步骤

（1）安装与配置 Flume。

（2）在本地系统中创建存储源数据信息和存储检查点文件的目录。

```
hadoop@master:~$ mkdir -p /home/hadoop/flume/logs
hadoop@master:~$ mkdir -p /home/hadoop/flume/position
```

（3）在 Flume 安装目录（/usr/local/flume）的 conf 文件夹中创建配置文件 taildir.conf，写入下列配置信息。

```
#配置组件名称
a1.sources = r1
a1.channels = c1
a1.sinks = k1
#配置Source组件
a1.sources.r1.type = TAILDIR
a1.sources.r1.channels = c1
a1.sources.r1.positionFile = /home/hadoop/flume/postion/taildir.json
a1.sources.r1.filegroups = f1
a1.sources.r1.filegroups.f1 = /home/hadoop/flume/logs/access.log
a1.sources.r1.fileHeader = true
#配置Channel组件
a1.channels.c1.type = memory
a1.channels.c1.capacity = 10000
a1.channels.c1.transactionCapacity = 100
a1.channels.c2.keep-alive=300
# 配置Sink组件
a1.sinks.k1.type = hdfs
a1.sinks.k1.hdfs.path = hdfs://master:9000/flume/logs/%Y-%m-%d/
a1.sinks.k1.hdfs.useLocalTimeStamp = true
a1.sinks.k1.hdfs.fileType = DataStream
a1.sinks.k1.hdfs.writeFormat = Text
a1.sinks.k1.hdfs.filePrefix=a1_%Y%m%d_%H
a1.sinks.k1.hdfs.fileSuffix = .log
a1.sinks.k1.hdfs.round = true
a1.sinks.k1.hdfs.roundValue = 10
a1.sinks.k1.hdfs.roundUnit = minute
# Source组件和Sink组件与Channel组件绑定
a1.sources.r1.channels = c1
a1.sinks.k1.channel = c1
```

（4）启动 Flume：

```
$ cd /usr/local/flume
```

```
$ flume-ng agent --name a1 --conf ./conf --conf-file ./conf/taildir.conf -Dflume.root.logger=INFO,console
```

（5）向/home/hadoop/logs/access.log 文件写入测试数据：

```
$ echo "hello everyone" >> /home/hadoop/flume/hdfslogs/access.log
```

（6）查看目的日志信息。在 HDFS 上可以查看生成的文件和采集到的数据，如图 8-12 所示。

```
hadoop@master:~$ hdfs dfs -ls hdfs://master:9000/flume/logs
ls: `hdfs://master:9000/flume/logs': No such file or directory
hadoop@master:~$ hdfs dfs -ls hdfs://master:9000/flume/logs
Found 1 items
drwxr-xr-x   - hadoop supergroup          0 2021-04-10 12:20 hdfs://master:9000/flume/logs/2021-04-10
hadoop@master:~$ hdfs dfs -ls hdfs://master:9000/flume/logs/2021-04-10
Found 1 items
-rw-r--r--   1 hadoop supergroup         21 2021-04-10 12:20 hdfs://master:9000/flume/logs/2021-04-10/a1_20210410_12.1618028399282.log
hadoop@master:~$ hdfs dfs -cat hdfs://master:9000/flume/logs/2021-04-10/a1*
"hello everyone"
```

图 8-12 采集本地文件数据到 HDFS

### 8.4.2 多源与多目的地数据采集

**1．案例说明**

有 3 个服务器，分别为 Flume1、Flume2 和 Flume3，其中 Flume1 和 Flume2 实时收集本地日志数据（/home/hadoop/flume/data 目录中的 access.log、nginx.log 和 web.log），并将收集到的日志数据发送到 Flume3。Flume3 中的 Source 组件采用多路 Channel 选择器，将 access 类型日志发送到 Channel1，并通过 HDFS Sink 组件将数据写入 HDFS 中；将 nginx 类型日志发送到 Channel2，并通过 File Roll Sink 组件将数据写入本地文件系统中；将 web 类型日志发送到 Channel3，并通过 HBase Sink 组件将数据写入 HBase 的 weblog 表中。Flume 组件的综合应用架构如图 8-13 所示。

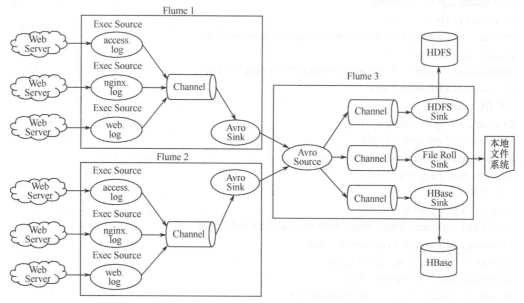

图 8-13 Flume 组件的综合应用架构

**2．操作步骤**

（1）安装与配置 Flume。

在 Flume1、Flume2 和 Flume3 三个服务器上分别安装与配置 Flume，并且 Flume3 处于 Hadoop 集群环境中。

（2）在 Flume1 和 Flume2 服务器上创建源数据文件所在的本地文件夹。

```
$ mkdir -p /home/hadoop/flume/data
```

（3）在 Flume3 服务器上创建本地文件夹、启动集群、启动 HBase 并创建表。

```
$ mkdir -p /home/hadoop/flume/logs/nginx
$ start-all.sh
$ start-hbase.sh
$ hbase shell
hbase(main):001:0> create 'weblog','log'
```

（4）在 Flume1 和 Flume2 服务器的 Flume 安装目录（/usr/local/flume）的 conf 文件夹中创建配置文件 exec_source_avro_sink.conf，进行 Flume 配置。

```
$ gedit /usr/local/flume/conf/exec_source_avro_sink.conf
```

写入下列配置信息：

```
#配置组件名称
a1.sources = r1 r2 r3
a1.sinks = k1
a1.channels = c1
#配置Source组件r1
a1.sources.r1.type = exec
a1.sources.r1.command = tail -F /home/hadoop/flume/data/access.log
#配置两个拦截器
a1.sources.r1.interceptors = i1 i2
#静态拦截器
a1.sources.r1.interceptors.i1.type = static
a1.sources.r1.interceptors.i1.key = type
a1.sources.r1.interceptors.i1.value = access
#主机拦截器
a1.sources.r1.interceptors.i2.type=host
#配置Source组件r2
a1.sources.r2.type = exec
a1.sources.r2.command = tail -F /home/hadoop/flume/data/nginx.log
a1.sources.r2.interceptors = i1 i2
a1.sources.r2.interceptors.i1.type = static
a1.sources.r2.interceptors.i1.key = type
a1.sources.r2.interceptors.i1.value = nginx
a1.sources.r2.interceptors.i2.type=host
#配置Source组件r3
a1.sources.r3.type = exec
a1.sources.r3.command = tail -F /home/hadoop/flume/data/web.log
a1.sources.r3.interceptors = i1 i2
a1.sources.r3.interceptors.i1.type = static
a1.sources.r3.interceptors.i1.key = type
a1.sources.r3.interceptors.i1.value = web
a1.sources.r3.interceptors.i2.type=host
# 配置Sink组件
a1.sinks.k1.type = avro
a1.sinks.k1.hostname =master
a1.sinks.k1.port =41414
# 配置Channel组件
a1.channels.c1.type = memory
a1.channels.c1.capacity = 2000
a1.channels.c1.transactionCapacity = 100
```

```
#Source组件和Sink组件与Channel组件绑定
a1.sources.r1.channels = c1
a1.sources.r2.channels = c1
a1.sources.r3.channels = c1
a1.sinks.k1.channel = c1
```

（5）在 Flume3 服务器的 Flume 安装目录（/usr/local/flume）的 conf 文件夹中创建配置文件 avro_source_multi_sink.conf，写入下列配置信息。

```
#配置组件名称
a1.sources = r1
a1.sinks = k1 k2 k3
a1.channels = c1 c2 c3
#配置Source组件
a1.sources.r1.type = avro
a1.sources.r1.bind =master
a1.sources.r1.port =41414
#添加时间戳拦截器
a1.sources.r1.interceptors = i1
a1.sources.r1.interceptors.i1.type =timestamp
#配置选择器
a1.sources.r1.selector.type=multiplexing
a1.sources.r1.selector.header=type;
a1.sources.r1.selector.mapping.access=c1
a1.sources.r1.selector.mapping.nginx=c2
a1.sources.r1.selector.mapping.web=c3
#配置Channel组件
a1.channels.c1.type = memory
a1.channels.c1.capacity = 2000
a1.channels.c1.transactionCapacity = 100
a1.channels.c2.type = memory
a1.channels.c2.capacity = 2000
a1.channels.c2.transactionCapacity = 100
a1.channels.c3.type = memory
a1.channels.c3.capacity = 2000
a1.channels.c3.transactionCapacity = 100
#配置HDFS Sink组件
a1.sinks.k1.type = hdfs
a1.sinks.k1.hdfs.path=hdfs://master:9000/flume/logs/%{type}/%Y%m%d
a1.sinks.k1.hdfs.filePrefix =%{type}
a1.sinks.k1.hdfs.filesuffix =.log
a1.sinks.k1.hdfs.fileType = DataStream
a1.sinks.k1.hdfs.writeFormat = Text
a1.sinks.k1.hdfs.rollCount = 0
a1.sinks.k1.hdfs.rollInterval = 30
a1.sinks.k1.hdfs.rollSize = 10485760
a1.sinks.k1.hdfs.batchSize = 100
a1.sinks.k1.hdfs.threadsPoolSize=10
a1.sinks.k1.hdfs.callTimeout=30000
#配置File Roll Sink组件
a1.sinks.k2.type=file_roll
```

```
a1.sinks.k2.sink.directory=/home/hadoop/flume/logs/nginx
a1.sinks.k2.sink.rollInterval=0
#配置HBase Sink组件
a1.sinks.k3.type = hbase2
a1.sinks.k3.table =weblog
a1.sinks.k3.columnFamily = log
a1.sinks.k1.serializer = org.apache.flume.sink.hbase2.RegexHBase2EventSerializer
#Source组件和Sink组件与Channel组件绑定
a1.sources.r1.channels = c1 c2 c3
a1.sinks.k1.channel = c1
a1.sinks.k2.channel = c2
a1.sinks.k3.channel = c3
```

（6）启动 Flume。首先启动 Flume3 服务器上的 Flume：

```
$ cd /usr/local/flume
$ flume-ng agent --name a1 --conf ./conf --conf-file ./conf/avro_source_multi_sink.conf -Dflume.root.logger=INFO,console
```

然后分别启动 Flume1 和 Flume2 服务器上的 Flume：

```
$ cd /usr/local/flume
$ flume-ng agent --name a1 --conf ./conf --conf-file ./conf/avro_source_avro_sink.conf -Dflume.root.logger=INFO,console
```

（7）分别向 Flume1 和 Flume2 服务器的/home/hadoop/flume/data 文件写入测试数据。

```
$ echo "this is an access log! " >> /home/hadoop/flume/data/access.log
$ echo "this is an nginx log!" >> /home/hadoop/flume/data/nginx.log
$ echo "this is an web log!" >> /home/hadoop/flume/data/web.log
```

（8）查看目的地日志信息，可以查看不同 Sink 对应目的地的数据采集信息。

# 本 章 小 结

本章首先介绍了 Flume 的功能特性、版本、架构及 Flume 安装与配置，然后详细介绍了 Flume 组件的配置，包括三大核心组件 Source、Channel、Sink，以及可选组件 Interceptor、Selector、Sink Processor，最后介绍了两个 Flume 数据采集案例及其实施。通过本章的学习，读者可以根据业务需要进行 Flume 数据采集方案设计与实施，将不同来源的数据采集并保存到不同的目的地，以便进行后续的数据分析与处理。

# 思考题与习题

1．简答题

（1）简述 Flume 不同数据流模型架构。
（2）简述 Flume 常用的组件及其作用。
（3）简述 Interceptor 组件的分类及其作用。
（4）简述 Sink Processor 组件的分类及其特征。

2．选择题

（1）关于 Flume 的描述不正确的是（　　）。
A．Flume 是日志采集系统

B．Flume 能够进行日志采集、聚合和传输

C．Flume 是一个分布式处理系统

D．Flume 的数据流模型为 Source-Sink-Channel

（2）关于 Flume 组件描述不正确的是（　　）。

A．event 是数据流的基本单位

B．Agent 负责将数据源的数据转发到目的地

C．Source 负责从外部读取数据并写入 Sink

D．写入 Channel 的消息会一直存在，直到 Sink 将其取走

（3）下列不是 Flume 的 Source 组件类型的是（　　）。

A．Avro Source　　　B．JOSN Source　　　C．Kafka Source　　　D．HTTP Source

（4）为了采集存储在文件夹中不断增减的文件数据，Flume Source 应采用哪种类型？（　　）

A．Avro Source　　　　　　　　　　B．Spooling Directory Source

C．Kafka Source　　　　　　　　　　D．HTTP Source

（5）关于 Channel 组件描述不正确的是（　　）。

A．是 Flume 采集的 event 暂存区

B．Channel 中 event 可以被多个 Sink 读取

C．Chanel 中 event 被 Sink 读取后删除

D．Channel 具有事务性，允许失败重写和失败重读

（6）下列不是 Flume 的 Channel 组件类型的是（　　）。

A．HBase Channel　　B．File Channel　　C．Kafka Channel　　D．Memory Channel

（7）关于 Sink 组件描述不正确的是（　　）。

A．Sink 组件从 Channel 组件消费 event

B．一个 Sink 只能读取一个 Channel 中的 event

C．Sink 可以将读取到的 event 存储或输入到下一个 Agent

D．Sink 读取 Channel 后，Channel 立即删除相应的 event

（8）下列不是 Flume 的 Sink 组件类型的是（　　）。

A．HDFS Sink　　　B．Mongo Sink　　　C．Kafka Sink　　　D．Avro Sink

（9）关于 Flume Interceptor 组件描述不正确的是（　　）。

A．Source 将 event 写入 Channel 之前调用拦截器

B．Channel 和 Sink 之间可以有多个拦截器

C．不同的拦截器使用不同的规则处理 event

D．通过实现 Interceptor 接口实现自定义的拦截器

（10）下列哪个不是 Flume 的内置 Interceptor？（　　）

A．Object Interceptor　　　　　　　　B．Timestamp Interceptor

C．Host Interceptor　　　　　　　　　D．Static Interceptor

3．实训题

（1）利用 Flume 将本地文件夹中新增的文件采集到 Hadoop 的 HDFS 中。

（2）利用 Flume 将本地文件中新追加的内容实时采集并上传到 HDFS 中。

（3）采集 3 个不同 Web 服务器（server1、server2、server3）上的日志数据并保存到 HDFS，根据日志信息来源的服务器名不同，分别将数据保存到 /flume/server1、/flume/server2 和 /flume/server3，子目录根据日志采集时间戳中的年、月、日分区。

# 第 9 章　网站日志分析

网站日志分析是大数据技术的一个重要应用，通过网页浏览量、新增 IP 数、跳出率等 KPI 指标分析，可以了解网站访客的访问规律、制订运行策略、调整板块规划等，为网站高效运营提供决策支持。本章将介绍网站日志分析项目的需求分析、方案设计、数据采集、数据分析、数据迁移与可视化等。

- 需求分析：网站日志分析的必要性、数据格式说明、KPI 指标。
- 方案设计：数据处理流程和组件使用。
- 数据采集：利用 Flume 将本地 "假脱机" 目录中的文件采集到 HDFS。
- 数据预处理：采用 MapReduce 对 HDFS 中采集的原始数据进行清洗。
- 数据分析：采用 Hive 工具对预处理后的数据进行各种 KPI 指标分析。
- 分析结果导出与可视化：将 Hive 数据分析结果利用 Sqoop 导出到 MySQL 数据库，并利用 PyEcharts 进行可视化。

## 9.1　需求分析

### 9.1.1　网站日志分析的必要性

网站日志分析可以实时对网站流量进行分析统计，帮助网站管理员、运营人员、推广人员等实时获取网站流量，并从流量来源、网站内容、网站访客特性等多方面提供网站分析的数据依据，从而提高网站流量，提升网站用户体验，让访客更多地沉淀下来变成会员或客户，通过更少的投入获取更大化的收益。

通过网站日志分析，可以获取指定时间段的网络流量数据，通过流量的趋势变化形态，可以帮助用户分析出网站访客的访问规律、网站发展状况等；了解网站质量和运营状况；掌握流量规律，制订运营策略；监控流量起伏，了解运营效果；及时发现异常，避免流量继续下跌。

### 9.1.2　网站日志数据说明

在网站日志中，每条日志通常代表着用户的一次访问行为，格式为：

$remote_addr - $remote_user [$time_local] $request $status $body_bytes_sent $http_referer $http_user_agent $http_x_forwarded_for

各部分参数的含义如下。

$remote_addr：记录客户端的 IP 地址。

$remote_user：记录客户端的用户名称。

$time_local：记录访问时间与时区。

$request：记录请求的 URL 与 HTTP 协议。

$status：记录请求状态，成功是 200。

$body_bytes_sent：记录发送给客户端文件主体内容大小。

$http_referer：记录从哪个页面链接访问过来的。

$http_user_agent：记录客户端浏览器的相关信息。

$http_x_forwarded_for：访问用户的真实 IP 地址。

例如，下面是一条网站日志记录：

> 115.239.212.132 - - [01/May/2019:04:21:01 +0800] "GET /wp-includes/js/wp-emoji-release.min.js HTTP/1.0" 200 16242 "http://alumni.neusoft.edu.cn/?portfolio-posts=%E9%97%AB%E6%98%8E" "Mozilla/5.0 (Linux; U; Android 4.1; en-us; GT-N7100 Build/JRO03C;Baiduspider-ads)AppleWebKit/534.30 (KHTML, like Gecko) Version/4.0 Mobile Safari/534.30" "-"

其中：

$remote_addr：115.239.212.132

$remote_user：-

$time_local：[01/May/2019:04:21:01 +0800]

$request："GET /wp-includes/js/wp-emoji-release.min.js HTTP/1.0"

$status：200

$body_bytes_sent：16242

$http_referer："http://alumni.neusoft.edu.cn/?portfolio-posts=%E9%97%AB%E6%98%8E"

$http_user_agent："Mozilla/5.0 (Linux; U; Android 4.1; en-us; GT-N7100 Build/JRO03C;Baiduspider-ads)AppleWebKit/534.30 (KHTML, like Gecko) Version/4.0 Mobile Safari/534.30"

$http_x_forwarded_for："-"

### 9.1.3 网站日志分析 KPI 指标

网站日志分析通常采用关键绩效指标考核法，即 KPI（Key Performance Indicator）绩效考核法。网站日志分析的主要任务就是通过对网站日志的分析，根据 KPI 指标设计，提取 KPI 数据。

网站日志分析的 KPI 指标包括如下几项。

① 页面浏览量 PV（PageView）统计：所有用户浏览页面的总和。

② 独立 IP 数 UIP（UniqueIP）统计：特定时间内访问网站的不重复 IP 总数。

③ 跳出率 BR（Bounce Rate）统计：只浏览了一个页面便离开了网站的浏览量占总的浏览量的百分比，即只浏览了一个页面的访问次数/全部的访问次数的汇总。

④ 新增 IP 数 NIP（NewIP）统计：一天内新增的 IP 总数。

⑤ 用户每小时 PV 的统计（Time）。

⑥ 用户来源域名的统计（Source）。

⑦ 用户的访问设备统计（Browser）。

## 9.2 方案设计

在本项目中，将保存在本地文件系统中的网站日志，通过 Flume 或 Shell 命令将日志数据上传到 HDFS，采用 MapReduce 对数据进行预处理后，使用 Hive 进行统计分析，分析结果导出到 MySQL 数据库中，最后基于 MySQL 数据库中的数据进行可视化，方案设计如图 9-1 所示。

① 将日志数据上传到 HDFS 中。如果数据量较小，可以通过 HDFS Shell 命令上传数据到 HDFS 中；如果数据量较大，可以使用 Flume 来完成日志收集工作。

② 使用 MapReduce 将 HDFS 中的日志数据进行预处理，即数据清洗。日志原始数据可能在格式上不满足要求，需要通过 MapReduce 程序来将 HDFS 中的数据进行清洗过滤，转换成

Hive 统计分析所需要的格式。

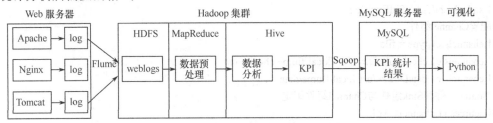

图 9-1　网站日志分析方案设计

③ 使用 Hive 对预处理后的数据进行统计分析。在 Hive 中建立一个外部分区表，关联到预处理后的 HDFS 目录，然后定期将预处理后的数据添加到 Hive 表的相应分区，最后执行 HiveQL 完成统计并保存结果。

④ 利用 Sqoop 将 Hive 统计分析结果导出到关系数据库 MySQL 中。

⑤ 利用 Python 程序设计，图形化显示 MySQL 中数据的分析结果。

## 9.3　数 据 采 集

在本项目中，将 Web 服务器生成的日志文件放入本地目录/home/hadoop/weblogs，通过 Flume，将新生成的日志文件采集存储到 HDFS 的/weblogs 目录中，根据数据采集日期的年、月进行分区。

（1）在本地 Linux 文件创建"假脱机"目录及 Channel 组件需要的目录。

$ mkdir -p /home/hadoop/flume/source/weblogs
$ mkdir -p /usr/local/flume/channel/checkpoint
$ mkdir -p /usr/local/flume/channel/data

（2）编写 Flume 配置文件：

$ gedit /usr/local/flume/conf/hdfsspool.conf

写入下列内容：

```
#配置组件名称
a1.sources = r1
a1.sinks = k1
a1.channels = c1
#配置Source名称
a1.sources.r1.type = spooldir
a1.sources.r1.spoolDir =/home/hadoop/flume/source/weblogs
a1.sources.source1.inputCharset =UTF-8
a1.sources.r1.fileHeader = true
a1.sources.r1.interceptors = i1
a1.sources.r1.interceptors.i1.type = timestamp
#配置Sink组件
a1.sinks.k1.type = hdfs
a1.sinks.k1.hdfs.path = hdfs://Master:9000/weblogs/%Y_%m
a1.sinks.k1.hdfs.writeFormat = Text
a1.sinks.k1.hdfs.fileType = DataStream
a1.sinks.k1.hdfs.rollInterval =60
a1.sinks.k1.hdfs.rollSize = 0
a1.sinks.k1.hdfs.rollCount = 0
```

```
a1.sinks.k1.hdfs.filePrefix =%Y_%m_%d_%H_%M_%S
a1.sinks.k1.hdfs.fileSuffix =.log
#配置Channel组件
a1.channels.c1.type = file
a1.channels.c1.checkpointDir =/usr/local/flume/checkpoint
a1.channels.c1.dataDirs = /usr/local/flume/data
# Source组件和Sink组件与Channel组件绑定
a1.sources.r1.channels = c1
a1.sinks.k1.channel = c1
```

（3）启动 Flume Agent a1 服务器：

$ flume-ng    agent --name a1 --conf /usr/local/flume/conf --conf-file /usr/local/flume/conf/hdfsspool.conf -Dflume.root.logger=DEBUG,consol

（4）将日志文件存入 Spooling Directory Source 中，即/home/hadoop/flume/source/weblogs。

（5）查看 HDFS 上收集到的日志信息，如图 9-2 所示。

```
hadoop@master:~$ hdfs dfs -ls -R /weblogs
drwxr-xr-x   - hadoop supergroup          0 2021-04-13 21:29 /weblogs/2021_04
-rw-r--r--   1 hadoop supergroup     308866 2021-04-13 21:29 /weblogs/2021_04/2021_04_13_21_28_23.1618320506796.log
-rw-r--r--   1 hadoop supergroup    6218105 2021-04-13 21:29 /weblogs/2021_04/2021_04_13_21_28_24.1618320510919.log
-rw-r--r--   1 hadoop supergroup    7851875 2021-04-13 21:29 /weblogs/2021_04/2021_04_13_21_28_25.1618320516381.log
-rw-r--r--   1 hadoop supergroup    6596093 2021-04-13 21:29 /weblogs/2021_04/2021_04_13_21_28_26.1618320522466.log
-rw-r--r--   1 hadoop supergroup    2924388 2021-04-13 21:29 /weblogs/2021_04/2021_04_13_21_28_27.1618320526885.log
```

图 9-2  HDFS 收集到的日志信息

## 9.4　数据预处理

为了方便进行日志数据分析，需要将输入的日志数据进行清洗，分离出日志文件中的客户端的 IP 地址、客户端用户名称、访问日期与时间、请求对象的 URL 与 HTTP 协议、请求状态、发送给客户端文件主体内容大小、访问链接、访问域名和客户端浏览器的相关信息等，然后将这些信息以格式化的方式存储到 HDFS 中，为后续的日志数据分析处理做好准备。

（1）在 Eclipse 中创建一个 Maven 工程（详见 4.4.2 节），并配置 Maven 工程（详见 4.4.3 节）。

（2）创建一个 KPI 类，用于封装从日志文件中提取出来的各部分日志信息。

```
package com.bigdata.example;
import java.text.ParseException;
import java.text.SimpleDateFormat;
import java.util.Date;
import java.util.Locale;
public class KPI {
    private String remote_addr;// 记录客户端的IP地址
    private String remote_user;// 记录客户端用户名称，忽略属性"-"
    private String time_local;// 记录访问时间与时区
    private String request;// 记录请求的URL与HTTP协议
    private String status;// 记录请求状态；成功是200
    private String body_bytes_sent;// 记录发送给客户端文件主体内容大小
    private String http_referer;//记录从哪个页面链接访问过来的
    private String http_user_agent;// 记录客户端浏览器的相关信息
    private boolean valid = true;// 判断数据是否合法
    private Date time_local_date; // 记录请求日期时间
    private String time_local_date_hour; // 记录请求日期和小时
    private String http_referer_domain; // 记录链接网页的域名
    public String getRemote_addr() {
```

```java
        return remote_addr;
    }
    public void setRemote_addr(String remote_addr) {
        this.remote_addr = remote_addr;
    }
    public String getRemote_user() {
        return remote_user;
    }
    public void setRemote_user(String remote_user) {
        this.remote_user = remote_user;
    }
    public String getTime_local() {
        return time_local;
    }
    public void setTime_local(String time_local) {
        this.time_local = time_local;
    }
    public String getRequest() {
        return request;
    }
    public void setRequest(String request) {
        this.request = request;
    }
    public String getStatus() {
        return status;
    }
    public void setStatus(String status) {
        this.status = status;
    }
    public String getBody_bytes_sent() {
        return body_bytes_sent;
    }
    public void setBody_bytes_sent(String body_bytes_sent) {
        this.body_bytes_sent = body_bytes_sent;
    }
    public String getHttp_referer() {
        return http_referer;
    }
    public void setHttp_referer(String http_referer) {
        this.http_referer = http_referer;
    }
    public String getHttp_user_agent() {
        return http_user_agent;
    }
    public void setHttp_user_agent(String http_user_agent) {
        this.http_user_agent = http_user_agent;
    }
    public boolean isValid() {
        return valid;
```

```java
    }
    public void setValid(boolean valid) {
        this.valid = valid;
    }
    public Date getTime_local_Date() throws ParseException {
        SimpleDateFormat df = new SimpleDateFormat("dd/MMM/yyyy:HH:mm:ss", Locale.US);
        return df.parse(this.time_local);
    }
    public String getTime_local_Date_hour() throws ParseException {
        SimpleDateFormat df = new SimpleDateFormat("yyyyMMddHH");
        return df.format(this.getTime_local_Date());
    }
    public String getHttp_referer_domain() {
        if (http_referer.length() < 8) {
            return http_referer;
        }
        String str = this.http_referer.replace("\"", "").replace("http://", "").replace("https://", "");
        return str.indexOf("/") > 0 ? str.substring(0, str.indexOf("/")) : str;
    }
    public static KPI parser(String line) {
        String ref = "";
        KPI kpi = new KPI();
        String[] arr = line.split(" ");
        if (arr.length > 11) {
            kpi.setRemote_addr(arr[0]);
            kpi.setRemote_user(arr[1]);
            kpi.setTime_local(arr[3].substring(1));
            kpi.setRequest(arr[6]);
            kpi.setStatus(arr[8]);
            kpi.setBody_bytes_sent(arr[9]);
            kpi.setHttp_referer(arr[10].replace("\"", ""));
            if (arr.length > 12) {
                for (int i = 11; i < arr.length - 1; i++) {
                    if (i != arr.length - 1)
                        ref = ref + arr[i] + " ";
                    else
                        ref = ref + arr[i];
                }
                kpi.setHttp_user_agent(ref.replace("\"", ""));
            } else {
                kpi.setHttp_user_agent(arr[11].replace("\"", ""));
            }
            if (Integer.parseInt(kpi.getStatus()) >= 400) {// 大于400,HTTP错误
                kpi.setValid(false);
            }
        } else {
            kpi.setValid(false);
        }
        return kpi;
    }
    public String formatLog() {
```

```java
        String logInfo = null;
        SimpleDateFormat df = new SimpleDateFormat("yyyy-MM-dd", Locale.US);
        if (this.isValid()) {
            try {
                logInfo = this.getRemote_addr() + "\t" + this.getTime_local() + "\t"
                    + df.format(this.getTime_local_Date()) + "\t" + this.getTime_local_Date_hour() + "\t"
                    + this.getRequest() + "\t" + this.getStatus() + "\t" + this.getBody_bytes_sent() + "\t"
                    + this.getHttp_referer() + "\t" + this.getHttp_referer_domain() + "\t"
                    + this.getHttp_user_agent();
            } catch (Exception e) {
                e.printStackTrace();
            }
        }
        return logInfo;
    }
}
```

(3) 编写一个 MapReduce 程序预处理日志文件,并将预处理后的数据保存到 HDFS 的 /cleanweblogs 目录中。

```java
package com.bigdata.example;
import java.io.IOException;
import org.apache.hadoop.conf.Configuration;
import org.apache.hadoop.fs.FileSystem;
import org.apache.hadoop.fs.Path;
import org.apache.hadoop.io.NullWritable;
import org.apache.hadoop.io.Text;
import org.apache.hadoop.mapreduce.Job;
import org.apache.hadoop.mapreduce.Mapper;
import org.apache.hadoop.mapreduce.Reducer;
import org.apache.hadoop.mapreduce.lib.input.FileInputFormat;
import org.apache.hadoop.mapreduce.lib.output.FileOutputFormat;
public class LogCleaner {
    public static class LogMapper extends Mapper<Object, Text, Text, NullWritable> {
        private Text word = new Text();
        public void map(Object key, Text value, Context context) throws IOException, InterruptedException {
            KPI kpi = KPI.parser(value.toString());
            if (kpi.isValid()) {
                String formatLogInfo = kpi.formatLog();
                word.set(formatLogInfo.trim());
                context.write(word, NullWritable.get());
            }
        }
    }
    public static class LogReducer extends Reducer<Text, NullWritable, Text, NullWritable> {
        public void reduce(Text key, Iterable<NullWritable> values, Context context)
                throws IOException, InterruptedException {
            for (NullWritable val : values) {
                context.write(key, NullWritable.get());
            }
        }
```

```java
        }
        public static void main(String[] args) throws Exception {
            Configuration conf = new Configuration();
            conf.set("fs.defaultFS", "hdfs://master:9000");
            String[] otherArgs = new String[] { "/weblogs/2021_04", "/cleanweblogs/2021_04" };
            FileSystem fileSystem = FileSystem.get(conf);
            if (fileSystem.exists(new Path(otherArgs[1]))) {
                fileSystem.delete(new Path(otherArgs[1]), true);
            }
            Job job = Job.getInstance();
            job.setJarByClass(LogCleaner.class);
            job.setMapperClass(LogMapper.class);
            job.setReducerClass(LogReducer.class);
            job.setOutputKeyClass(Text.class);
            job.setOutputValueClass(NullWritable.class);
            FileInputFormat.addInputPath(job, new Path(otherArgs[0]));
            FileOutputFormat.setOutputPath(job, new Path(otherArgs[1]));
            System.exit(job.waitForCompletion(true) ? 0 : 1);
        }
    }
```

（4）查看 HDFS 的/cleanweblogs 中生成的目录与文件，如图 9-3 所示。

```
hadoop@master:~$ hdfs dfs -ls -R /cleanweblogs
drwxr-xr-x   - hadoop supergroup          0 2021-04-13 21:32 /cleanweblogs/2021_04
-rw-r--r--   1 hadoop supergroup          0 2021-04-13 21:32 /cleanweblogs/2021_04/_SUCCESS
-rw-r--r--   1 hadoop supergroup   24186212 2021-04-13 21:32 /cleanweblogs/2021_04/part-r-00000
```

图 9-3 预处理后的文件

## 9.5 数 据 分 析

为了能够使用 Hive 进行统计分析，首先需要在 Hive 中创建一个表，然后将预处理后的数据导入该表中。本项目使用分区表，以日期中的年、月作为分区的字段，与预处理后的日志存储文件目录相关联。

### 1. 数据分析准备
（1）创建数据库及分区表

```sql
CREATE DATABASE IF NOT EXISTS logdb;
USE logdb;
CREATE EXTERNAL TABLE IF NOT EXISTS logs (
  ip string,time_local string,time_local_date string,time_local_date_hour string, url string,status string,
  quantity string,referer string,referer_domain string,user_agent string)
partitioned by (logdate string)
row format delimited fields terminated by '\t'
location '/cleanweblogs';
```

（2）创建一个分区并将已有数据加载到分区表中

```sql
ALTER TABLE logs ADD PARTITION(logdate='2021_04') LOCATION '/cleanweblogs/2021_04';
```

### 2. 利用 Hive 进行数据分析
（1）页面浏览量 PV 统计

页面浏览量是指所有用户浏览页面的总和，一个独立用户每打开一个页面就被记录一次。因此只需要统计日志中的记录个数即可。HiveQL 代码如下：

```
CREATE TABLE IF NOT EXISTS log_pv_2021_04
AS
SELECT time_local_date ,count(*)   AS quantity   FROM logs WHERE logdate='2021_04'
GROUP BY time_local_date;
```

（2）独立 IP 数 UIP 统计

一天之内，访问网站的不同独立 IP 个数的和。其中，同一 IP 无论访问了几个页面，独立 IP 数均为 1。因此，只需要统计日志中处理的独立 IP 数即可。HiveQL 代码如下：

```
CREATE TABLE IF NOT EXISTS log_ip_2021_04
AS
SELECT time_local_date,count(DISTINCT ip)   AS quantity   FROM logs WHERE logdate='2021_04'
GROUP BY time_local_date;
```

（3）跳出率 BR 统计

只浏览了一个页面便离开了网站的访问次数，即只浏览了一个页面的访问次数／全部的访问次数汇总。通过用户的 IP 进行分组，如果分组后的记录数只有一条，那么即为跳出用户。将这些用户的数量相加，就得出了跳出用户数。HiveQL 代码如下：

```
CREATE TABLE IF NOT EXISTS log_ip_br_2021_04
AS
SELECT br_ip_tab.time_local_date,round(br_count/vv_count,4) bounceRate FROM
(SELECT time_local_date,count(ip) as br_count FROM (
SELECT time_local_date,ip FROM logs WHERE logdate='2021_04'
GROUP BY time_local_date,ip HAVING count(ip)=1 ) as br_tab
GROUP BY time_local_date ) as br_ip_tab
JOIN
(SELECT time_local_date,count(ip) as vv_count FROM logs WHERE logdate='2021_04'
GROUP BY time_local_date) all_ip_tab
ON br_ip_tab.time_local_date=all_ip_tab.time_local_date;
```

（4）新增 IP 数 NIP 统计

一天内访问网站的所有 IP 去重后，检查有多少是在历史数据中未出现过的，这些数量指的就是新增的 IP 总数。HiveQL 代码如下：

```
CREATE TABLE IF NOT EXISTS log_newip_2021_04
AS
select time_local_date ,count(ip) quantity from (
select ip,time_local_date,row_number()over(partition by ip order by time_local_date asc) flag
from logs WHERE logdate='2021_04') ips
where flag=1 group by time_local_date;
```

（5）用户每小时 PV 的统计（Time）

```
CREATE TABLE IF NOT EXISTS log_time_2021_04
AS
SELECT time_local_date,substr(time_local_date_hour,9) time_local_hour ,count(ip) quantity
FROM logs WHERE logdate='2021_04' group by time_local_date,substr(time_local_date_hour,9);
```

（6）用户来源域名的统计（Source）

```
CREATE TABLE IF NOT EXISTS log_source_2021_04
AS
SELECT time_local_date,referer_domain,count(referer_domain) quantity FROM logs
WHERE logdate='2021_04' GROUP BY time_local_date,referer_domain;
```

**（7）用户的访问设备统计（Browser）**

```
CREATE TABLE IF NOT EXISTS log_browser_2021_04
AS
SELECT time_local_date,user_agent,count(user_agent) quantity FROM logs
WHERE logdate='2021_04' GROUP BY time_local_date,user_agent;
```

## 9.6 数据分析结果导出及可视化

为了充分利用日志分析结果，使用 Sqoop 将 Hive 分析结果导出到 MySQL 数据库中。

### 1. 在 MySQL 中创建接收数据的表

```
CREATE DATABASE IF NOT EXISTS logdb;
USE logdb;
CREATE TABLE IF NOT EXISTS log_pv_2021_04 (time_local_date char(50),quantity bigint);
CREATE TABLEIF NOT EXISTS log_ip_2021_04 (time_local_date char(50),quantity bigint);
CREATE TABLEIF NOT EXISTS log_ip_br_2021_04(time_local_date char(50),bounceRate float);
CREATE TABLE IF NOT EXISTS log_newip_2021_04(time_local_date char(50),quantity bigint);
CREATE TABLE IF NOT EXISTS log_time_2021_04 (local_date char(50), local_hour char(10),
quantity bigint ) ;
CREATE TABLE IF NOT EXISTS log_source_2021_04 (local_date char(50),referer varchar(2000),
quantity bigint );
CREATE TABLE IF NOT EXISTS log_browser_2021_04 ( local_date char(50),agent varchar(2000),
quantity bigint );
```

### 2. 利用 Sqoop 将数据导出到 MySQL 数据库中

**（1）导出页面浏览量 PV 统计结果**

```
$ sqoop export --connect jdbc:mysql://localhost:3306/logdb --username root --password root --table log_pv_2021_04 --export-dir /user/hive/warehouse/logdb.db/log_pv_2021_04 --input-fields-terminated-by '\001' -m 1
```

**（2）导出独立 IP 数 UIP 统计结果**

```
$ sqoop export --connect jdbc:mysql://localhost:3306/logdb --username root --password root --table log_ip_2021_04 --export-dir /user/hive/warehouse/logdb.db/log_ip_2021_04 --input-fields-terminated-by '\001' -m 1
```

**（3）导出跳出率 BR 统计结果**

```
$ sqoop export --connect jdbc:mysql://localhost:3306/logdb --username root --password root --table log_ip_br_2021_04 --export-dir /user/hive/warehouse/logdb.db/log_ip_br_2021_04 --input-fields-terminated-by '\001'-m 1
```

**（4）导出新增 IP 数 NIP 统计结果**

```
$ sqoop export --connect jdbc:mysql://localhost:3306/logdb --username root --password root --table log_newip_2021_04 --export-dir /user/hive/warehouse/logdb.db/log_newip_2021_04 --input-fields-terminated-by '\001' -m 1
```

**（5）导出每天浏览量统计结果**

```
$ sqoop export --connect jdbc:mysql://localhost:3306/logdb --username root --password root --table log_time_2021_04 --export-dir /user/hive/warehouse/logdb.db/log_time_2021_04 --input-fields-terminated-by '\001' -m 1
```

**（6）导出每天来源域数量统计结果**

```
$ sqoop export --connect jdbc:mysql://localhost:3306/logdb --username root --password root --table log_source_2021_04 --export-dir /user/hive/warehouse/logdb.db/log_source_2021_04 --input-fields-terminated-by '\001' -m 1
```

（7）导出每天客户端不同浏览器数量统计结果

$ sqoop export --connect jdbc:mysql://localhost:3306/logdb --username root --password root --table log_browser_2021_04 --export-dir /user/hive/warehouse/logdb.db/log_browser_2021_04 --input-fields-terminated-by '\001' -m 1

日志分析结果从 Hive 表导出到 MySQL 数据库后，可以查看导出结果，如图 9-4 至图 9-9 所示。

```
mysql> SELECT * FROM log_pv_2021_04 limit 5;
+-----------------+----------+
| time_local_date | quantity |
+-----------------+----------+
| 2021-04-01      |     4110 |
| 2021-04-02      |     7284 |
| 2021-04-03      |     5910 |
| 2021-04-04      |     4184 |
| 2021-04-05      |     3221 |
+-----------------+----------+
```

图 9-4　页面浏览量 PV 统计结果

```
mysql> SELECT * FROM log_ip_2021_04 LIMIT 5;
+-----------------+----------+
| time_local_date | quantity |
+-----------------+----------+
| 2021-04-01      |      148 |
| 2021-04-02      |      289 |
| 2021-04-03      |      227 |
| 2021-04-04      |      176 |
| 2021-04-05      |      109 |
+-----------------+----------+
```

图 9-5　独立 IP 数 UIP 统计结果

```
mysql> SELECT * FROM log_ip_br_2021_04 limit 5;
+-----------------+------------+
| time_local_date | bounceRate |
+-----------------+------------+
| 2021-04-01      |     0.0085 |
| 2021-04-02      |     0.0097 |
| 2021-04-03      |     0.0069 |
| 2021-04-04      |     0.0093 |
| 2021-04-05      |     0.0078 |
+-----------------+------------+
```

图 9-6　跳出率 BR 统计结果

```
mysql> SELECT * FROM log_newip_2021_04 LIMIT 5;
+-----------------+----------+
| time_local_date | quantity |
+-----------------+----------+
| 2021-04-01      |      148 |
| 2021-04-02      |      279 |
| 2021-04-03      |      218 |
| 2021-04-04      |      164 |
| 2021-04-05      |      105 |
+-----------------+----------+
```

图 9-7　新增 IP 数 NIP 统计结果

```
mysql> SELECT * FROM log_time_2021_04 limit 5;
+------------+------------+----------+
| local_date | local_hour | quantity |
+------------+------------+----------+
| 2021-04-01 | 04         |        1 |
| 2021-04-01 | 07         |       15 |
| 2021-04-01 | 08         |       97 |
| 2021-04-01 | 09         |      110 |
| 2021-04-01 | 10         |      189 |
+------------+------------+----------+
```

图 9-8　每天浏览量统计结果

```
mysql> SELECT * FROM log_source_2021_04 limit 5;
+------------+------------------+----------+
| local_date | referer          | quantity |
+------------+------------------+----------+
| 2021-04-01 | -                |      360 |
| 2021-04-01 | 172.24.4.42      |        2 |
| 2021-04-01 | 42.neusoft.edu.cn|     3584 |
| 2021-04-01 | blank            |        3 |
| 2021-04-01 | blog.sina.com.cn |        3 |
+------------+------------------+----------+
```

图 9-9　客户端不同浏览器数量统计结果

### 3．可视化

可以采用多种可视化技术将日志分析结果可视化，本项目采用 PyEcharts 对统计分析结果进行可视化。

（1）利用折线图实现页面浏览量 PV 可视化

```
import pymysql
import pyecharts.options as opts
from pyecharts.charts import Bar,Line
import pandas as pd
date='2021-04'
conn = pymysql.connect(host='192.168.189.128',user='root',passwd='root',db='logdb')
sql='select distinct time_local_date,quantity from log_pv_2021_04 where substr(time_local_date,1,7)="'+date+'"order by time_local_date'
df = pd.read_sql(sql,conn)
dt=df['time_local_date'].str.split("-").str[2].tolist()
year,month=date.split("-")[0],date.split("-")[1]
pagetitle=year+"年"+month+"月日浏览量"
pv=df['quantity'].tolist()
line =Line();
line.add_xaxis(dt)
```

```
line.add_yaxis("浏览量PV",pv)
line.set_global_opts(title_opts=opts.TitleOpts(title=pagetitle))
line.render("./linepv.html")
```

可视化结果如图 9-10 所示。

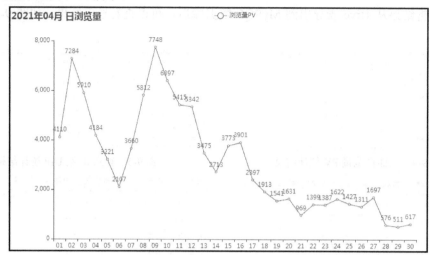

图 9-10　利用折线图实现页面浏览量 PV 可视化

（2）利用柱状图实现页面浏览量 PV 可视化

```
import pymysql
import pyecharts.options as opts
from pyecharts.charts import Line,Bar
import pandas as pd
date='2021-04'
conn = pymysql.connect(host='192.168.189.128',user='root',passwd='root',db='logdb')
sql='select distinct time_local_date,quantity from log_pv_2021_04 where substr(time_local_date,1,7)="'+date+'" order by time_local_date'
df = pd.read_sql(sql,conn)
dt=df['time_local_date'].str.split("-").str[2].tolist()
year,month=date.split("-")[0],date.split("-")[1]
pagetitle=year+"年"+month+"月日浏览量"
pv=df['quantity'].tolist()
bar =Bar();
bar.add_xaxis(dt)
bar.add_yaxis("浏览量PV",pv)
bar.set_global_opts(title_opts=opts.TitleOpts(title=pagetitle))
bar.render("./barpv.html")
```

可视化结果如图 9-11 所示。

（3）利用复杂折线图实现页面浏览量 PV 可视化

```
import pymysql
import pyecharts.options as opts
from pyecharts.charts import Line
import pandas as pd
date='2021-04'
conn = pymysql.connect(host='192.168.189.128',user='root',passwd='root',db='logdb')
sql='select distinct time_local_date,quantity from log_pv_2021_04
```

```
where substr(time_local_date,1,7)="'+date+ "'order by time_local_date'
df = pd.read_sql(sql,conn)
dt=df['time_local_date'].str.split("-").str[2].tolist()
year,month=date.split("-")[0],date.split("-")[1]
pagetitle=year+"年"+month+"月日浏览量"
pv=df['quantity'].tolist()
line =Line();
line.add_xaxis(dt)
line.add_yaxis("浏览量PV", pv)
line.set_global_opts(
title_opts=opts.TitleOpts(title=pagetitle,subtitle="此处是副标题",pos_left="10%"),
toolbox_opts=opts.ToolboxOpts(),legend_opts=opts.LegendOpts(pos_left="40%") )
line.set_series_opts(label_opts=opts.LabelOpts(is_show=True),markpoint_opts=opts.MarkPointOpts(
data=[ opts.MarkPointItem(type_="max",name="最大值",symbol="triangle",symbol_size=[80,60]),
opts.MarkPointItem(type_="min",name="最小值"),opts.MarkPointItem(type_="average",name="平均值
")]),markline_opts=opts.MarkLineOpts(data=[ opts.MarkLineItem(type_="min",name="最小等位线"),
opts.MarkLineItem(type_="average",name="平均等位线")])
)
line.render("./complexpv.html")
```

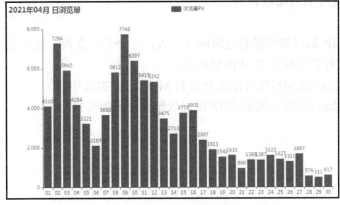

图 9-11　利用柱状图实现页面浏览量 PV 可视化

可视化结果如图 9-12 所示。

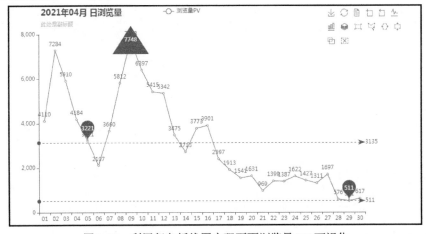

图 9-12　利用复杂折线图实现页面浏览量 PV 可视化

可以采用类似的程序设计进行独立 IP 数、跳出率、新增 IP 数等 KPI 指标的可视化。

·271·

# 本 章 小 结

本章首先介绍了网站日志分析项目需求分析与方案设计，然后按照大数据业务处理流程，分别介绍了利用 Flume 进行网站日志采集、利用 MapReduce 进行网站日志数据预处理、利用 Hive 进行网站日志 KPI 指标分析、利用 Sqoop 将分析结果导出到 MySQL 数据库，最后采用 PyEcharts 工具进行结果可视化。通过本章学习，读者可以掌握大数据分析业务的处理流程，掌握不同 Hadoop 组件在业务处理不同环节的应用，为独立完成大数据项目开发奠定基础。

## 思考题与习题

### 1. 简答题
（1）简述大数据分析业务的处理流程。
（2）简述网站日志分析的作用。
（3）数据预处理的主要任务有哪些？
（4）数据分析技术有哪些？有何差异？
（5）数据可视化工具有哪些？

### 2. 实训题
（1）页面独立 IP 访问量是指特定时间（一天、一个月）之内，访问某个页面的不重复 IP 总数。利用 Hive 进行页面独立 IP 访问量统计。
（2）将页面独立 IP 访问量统计结果导出到 MySQL 数据库中。
（3）利用 PyEcharts 进行页面独立 IP 访问量统计结果的可视化。

# 第10章  Hadoop 与 HBase 分布式集群安装与配置

前面章节主要介绍了基于 Hadoop 伪分布式环境下的各种组件的管理与开发，但在实际生产环境中，大数据的管理与开发都是基于 Hadoop 分布式环境进行的。本章将介绍 Hadoop 分布式集群和 Hbase 分布式集群的安装与配置。

- Hadoop 分布式集群安装与配置：包括 4 个服务器节点的 Hadoop 分布式集群安装与配置。
- HBase 分布式集群安装与配置：基于 Hadoop 分布式集群的 HBase 分布式集群安装与配置。

## 10.1  Hadoop 分布式集群安装与配置

在实际生产环境中，Hadoop 集群都是在多个服务器上安装与配置的，本章将使用 VMware 虚拟出 4 个服务器，操作系统为 Ubuntu18，Hadoop 版本为 Hadoop 3.1.2，构建一个 Hadoop 分布式集群。集群各个服务器需要安装的软件和需要运行的服务即 Hadoop 分布式集群节点规划如表 10-1 所示，其中 master 服务器作为主节点、其他 3 个服务器（slave1、slave2、slave3）作为子节点。

表 10-1  Hadoop 分布式集群节点规划

| 服务 \ 主机名 | master | slave1 | slave2 | slave3 |
| --- | --- | --- | --- | --- |
| IP | 192.168.189.128 | 192.168.189.129 | 192.168.189.130 | 192.168.189.131 |
| JDK | ✔ | ✔ | ✔ | ✔ |
| NameNode | ✔ | | | |
| ResourceManager | ✔ | | | |
| SecondaryNameNode | ✔ | | | |
| NodeManager | | ✔ | ✔ | ✔ |
| DataNode | | ✔ | ✔ | ✔ |

### 1. 设置各个服务器的静态 IP 地址

在集群的各个服务器上分别编辑网卡配置文件/etc/netplan/01-network-manager-all.yaml。

```
$ sudo gedit /etc/netplan/01-network-manager-all.yaml
```

写入下列内容（注意：各个服务器 IP 地址不同）。

```
network:
  version: 2
  renderer: NetworkManager
  ethernets:
    ens33:
      dhcp4: no
      dhcp6: no
      addresses: [192.168.189.128/24]
      gateway4 : 192.168.189.2
```

```
    nameservers:
        addresses: [114.114.114.114,172.17.1.7]
```

配置完成后,保存并退出。执行下列命令让静态 IP 地址配置立即生效:

`$ sudo netplan apply`

### 2. 在各个服务器上创建用户 hadoop

(1) 创建用户 hadoop:

`$ sudo useradd -m hadoop -s /bin/bash`

设置用户 hadoop 密码为 hadoop(注意:密码不回显):

`$ sudopasswdhadoop`

(2) 将用户 hadoop 添加到 sudo 组,为用户授权。

`$ sudo adduser hadoop sudo`

### 3. 在各个服务器上修改主机名与域名映射

(1) 修改服务器的主机名称。

编辑/etc/hostname 文件,写入服务器主机名称(注意:各个服务器主机名称不同)。

`$ sudogedit/etc/hostname`

在 master 服务器中写入"master",在 slave1 服务器中写入"slave1",在 slave2 服务器中写入"slave2",在 slave3 服务器中写入"slave3"。

(2) 将所有的服务器 IP 地址与主机名称写入/etc/hosts 中,完成域名映射的添加。

`$ sudo gedit /etc/hosts`

在文件中添加下列信息:

```
192.168.189.128 master
192.168.189.129 slave1
192.168.189.130 slave2
192.168.189.131 slave3
```

(3) 重新启动服务器,以用户 hadoop 登录系统。

`$ sudo reboot`

### 4. 在各个服务器上进行 SSH 免密码登录设置

为了实现 Hadoop 集群服务器之间 SSH 免密码登录,需要进行 SSH 免密码登录设置。

(1) 在服务器上安装 SSH:

`$ sudo apt-get update`

`$ sudo  apt-get  install openssh-server`

(2) 在服务器上生成公钥和私钥:

`$ ssh-keygen -t  rsa`

生成过程如图 10-1 所示(确认环节按回车键即可)。

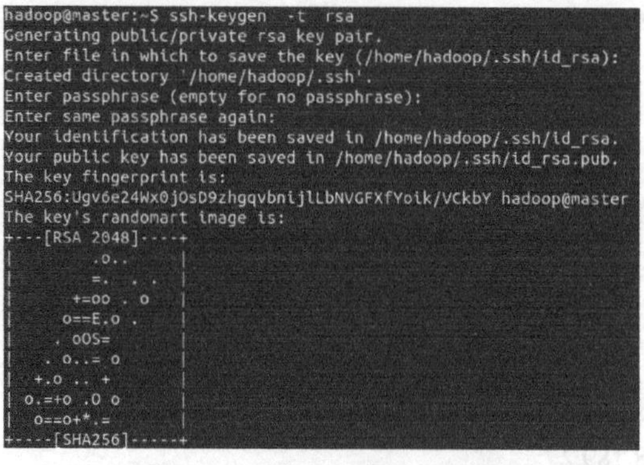

图 10-1 公钥与私钥生成过程

执行完成后，在~/目录下（/home/hadoop）自动创建目录.ssh，内部创建 id_rsa（私钥）、id_rsa.pub（公钥）两个文件。

（3）将服务器的公钥发送到集群所有服务器的.ssh/authorized_keys 文件中。

```
$ cd   ~/.ssh
$ ssh-copy-id -i master
$ ssh-copy-id -i slave1
$ ssh-copy-id -i slave2
$ ssh-copy-id -i slave3
```

执行过程如图 10-2 所示。

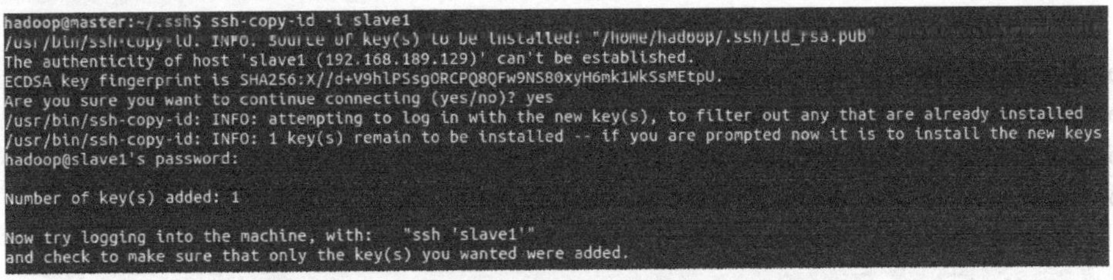

图 10-2　公钥发送到集群所有服务器的执行过程

所有服务器执行完上述操作后，查看服务器的~/.ssh/authorized_keys 文件，可以看到相应的客户端节点公钥信息，如图 10-3 所示。

图 10-3　authorized_keys 文件中的客户端节点公钥信息

（4）测试 SSH 免密码登录。

在每个服务器上执行下列操作，测试 SSH 免密码登录配置是否成功。

```
$ ssh master
$ ssh slave1
$ ssh slave2
$ ssh slave3
```

测试成功后，必须执行 exit 命令结束远程登录，如图 10-4 所示。

图 10-4　SSH 免密码登录测试

### 5．在各个服务器上安装 Java 环境

由于 Hadoop 是使用 Java 语言开发的，因此 Hadoop 平台运行需要 Java 环境。

（1）在目录/usr/lib 中创建 jvm 目录，并将目录所有者修改为用户 hadoop。
$ sudo mkdir /usr/lib/jvm/
$ sudochown -R hadoop /usr/lib/jvm

（2）使用 tar 命令解压安装 jdk-8u121-linux-x64.tar.gz 文件到目录/usr/lib/jvm。
$ cd ~/Downloads    #进入jdk-8u121-linux-x64.tar.gz文件所在目录
$ sudo tar -zxvf jdk-8u231-linux-x64.tar.gz -C /usr/lib/jvm    #解压文件到/usr/lib/jvm

（3）配置 JDK 环境变量，并使其生效。
① 使用 gedit 命令打开用户的配置文件.bashrc。
$ gedit    ~/.bashrc
② 在文件中加入下列内容：
export JAVA_HOME=/usr/lib/jvm/jdk1.8.0_231
export JRE_HOME=$JAVA_HOME/jre
export PATH=$JAVA_HOME/bin:$JAVA_HOME/jre/bin:$PATH
export CLASSPATH=$CLASSPATH:.:$JAVA_HOME/lib:$JAVA_HOME/jre/lib
③ 使环境变量生效。
$ source    ~/.bashrc
④ 验证 JDK 是否安装成功。
$ java    -version
如果打印出 Java 版本信息，则 Java 环境安装成功，如图 10-5 所示。

```
hadoop@master:~/Downloads$ java -version
java version "1.8.0_231"
Java(TM) SE Runtime Environment (build 1.8.0_231-b11)
Java HotSpot(TM) 64-Bit Server VM (build 25.231-b11, mixed mode)
```

图 10-5    Java 版本信息

## 6．Hadoop 分布式集群安装与配置

（1）在 master 服务器上，使用 tar 命令解压安装 hadoop-3.1.2.tar.gz 文件到目录/usr/local，并重命名为 hadoop。
$ cd ~/Downloads
$ sudo tar -zxvf hadoop-3.1.2.tar.gz -C /usr/local
$ cd /usr/local
$ sudo mv    hadoop-3.1.2    hadoop

（2）在 master 服务器上，将目录/usr/local/hadoop 的所有者修改为用户 hadoop。
$ sudo chown -R hadoop /usr/local/hadoop

（3）在 master 服务器上，配置环境变量，并使其生效。
① 使用 gedit 命令打开用户的配置文件.bashrc。
$ gedit    ~/.bashrc
② 在文件中加入下列内容：
export HADOOP_HOME=/usr/local/hadoop
export PATH=$HADOOP_HOME/bin:$HADOOP_HOME/sbin:$PATH
③ 使环境变量生效。
$ source    ~/.bashrc

（4）在 master 服务器上，配置 Hadoop 文件。
在 Hadoop 分布式集群配置中，需要配置/usr/local/hadoop/etc/hadoop 目录中的 6 个配置文件，分别为 hadoop-env.sh、core-site.xml 和 hdfs-site.xml、yarn-site.xml、mapred-site.xml 和 workers，

属性可以采用只读配置文件中的默认设置。

① 配置 hadoop-env.sh 文件。用 gedit 命令编辑 hadoop-env.sh 文件：

$ gedit /usr/local/hadoop/etc/hadoop/hadoop-env.sh

在文件中添加下列代码：

export JAVA_HOME=/usr/lib/jvm/jdk1.8.0_231

② 配置 core-site.xml 文件。用 gedit 命令编辑 core-site.xml 文件：

$ gedit /usr/local/hadoop/etc/hadoop/core-site.xml

在<configuration>和</configuration>标记之间写入下列内容：

```xml
<property>
    <name>fs.defaultFS</name>
    <value>hdfs://master:9000/</value>
</property>
<property>
    <name>hadoop.tmp.dir</name>
    <value>file:/usr/local/hadoop/tmp</value>
</property>
```

③ 配置 hdfs-site.xml 文件。用 gedit 命令编辑 hdfs-site.xml 文件：

$ gedit /usr/local/hadoop/etc/hadoop/hdfs-site.xml

在<configuration>和</configuration>标记之间写入下列内容：

```xml
<property>
    <name>dfs.namenode.name.dir</name>
    <value>file:/usr/local/hadoop/dfs/name</value>
</property>
<property>
    <name>dfs.datanode.data.dir</name>
    <value>file:/usr/local/hadoop/dfs/data</value>
</property>
<property>
    <name>dfs.replication</name>
    <value>3</value>
</property>
<property>
    <name>dfs.namenode.secondary.http-address</name>
    <value>master:50090</value>
</property>
```

④ 配置 yarn-site.xml 文件。用 gedit 命令编辑 yarn-site.xml 文件：

$ gedit /usr/local/hadoop/etc/hadoop/yarn-site.xml

在<configuration>和</configuration>标记之间写入下列内容：

```xml
<property>
    <name>yarn.resourcemanager.hostname</name>
    <value>master</value>
</property>
<property>
    <name>yarn.nodemanager.aux-services</name>
    <value>mapreduce_shuffle</value>
</property>
```

⑤ 配置 mapred-site.xml 文件。用 gedit 命令编辑 mapred-site.xml 文件：

```
$ gedit /usr/local/hadoop/etc/hadoop/mapred-site.xml
```
在<configuration>和</configuration>标记之间写入下列内容：
```
<property>
    <name>mapreduce.framework.name</name>
    <value>yarn</value>
</property>
<property>
    <name>mapreduce.jobhistory.address</name>
    <value>master:10020</value>
</property>
<property>
    <name>mapreduce.jobhistory.webapp.address</name>
    <value>master:19888</value>
</property>
<property>
    <name>yarn.app.mapreduce.am.env</name>
    <value>HADOOP_MAPRED_HOME=/usr/local/hadoop</value>
</property>
<property>
    <name>mapreduce.map.env</name>
    <value>HADOOP_MAPRED_HOME=/usr/local/hadoop</value>
</property>
<property>
    <name>mapreduce.reduce.env</name>
    <value>HADOOP_MAPRED_HOME=/usr/local/hadoop</value>
</property>
```
⑥ 配置 workers 文件。用 gedit 命令编辑 workers 文件：
```
$ gedit /usr/local/hadoop/etc/hadoop/workers
```
删除 workers 文件中原有的 localhost，将集群中所有子节点服务器主机名或 IP 地址写入文件，每行一个主机名或 IP 地址。
```
slave1
slave2
slave3
```
（5）将 master 服务器上配置好的 hadoop 目录发送到其他子节点服务器。由于权限问题，需要先将 hadoop 文件夹发送到其他服务器的宿主目录：
```
$ sudo scp -r /usr/local/hadoop hadoop@slave1:~/
$ sudo scp -r /usr/local/hadoop hadoop@slave2:~/
$ sudo scp -r /usr/local/hadoop hadoop@slave3:~/
```
（6）在各个子节点服务器进行 hadoop 目录移植，并修改所有者。

① 将 hadoop 目录移动到/usr/local/目录下：
```
$ sudo mv   ~/hadoop   /usr/local/
```
② 将 hadoop 目录的所有者修改为用户 hadoop：
```
$ sudo chown   -R hadoop:hadoop   /usr/local/hadoop
```
（7）将 master 服务器上的配置文件~/.bashrc 发送到其他子节点服务器：
```
$ sudo scp ~/.bashrc hadoop@slave1:~/
$ sudo scp ~/.bashrc hadoop@slave2:~/
$ sudo scp ~/.bashrc hadoop@slave3:~/
```

(8)在各个子节点服务器上使环境变量立即生效:

$ source ~/.bashrc

(9)在 master 服务器上进行集群格式化:

$ hdfs namenode -format

注意:如果格式化之后,重新修改了系统配置,或者 Hadoop 启动不了,可能需要重新格式化操作。重新格式化,需要首先停止 Hadoop 运行,然后删除 tmp、dfs、logs 文件夹。例如:

$ stop-all.sh
$ cd /usr/local/hadoop
$ rm -r dfs/ logs/ tmp/
$ hdfs namenode -format

(10)启动 Hadoop 服务:

$ start-all.sh

### 7. 验证安装结果

(1)查看 Hadoop 进程。Hadoop 集群启动后,可以使用 jps 命令查看 Hadoop 分布式集群中不同节点运行的进程,如图 10-6 所示。

图 10-6 Hadoop 集群不同节点运行的进程

(2)浏览器访问。通过 http://master:8088/cluster 可以访问 YARN 管理界面,如图 10-7 所示。

图 10-7 YARN 管理界面

通过 http://master:9870 可以查看 HDFS 管理界面,如图 10-8 所示。

图 10-8 HDFS 管理界面

## 10.2 HBase 分布式集群安装与配置

Hadoop 分布式集群安装并成功启动后,就可以在 Hadoop 集群上安装 HBase 分布式集群。将 Hadoop 集群中的 master 服务器作为主节点(HMaster)、其他 3 台服务器(slave1、slave2、slave3)作为子节点(HRegionServer),内置 Zookeeper 也运行在 3 个子节点上。HBase 分布式集群节点规划如表 10-2 所示。

表 10-2 HBase 分布式集群节点规划

| 服务\主机名 | master | slave1 | slave2 | slave3 |
|---|---|---|---|---|
| IP | 192.168.189.128 | 192.168.189.129 | 192.168.189.130 | 192.168.189.131 |
| HMaster | ✔ | | | |
| HRegionServer | | ✔ | ✔ | ✔ |
| QuorumPeerMain | | ✔ | ✔ | ✔ |

HBase 分布式集群配置较为简单,只需在主节点进行配置,然后将配置后的目录发送到各个子节点就可以了。

### 1. 在 master 服务器上安装 HBase

(1)使用 tar 命令解压安装包 hbase-2.2.6-bin.tar.gz 至路径/usr/local。

$ cd ~/Downloads
$ sudo tar -zxvf hbase-2.2.6-bin.tar.gz -C /usr/local

(2)将解压的文件名 hbase-2.2.6 重命名为 hbase。

$ sudo mv /usr/local/hbase-2.2.6 /usr/local/hbase

(3)将目录/usr/local/hbas 目录所有者改为用户 hadoop。

$ sudo chown -R hadoop /usr/local/hbase

(4)配置环境变量,并使其生效。

① 使用 gedit 命令打开用户的配置文件.bashrc:

$ gedit ~/.bashrc

② 在文件中加入下列内容:

export HBASE_HOME=/usr/local/hbase
export PATH=$PATH:$HBASE_HOME/bin

③ 使环境变量生效:

$ source ~/.bashrc

(5)查看 HBase 版本,确定 HBase 安装成功。

$ hbase version

命令执行后,输出版本信息如图 10-9 所示。

```
hadoop@master:~/Downloads$ hbase version
HBase 2.2.6
Source code repository git://or1-hadoop-build02.awsus/home/zhangguanghao1/code/hbase revision=88c9a386176e2c2b5fd9915d0e9d3ce17d0e456e
Compiled by zhangguanghao1 on Tue Sep 15 17:36:14 CST 2020
```

图 10-9 HBase 版本信息

### 2. 在 master 服务器上进行 HBase 集群分布式配置

HBase 已经安装成功,就可以进行 HBase 分布式集群配置。

(1)配置 hbase-env.sh 文件。使用 gedit 命令编辑 hbase-env.sh 文件:

$ gedit /usr/local/hbase/conf/hbase-env.sh

在文件中写入下列内容：
export JAVA_HOME=/usr/lib/jvm/jdk1.8.0_231
export HBASE_CLASSPATH=/usr/local/hadoop/etc/hadoop
export HBASE_MANAGES_ZK=true

注意：HBASE_MANAGES_ZK=true 指定采用 HBase 自带的 Zookeeper。

（2）配置 hbase-site.xml 文件。修改 hbase.rootdir，指定 HBase 数据在 HDFS 上的存储路径；将 hbase.cluster.distributed 设置为 true，设置集群处于分布式模式；hbase.zookeeper.quorum 设置 Zookeeper 运行的节点，其数量为奇数。

用 gedit 命令编辑 hbase-site.xml 文件：

$ gedit /usr/local/hbase/conf/hbase-site.xml

修改后<configuration>和</configuration>标记之间内容为：

```
<property>
    <name>hbase.rootdir</name>
    <value>hdfs://master:9000/hbase</value>
</property>
<property>
    <name>hbase.cluster.distributed</name>
    <value>true</value>
</property>
<property>
    <name>hbase.zookeeper.quorum</name>
    <value>slave1,slave2,slave3</value>
</property>
<property>
    <name>hbase.unsafe.stream.capability.enforce</name>
    <value>false</value>
</property>
```

（3）配置 HRegionServer 文件。用 gedit 命令编辑 HRegionServer 文件：

$ gedit /usr/local/hadoop/etc/hadoop/workers

删除 HRegionServer 文件中原有的 localhost，将集群中所有 HRegionServer 服务器主机名或 IP 地址写入文件，每行一个主机名或 IP 地址。

slave1
slave2
slave3

（4）将 master 服务器上配置好的 hbase 目录发送到所有 HRegionServer 服务器。

由于权限问题，需要先将 hbase 目录发送到 HRegionServer 服务器的宿主目录。

$ sudo scp -r /usr/local/hbase hadoop@slave1:~/
$ sudo scp -r /usr/local/hbase hadoop@slave2:~/
$ sudo scp -r /usr/local/hbase hadoop@slave3:~/

3. 在各个 HRegionServer 服务器上进行 hbase 目录移植并修改所有者

① 将 hbase 目录移动到/usr/local/目录下

$ sudo mv   ~/hbase   /usr/local/

② 将 hbase 目录的所有者修改为用户 hadoop

$ sudo chown   -R hadoop:hadoop   /usr/local/hbase

4. 在 master 服务器上启动与关闭 HBase

由于 HBase 分布式集群是运行在 Hadoop 集群之上的，因此启动 HBase 之前，需要先启动

Hadoop 集群。HBase 启动后，可以通过 Shell、Web 方式验证 HBase 配置结果。

（1）在 master 服务器上启动 Hadoop：

$ start-all.sh

（2）在 master 服务器上启动 Hbase：

$ start-hbase.sh

（3）查看 HBase 进程。可以通过运行 jps 命令查看 HBase 启动后启动的相关 Java 进程。如图 10-10 所示为 HMaster 服务器上运行的进程，如图 10-11 所示为 HRegionServer 服务器上运行的进程。

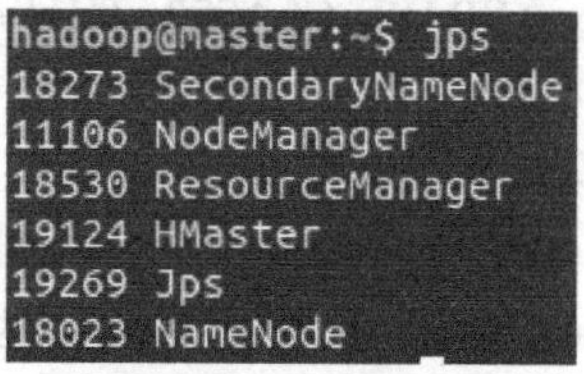

图 10-10　HMaster 服务器上运行的进程

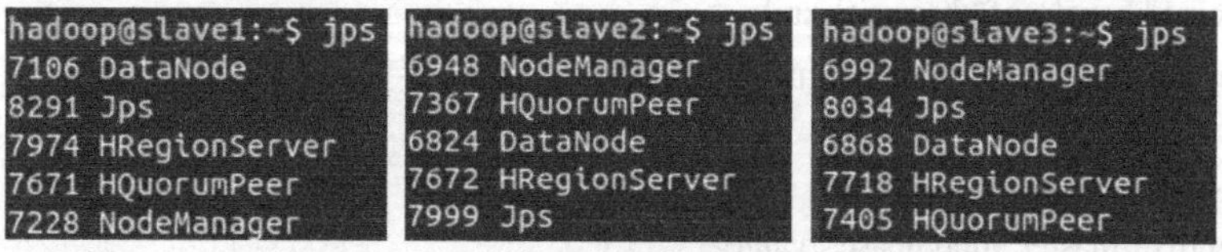

图 10-11　HRegionServer 服务器上运行的进程

（4）在 master 服务器上打开 HBase 命令行窗口。输入 hbase shell 命令后，进入 HBase 命令行窗口，如图 10-12 所示。

图 10-12　HBase 命令行窗口

（5）通过浏览器查看 HBase 运行状态。打开浏览器，输入网址 http://master:16010，可以查看 HBase 运行状态，如图 10-13 所示。

图 10-13　查看 HBase 运行状态

（6）在 master 服务器上关闭 Hbase：

$ stop-hbase.sh

## 本 章 小 结

本章首先介绍了利用 VMware 虚拟 4 个服务器节点，进行 Hadoop 分布式集群安装与配置，然后介绍了基于 Hadoop 分布式集群的 HBase 分布式集群的安装与配置。为了与前面章节中 Hadoop 伪分布式环境的衔接，本章中配置的 Hadoop 分布式集群的访问方式与伪分布式访问方式相同，都是 hdfs://master:9000。通过本章的学习，读者可以掌握 Hadoop 分布式集群的安装与配置，以及基于 Hadoop 分布式集群的 HBase 分布式集群的安装与配置，为将来在生产环境中的实际应用开发奠定基础。

## 思考题与习题

1. 简答题

（1）简述 Hadoop 分布式集群安装需要进行的准备工作。
（2）简述 Hadoop 分布式集群安装与配置步骤。
（3）简述 HBase 分布式集群安装与配置步骤。

2. 实训题

（1）完成一个由 5 个服务器节点构成的 Hadoop 分布式集群的安装与配置。
（2）完成基于 5 个服务器节点的 HBase 分布式集群的安装与配置。

# 参考文献

[1] hadoop 官网：https://hadoop.apache.org/

[2] hadoop 开发文档：https://hadoop.apache.org/docs/r3.1.2/

[3] NoSQL 数据库：http://nosql-database.org/

[4] hbase 官网：http://hbase.apache.org/

[5] hbase 开发文档：http://hbase.apache.org/book.html

[6] hive 官网：http://hive.apache.org/

[7] hive 开发文档：https://cwiki.apache.org/confluence/display/Hive/LanguageManual

[8] sqoop 官网：http://sqoop.apache.org/

[9] sqoop 开发文档：http://sqoop.apache.org/docs/1.99.7/index.html

[10] Flume 官网：http://flume.apache.org/

[11] Flume 开发文档：http://flume.apache.org/releases/content/1.9.0/FlumeDeveloperGuide.html

[12] Kafka 官网：http://kafka.apache.org/

[13] Kafka 开发文档：http://kafka.apache.org/documentation/

[14] Zookeeper 官网：https://zookeeper.apache.org/doc/current/index.html

[15] Zookeeper 开发文档：https://zookeeper.apache.org/doc/current/api/index.html

[16] eclipse 官网：https://www.eclipse.org/

[17] maven 官网：http://maven.apache.org/